Integrated Control of Citrus Pests in the Mediterranean Region

Editor

Vincenzo Vacante

Mediterranean University of Reggio Calabria
Italy

Co-Editor

Uri Gerson

The Hebrew University of Jerusalem
Israel

CONTENTS

PREFACE

Citrus originated from Southeast Asia, whence over the centuries it gradually spread to various regions of the world by human migrations and trade. During the Roman Empire citrus fruits were known in southern Europe, and their consumption began through the Middle Ages in the remaining parts of Europe. In the Mediterranean region citrus has been known for centuries and had intensively influenced human history and customs. This is evident from the part that citrus still plays in the religious rituals of some peoples, in citrus flowers being used as symbols (of purity, etc.), in fruits and leaves being utilized in culinary preparations, and along with other plants (cultivated and wild) being a characterizing element of the Mediterranean landscape. The most important aspect of citrus is its use as a nutrient, either fresh or as juice and/or derivatives, and also in pharmaceutics and/or cosmetics. Directly and/or indirectly, Mediterranean citriculture currently affects about 300 million people and in 2007 the FAO estimated a total harvested area of 8,322,605 ha. Citriculture thus represents an important agricultural sector, characterized by a heterogeneous structure, resulting from various geographical and climatic constraints and the different history of the settlements. This is reflected in several aspects of the industry, such as qualitative and quantitative standards of production, and emphasizes the need for planning a commonly shared modern approach. This becomes critical upon considering plant protection aspects, an area where various technical choices and patterns of behavior are significantly different between the various regions. That is the area in which this eBook strives to contribute, presenting an update of knowledge on the defense against citrus pests in the region, being primarily concerned with promoting the strategy of Integrated Pest Management (IPM). This strategy has been widely implemented in the North American citriculture and is providing economic, ecological and toxicological benefits to individual farms and to the community there. Unfortunately, at present the use of IPM in the Mediterranean region is, at best, either at an initial state, or uncommon, except in some countries like Israel.

This eBook aims to fill that gap and to address all personnel (college and university students, technicians and traders) who are concerned with citrus in the Mediterranean region, advocating the possibilities of IPM programs by using simple language which is not however devoid of scientific rigor. In order to meet this requirement a strict editorial policy was imposed, based on submitting "essential" information, focusing only on key topics. In addition, we avoided, as much as possible, a "too specialized" approach, which although essential in a strictly entomological context, would have made the text more complex and might have detracted from comprehending the whole discussion. The reader will therefore find in the volume useful information for implementing IPM principles in citriculture, including the use of decision tools and other available practical relevant means, which at this time cannot be found elsewhere. Topics of more specialist interests (*e.g.* systematics of natural enemies, ecology of populations, etc.), although essential in other contexts, are not required for the purposes the present project.

The table of contents is in two parts, the first discusses the agronomic aspects of citriculture, whereas the second is about the pests. The first chapter of this second part briefly describes the history of citriculture IPM in the Mediterranean region. The second deals with the implementation of IPM, highlighting the objectives of the strategy, its benefits, the decision tools and the relevant methodology used. A subsequent chapter presents basic systematics of the various groups of key pests, including a key for their recognition. The discussion of key pests is developed in 18 chapters, each of which illustrates the fundamental aspects of morphology and bio-ecology (including symptoms of damage), the main limiting natural factors, the endemic and emerging vector-borne Mediterranean citrus diseases and IPM strategies.

The complexity of the topic and the never-ending increases in the published information render our framework highly dynamic; not all available knowledge could thus be included. With that in mind, we extend an invitation to those who will follow us to fill the inevitable many gaps and to correct errors.

Finally, we want to thank Bentham Sciences Publishers for believing in our proposal and in helping make it happen. We also sincerely thank all colleagues who joined our initiative, and provided their contributions to this eBook with a singular spirit of Mediterranean cooperation, which makes it unique.

Vincenzo Vacante
Mediterranean University of Reggio Calabria
Italy

Uri Gerson
The Hebrew University of Jerusalem
Israel

List of Contributors

Yael Argov

Israel Cohen Institute for Biological Control, Plant Production and Marketing Board, Citrus Division, POB 80, Bet Dagan 50250, Israel

Moshe Bar-Joseph

Department of Plant Pathology and Weed Research ARO, (S. Tolkowsky Laboratory for Citrus Disease Research) The Volcani Center, Bet Dagan 50250, Israel

Hüseyin Başpınar

Department of Plant Protection, Adnan Menderes University, Agricultural Faculty, 09100 Aydın, Turkey

Carmelo Peter Bonsignore

Dipartimento Patrimomio Architettonico ed Urbanistico (PAU), Mediterranean University of Reggio Calabria, Via Melissari, Località Feo di Vito, 89060 Reggio Calabria, Italy

Antonino Catara

Department of Phytosanitary Sciences and Technologies, University of Catania, via S. Sofia 100, 95123 Catania, Italy

Arturo Cocco

Department of Plant Protection, University of Sassari, via E. De Nicola 1, 07100 Sassari, Italy

Alessandra De Grazia

Department of Agricultural and Forest Management Systems (GESAF) (Section of Entomology), Mediterranean University of Reggio Calabria, Località Feo di Vito, 89060 Reggio Calabria, Italy

Gavino Delrio

Department of Plant Protection, University of Sassari, via E. De Nicola 1, 07100 Sassari, Italy

Uri Gerson

Department of Entomology, Faculty of Agricultural, Food and Environmental Quality Sciences, The Hebrew University of Jerusalem, P.O. Box 12, Rehovot, 76-100, Israel

Alfonso Hermoso de Mendoza

Institut Valencià d'Investigacions Agràries (IVIA), Carretera de Nàquera km 5, 46113 Montcada (València), Spain

Paolo Inglese

Dipartimento di Colture Arboree, Università degli Studi di Palermo, Viale delle Scienze, 90128 Palermo, Italy

Andrea Lentini

Department of Plant Protection, Faculty of Agricultural Sciences, University of Sassari, via E. De Nicola 1, 07100 Sassari, Italy

Apostolos Kapranas

Department of Crop Production, Technological Educational Institute, 41110 Larissa, Greece

Rita Marullo

Department of Agricultural and Forest Management Systems (GESAF) (Section of Entomology), Mediterranean University of Reggio Calabria, Località Feo di Vito, 89060 Reggio Calabria, Italy

Giovanni Pensabene Bellavia

Institut Valencià d'Investigacions Agràries (IVIA), Carretera de Nàquera km 5, 46113 Montcada (València), Spain

Ezio Peri

Department SENFIMIZO, University of Palermo, Viale delle Scienze, 90128 Palermo, Italy

Francesco Porcelli

Dipartimento di Biologia e Chimica Agroforestale ed Ambientale (DiBCA), Sez. Entomologia e Zoologia, Facoltà di Agraria, Università degli Studi di Bari, Via Amendola 165/a, 70126 Bari, Italy

Nedim Uygun

Department of Plant Protection, Çukurova University, Agricultural Faculty, 01100 Adana, Turkey

Vincenzo Vacante

Dipartimento Patrimonio Architettonico ed Urbanistico (PAU), Mediterranean University of Reggio Calabria, Via Melissari, Località Feo di Vito, 89060 Reggio Calabria, Italy

Marcello Verdinelli

Institute of Ecosystem Study, National Research Council, Traversa La Crucca 3, 07100 Sassari, Italy

CHAPTER 1

The Citrus in the Mediterranean Region

Paolo Inglese[1*] and Giovanni Pensabene Bellavia[2]

[1]*Dipartimento di Colture Arboree, Università degli Studi di Palermo, Viale delle Scienze, 90128, Palermo, Italy and* [2]*Institut Valencià d'Investigacions Agràries (IVIA), Carretera de Nàquera km 5, 46113 Montcada (València), Spain*

Abstract: The origin and spread of citrus in the Mediterranean Region are briefly discussed, together with the fundamental characteristics of the main species, cultivars and rootstocks grown in the Region. The choice of the genotypes, the plant propagation, the sanitary control, and the cultural management (planting systems, nutrition and fertilization, irrigation and management of water, soil and ground soil management, harvesting and fruit quality) are presented. In conclusion, the future perspectives and genetic improvement for the Mediterranean citriculture are discussed.

Keywords: Citriculture, Mediterranean Region, fruit, rootstock, orchard management.

1. INTRODUCTION

1. Citrus Origin and Spread in the Mediterranean Region

Citrus spp. are the most extensively cultivated tree fruit crop in the world. Their origin is still controversial, as South China, Northeast India [1], the Indo-Chinese peninsula and Burma [2] have been considered as their areas of origin. More recent studies propose the Yunnan province of China as probably the center of origin, because of the great variety of species discovered there [3]. *Citrus* spp. were domesticated in Southeast Asia several thousand years ago and have then spread throughout the world.

Alexander the Great probably introduced citron (*Citrus medica* Linnaeus) into the Mediterranean Region in 300 B.C., and his military expeditions contributed by bringing citrons from India to Persia and Greece. Citron became a feature in the Jewish religion, in which it played a symbolic role, as proved by images discovered in synagogues [4].

Romans probably knew citron fruits, although there is no proof that they were cultivated. Only in 300 A.D. did writers of the Christian era describe the cultivation of citrus fruits in Italy.

The Arabs spread citrus throughout Europe and North Africa with the expansion of their empire. From the 8th until the end of 15th century they occupied the Iberian Peninsula where they developed a very important citriculture. Citron, sour orange (*C. aurantium* Linnaeus), lemons (*C. limon* Burn), limes and pummelos were probably the only citrus fruits cultivated in the Mediterranean Region during the 11th and 12th centuries, as described in numerous books from Spain [4].

Other citrus species, like sweet orange, were introduced into Europe by the Genoese in the 15th century; during their extensive commercial activities they brought back the so called 'Cajel' orange from East Asia. The Portuguese introduced more selected types of sweet orange in the early 16th century. These genotypes were the real sweet orange [*C. sinensis* (Linnaeus) Osbek], initially known as 'Portugal Orange' or 'Lisboa Orange'.

Mandarins were cultivated in Southeast Asia from ancient times, but were not introduced into Europe until the early 19th century. The presence of mandarins and mandarin-like cultivars is nowadays rapidly expanding due to much interest on easy-peeling fruits and the continuous delivery of new cultivars, with different characteristics and in terms of ripening time. The clementine is an outstanding production that is typical of the area. The relative importance of lemons has decreased in the last decades, due to strong competition from Argentina and the USA, whereas blood oranges still represent a unique product of the area, particularly in Italy.

***Address correspondence to Paolo Inglese:** Dipartimento di Colture Arboree, Università degli Studi di Palermo, Viale delle Scienze, 90128, Palermo, Italy; Tel: +3909123861234; Fax: +390917049025; E-mail: pinglese@unipa.it

The citrus industry in the Mediterranean area nowadays accounts for 20% of the world citrus industry. Spain is the largest orange exporter in the world, accounting for almost 30% of the citrus trade, whereas Italy contributes less than 5%. The citrus industry in the Mediterranean Region mainly produces fruit for fresh consumption, exports to other countries accounting for 35% of the product, representing 60% of the world citrus trade of fresh fruit. The orange season in the Mediterranean area and in Europe begins with the early ripening navel cultivars *Naveline* and *Newhall*, continuing with *Washington Navel* and red oranges (mainly *Tarocco*, in Italy) during winter (December-March). It lasts throughout June, with late navel (*Navelate* and *Lanelate*) and *Valencia late* or *Ovale Calabrese* (which are of residual production in Southern Italy). Mandarins and mandarin-like hybrids harvest begins in late September or early October with Satsumas, continuing, with different cultivars and hybrids in the different countries throughout April when the late *Ortanique* tangor ripens in Italy and Spain.

2. SPECIES AND CULTIVARS

The main use of citrus in the Mediterranean Region is as fresh fruit, of which the producing countries export an important part of their harvest. Fruit quality is therefore the main criterion for variety choice. Other criteria are productivity and adaptation to the specific environmental conditions of the different areas.

Climate is probably the most important factor that determines the distribution of the different citrus varieties. Temperature is a limiting condition for citriculture. In areas with prevailing high temperatures and low freezing frequency, both early and late varieties can be cultivated. High temperatures speed up fruit ripening, thus it is profitable for early ripening cultivars. In areas where low winter temperatures cause freezing damage, the use of late ripening cultivars may reduce this risk; on the other hand, if low temperatures slow down fruit ripening, early varieties could lose their commercial value.

The occurrence of rainfall during the season is an important factor when considering the choice of scions; in areas of heavy rainfall during harvest there is a risk of increased fungus infections, making fruit picking difficult and consequently reducing fruit quality. Wind is another important climatic factor because it can damage varieties with large fruits, such as navel oranges or grapefruit and thorn varieties.

Soil conditions strongly affect fruit characteristics and quality; sandy soils are normally used for early varieties, and require particular care in fertilization and irrigation strategies. Clay-loam soils are used for late varieties and for those with attractive action.

Farm structure is an important commercial factor that must be considered in variety choice. In large farms a staggered crop (of early and late varieties) can be produced during the entire season. This facilitates fruit picking, limits biotic and/or abiotic factors that might damage certain citrus genotypes and increases the value of the commercial product.

Small farms must obtain high quality product in order to receive the best price in the market, and their choices, in terms of cultivated species and varieties, will depend on the markets that they wish to address.

2.1. Lemon

Lemons are the third most important citrus fruit in the world. It was one of the first citrus fruit introduced into the Mediterranean Region. Spain is the primary producing country, but lemons are also cultivated in other Mediterranean areas. Lemons are very sensitive to low temperatures and show much adaptability to all kinds of soils. The Mediterranean Region has the best climate for these trees. In fact, lemons are grown in particularly moderate climatic Regions, like the Middle East, Murcia (Spain), Sicily (Italy), Israel, Southern Greece and Turkey.

In subtropical Regions lemons are of very low quality and for this reason they were replaced by Lima fruit.

The most important lemon cultivars in the Mediterranean Region are:

Fino. The origin of Fino is unknown but it possibly came from an older Spanish variety; it is also called *Mesero* or *Primofiori*. *Fino* is a very vigorous and thorny tree with high productivity. The fruits are small to medium sized lemons, spherical to oval in shape with a smooth thin rind. *Fino* has high fruit quality with high juice content and acidity. It is a winter producing lemon, with one main crop between October and February, when prices are at a premium.

Verna. This variety accounts for a great part of Spain's annual lemon crop and is also cultivated in Algeria and Morocco. *Verna* lemon is highly susceptible to the "Mal secco" [*Phoma tracheiphila* (Petri) L.A. Kantsch. and Gikaschvili] disease and for this reason its cultivation is limited in Italy and in other affected Mediterranean Regions. Under Mediterranean condition *Verna* usually flowers twice a year and in some years a third crop is produced. The fruit is usually large, with a pronounced nipple and an intense yellow. The juice content is not as high as in other lemon varieties. *Verna* trees are vigorous and productive; the delayed harvesting of fruit has a tendency to induce alternate bearing.

Eureka. This variety, of Californian origin, is probably the most widely spread variety in the world. In the Mediterranean *Region Eureka* is cultivated only in Israel, probably because its harvest period is the same as that of *Fino*, which has better quality and productivity. *Eureka* fruits are usually of medium size and elliptical, with a short neck and a medium nipple. The fruit has a very high juice content and high acidity. The tree possesses medium vigor and a spreading habit, with sparse foliage.

Interdonato. This is considered a lemon x citron hybrid, commercially grown in Italy (particularly in the Messina province), Greece and Turkey. The trees are thorn-less and vigorous; the fruits are oblong to cylindrical, with a nipple and few seeds. It is a juicy and acid variety, with early and large fruits. *Interdonato* is considered to be resistant to "Mal secco", a trait that contributes to its spread in disease-affected zones.

Femminello. Different clones of *Femminello* are currently cultivated in Italy (*F. comune*, *F. Siracusano*, *F. Zagara Bianca*). Their fruits are medium-sized, seeded, elliptic, have blunt nipples and they ship and store well. Juice content is not high but fairly acidic. *Femminello* has four annual crops: *Primofiore* (harvested from September to November), *Limoni* (December-May), *Bianchetti* (April-June) and *Verdelli* (May-July). The cultivar *Adamopoulou* is primarily cultivated in Greece. Is a seeded variety with a 31% juice content. Its cultivation has decreased due to "Mal secco".

2.2. Orange

Although oranges were introduced to the Mediterranean Region relatively late, this area is the principal center of diversification of the modern sweet orange [5].

Three sub-areas of diversification exist. The main one, the Iberian Peninsula, is associated with blond sweet oranges. The *Cadenera* orange was most important in the development of the Spanish citrus industry during the second half of the 19th century. Concurrently, a late Portuguese variety, *Don Joao*, was introduced into the Azores and then to the United States, where it was renamed *Valencia Late*. Today this variety is cultivated for both the fresh and juice markets. Similarly, 'Navel' oranges (*Umbigo*) were initially cultivated in Portugal and Spain. After the Portuguese introduced the sweet orange into Bahia (Brazil), the navel variety *Bahia* (*Washington Navel*) originated as a budsport mutation of the cultivar *Selecta* during the earliest decades of the 19th century. After being imported into the USA in 1870 by the Washington State Department of Agriculture, it became the leading world variety for fresh fruit consumption [6].

The second area of diversification of sweet orange consists of Tunisia, Malta and Sicily, from where the blood and half-blood oranges originated. These oranges consist of three groups. The *Moro* group originated in Liguria and then diversified in Sicily. *Tarocco* is currently the most widely cultivated variety in this group. The second group is *Maltese*, largely cultivated in Tunisia. Different early cultivars, like *Bokobza*, or late as *Barlerin,* were selected from within this group. The last group is the *Doubles fines*, which also diversified in Spain, where the blood 'Washington' variety was selected. The Near East constitutes the third center of sweet orange diversification, whence blond oranges, such as *Shamouti* and *Beladi* were selected.

Nowadays sweet oranges are classified in three groups: Navel oranges, Common oranges (or Blond oranges) and Blood oranges.

Navel oranges. This group is characterized by the presence of a second, smaller fruit embedded inside the principal fruit. The smaller fruit originates from the development of carpels in a second whorl. Navel fruits are seedless and the propagation of its trees is always via cuttings or grafting. This does not allow the usual improvement methodologies, and new varieties had originated only from mutations of other genotypes. The most interesting cultivars of this group are:

Washington Navel. This variety is probably the most popular in the navel group, and the leading variety in the world for fresh fruit consumption since the end of the 19th century. Fruits have a good organoleptic quality, are large and show a very rich orange colour, especially in Mediterranean countries. *Washington Navel* is considered a mid-season fruit, harvested from December to February. The peel is particularly sensitive and thus requires regular treatment to keep the fruit on the trees in order to delay ripening.

Navelina. This is an early ripening variety from California, whose fruit is generally ready for harvesting at the end of October. The *Navelina* tree is vigorous and in some years produces only alternately. The fruit has a good quality and an intensive. Its two main selections are known as *ISA 315* and *VCR.*

Newhall. It is a *Washington Navel* mutation, which in the Mediterranean climate is practically indistinguishable from *Naveline*, except for its earlier ripening.

Navelate. This orange tree is a spontaneous mutation from *Washington Navel*, although slightly inferior in fruit quality and less vigorous as a tree. The fruit matures later, its peel is yellow-orange, and it has high juice content. The fruit can remain on the trees for months in good condition.

Lanelate. Another variety that had originated from a mutation of *Washington Navel*, but the tree characteristics are indistinguishable from its source. The fruit can be harvested from January to May; it has good organoleptic qualities with high juice content and a good sugar/acid ratio.

Other promising cultivars for the Mediterranean Region are *Powell* and *Chislett,* which had originated in Australia. Their fruit may remain on the trees after harvest for a long time.

Common oranges. The fruits of this group are characterized by the absence of a navel and are less acidic in comparison to other oranges. They were usually meant for the industrial production of juice, but nowadays citrus cultivation in the Mediterranean Region is aimed mostly towards the fresh fruit market. For this reason the cultivation of these "common oranges" has drastically declined. Only cultivars with good fruit and organoleptic qualities are cultivated now. The most important cultivars of this group are:

Salustiana. It is a vigorous and very productive tree with a tendency to produce vertical branches. Fruits ripen between December and March and when mature can remain on the trees.

Valencia late. This variety is very adaptable to different climates and soils. The fruit is spherical and not very colored, with a high juice content and acidity. *Valencia* is a late ripening variety that can remain on the tree several months, even if it turns green under high temperatures.

Shamouti. Also known as 'Jaffa' orange, it is an important variety in the eastern Mediterranean Region, being cultivated in Israel, Turkey, Syria, Lebanon and Cyprus. *Shamouti* is clearly not adapted to hot and arid climates and is therefore restricted to the favorable parts of this area. Under advantageous conditions the trees produce seedless and red fruit, which mature in mid-season.

Other varieties of limited cultivation are *Verna* in Spain, *Hamlin* and *Cadenera* in North Africa (Algeria and Morocco) and *Ovale Calabrese* in Italy.

Blood oranges. Genotypes belonging to this group are thus called because of the red color of their pulp, which is due to the presence of anthocyanins in the sap of the vesicles. These oranges are cultivated in areas with prevailing low temperatures that allows an adequate presence of anthocyanins. The best temperature range for the red pigmentation is 13-14°C during ripening. Lower temperatures do not allow regular fruit ripening, whereas higher ones reduce anthocyanin synthesis. The fruit is seedless and commonly medium to small in size. Harvest starts in January. Sicily is the main producer of these oranges. The most important varieties are:

Moro. It originated at the beginning of the 19th century in Sicily. The fruit pulp and juice are a very intensive purple, but the external peel is not strongly colored. *Moro* is a productive genotype, whose fruit ripens just before the other blood oranges.

Tarocco. The fruits of this variety are medium sized, with good peel and pulp. It is popular for being sweet and juicy, and considered the best tasting of all blood oranges. Different clones, such as *T. Scirè*, *T. Gallo*, *T. Ippolito*, *T. Meli*, *T. Sant'Alfio*, *T. dal Muso*, are used. They differ in ripening times and the degree of fruit pigmentation.

2.3. Mandarins and Hybrids

The botanical classification of this citrus group is complex and confused; according to Swingle three different species exist within the "mandarin group", whereas Tanaka believed that 36 exist. To simplify this aspect we will consider only four groups: Satsuma (*C. unshiu* Marc.), Clementine [*C. reticulata* Blanco; Tangor (*C. reticulata* x *C. sinensis* or *C. aurantium*)], Mediterranean Mandarin (*C. deliciosa* Ten.) and other hybrids.

Satsuma. Early ripening is the principal characteristic of this group. Fruit quality is not very good, but it ripens in September – October, when no other citrus fruit is on thc fresh market. The fruit cannot remain on the tree for long, because it rapidly loses its juiciness and acidity. Post-harvest degreening treatments are often necessary to change its color. Spain, Morocco, Turkey, Italy and Israel are the principal producers of Satsuma in the Mediterranean Region. The most important varieties are:

Clausellina. This is a spontaneous mutation of Satsuma "*Owari*", maintaining similarities with its origin. It is a late flowering and early maturing tree, a trait that reduces the risk of low temperature damage. The fruit is not of high quality due to its low content in sugars and acids, but is a very productive variety.

Hashimoto. This belongs to the 'Wase' group, known as the earliest maturing Satsuma. It is a little tree that produces small fruits. Its maturity index is high because of a higher sugar content and lower acidity. Nowadays is not very common but could serve as an interesting genotype in the future.

Okitsu. This is a late flowering genotype that can be harvested in the last week of September. Because of its early ripening, *Okitsu* is often attacked by the Mediterranean fruit fly [*Ceratitis capitata* (Wiedemann)]; in addition, the fruit may get sunburned if exposed to the south.

Clementine. These fruits are nowadays the most economically important in the mandarin group, due to their attractiveness, pleasant taste and easy peeling characteristics. Fruits are normally seedless because of self-incompatibility, although their ovules and pollen are viable. All Clementine varieties are considered to be spontaneous mutations that had originated from a unique mutation genotype. The first Clementine (*Fina*) was obtained in North Africa by Father Clement Rodier; and the other varieties had originated from that genotype. Various clones are commercially very important for the export trade of many Mediterranean countries (*e.g.* Spain and North Africa); some of these clones are:

Loretina. This variety is a spontaneous mutation of 'Marisol' whose harvest normally starts a few days before *Marisol*. The fruit is small, but with good organoleptic characteristics, an intensive and a slightly rough peel.

Marisol. It is harvested at the same time as *Oronules*; the fruit is slightly acidic with a tender pulp and good juice content. The tree is vigorous with compact foliage and vertical branching.

Clemenrubi. This is a new variety that originated in Spain in 1996. It is a medium vigorous plant with short interknots and dense foliage. *Clemenrubi* has good productivity and is harvested early, in the second part of September.

Oronules. This is an early maturing genotype that is harvested in the first week of October. It is a small fruit with a very attractive color and good fruit quality. Fruit bearing is in alternate years.

Arrufatina. This is a vigorous, early maturing tree whose fruit ripens three weeks before '*Clemenules'*. The fruit has a good size and a slightly granulated pulp.

Monreal. This variety was first found in 1940 in the orchard of Vincent Monreal at Perregaux, Oran. It is one of the most important mandarin varieties in the Mediterranean Region, because it ripens early, in October. *Monreal* is self-compatible and without cross-pollination the fruit is regularly seedy. It is one of the sweetest and easiest to peel clementine in the mandarin group.

Clemenules. This is probably the most widely distributed Clementine variety in the Mediterranean Region. It is a vigorous variety that produces medium-sized fruit of good quality. It is seedless and easy to peel, with good juice content. As its harvest is between November and January; it is considered a mid-season maturing Clementine. The fruits rapidly lose their juiciness and thus cannot remain on the trees for long (only until January). Flowering is staggered, which causes a staggered harvesting, that may be completed in two or three times.

Algerian. Also known as *Comune*. *Comune* and *Monreal* are the two original Clementine clones from Algeria. The trees are of medium vigor and size and are nearly thornless, whereas the foliage is dense. Regular satisfactory bearing in most locations can be assured only by cross-pollination. The trees are very cold-resistant.

Sidi Aissa. This is a mutation of '*Fina'* Clementine, discovered in Morocco, where it is valued because of its large fruit size and durable peel. '*Sidi Aissa*' is very productive. It is said to have the same maturity period as Clemenules but better internal color and flavor.

Nour. This variety is an important Moroccan Clementine. The fruit has a good color but the rind texture is often coarse. The pulp is tender, juicy and of very good flavor. Harvest is from mid- to late- January. The fruit can stay on the trees in good condition until March.

Hernandina. This variety is a bud mutation of '*Fina'*, bearing small fruits with good organoleptic characteristics. It is a late (January-February) maturing cultivar; at times its fruits are not completely red, remaining green at the stylar end. It is suitable for cultivation in zones of late harvest with no risk of frost.

Carte noir. This Moroccan variety is also known as *'Bekria'*. It is a low-acidity, early ripening Clementine. The pulp is fairly juicy, seedless, soft-textured and with a pale orange color. It is much appreciated by the North African community in France. *Carte noir* is one of the earliest citrus fruits to mature in the Mediterranean Region, being available from mid-September to mid-October.

Mediterranean Mandarins. This group includes cultivars grown only in specific areas and with a restricted economic importance, probably due to the presence of too many seeds in the segments. Fruits have a remarkable characteristic flavor.

Common mandarin. These fruits are medium in size, moderately oblate, frequently slightly lobed. Usually they are with a low collar and a strongly furrowed neck; apex depressed and commonly slightly wrinkled; areole lacking and a small Navel-like structure is fairly common. These fruits are small, round, plump, and highly polyembryonic, with light green cotyledons and numerous seeds. Rind thin, not leathery, loosely adherent; surface smooth and glossy with large, deeply ed oil glands; yellowish-orange at maturity. With 10-12 segments, very loosely adherent; axis hollow. The light orange flesh is tender, juicy, and sweet, the flavor distinctly and pleasantly aromatic. The fruits are moderately early to early midseason in maturing. This cultivar bears fruit in a typical alternate pattern, which can be reduced with adequate pruning.

Avana. This is another variety cultivated in Sicily since the 19th century, when it was introduced by the Botanic Garden of Palermo. The fruits are oblate, with a yellow-orange peel and ripen from November to January.

Tardivo di Ciaculli. This is an important commercial variety in Italy, being a bud mutation of '*Avana'*, described in 1954 [7]. It is juicy with a fragrant flavor, ripening in February-March.

Hybrids. Hybrids resembling mandarins were introduced into the Mediterranean *Region* to extend the harvest period by producing late maturing fruit. The international citrus market demands late maturing fruit because Clementine varieties are produced only until February. Most of the cultivated hybrids are self-incompatible but possess viable pollen and ovules. This has caused serious problems in the introduction of such genotypes into the Mediterranean citriculture. Inter-pollination with Clementine may produce seedy fruit in mixed plantings, especially if pollinating insects are present. For this reason the principal goal for citrus breeding programs is to obtain new seedless and late maturing genotypes.

Ellendale (*C. reticulata* x *C. sinensis*)**.** This is a natural hybrid discovered in Australia, where it attains maximum fruit size and has excellent quality. In the Mediterranean *Region* it is less productive but maintains its quality. Fruit can be stored on the tree for an extended period, to be harvested in February-March.

Encore (*C. nobilis* x *C. deliciosa*). This Californian hybrid produces seedy fruit with a yellow-orange peel that completely ripens in March. Its flavor is similar to that of the common mandarin.

Fortune (*C. clementina* x *C. tangerina*). This hybrid was mainly cultivated in Spain because it ripens late (March-April) and due to its good organoleptic quality. Nowadays it has been replaced by other varieties because the fruits may contain seeds and because of its susceptibility to the fungus *Alternaria alternata* pv. *citri*.

Moncada (*C. clementina* x 'Kara' hybrid). A variety obtained in Spain in 1980. It produces mid-large sized fruit of good quality. The harvest period starts in January but the fruit does not lose its quality even if left on the tree for several months, to be harvested even in May or June. It has a slight tendency to alternate bearing.

Kara [*C. clementina* x (*C. unshiu* x *C. nobilis*)]. This late season fruit, available in March-April, has a rich and tart-sweet flavor and is slightly acid. It is a cross between *King* and *Satsuma Mandarin*, has seeds and is easy to peel. Nowadays its cultivation is decreasing but this genotype serves as a crossing parent in improvement programs.

Nadorcot ('Murcott' hybrid x unknown). Also known as *Afourer*, this variety is very productive, with excellent fruit quality. It is a late maturing genotype, whose harvest starts in February. *Nadorcot* was patented, which increases the costs for its propagation and cultivation.

Nova [*C. clementina* x (*C. paradisi* x *C. tangerina*)]. This variety has a sweet flavor and its peel has an attractive color. Commercially it is important in Spain and in Israel, where good quality fruit is obtained due to the prevailing climate. On the other hand it is susceptible to the *Alternaria* fungus and its harvest period overlaps that of other Clementine varieties (December-January).

Ortanique (unknown mandarin x *C. sinensis*). This is a late season ripening genotype that may be harvested from March to May and can be stored for a short time. The fruit is juicy with excellent organoleptic characteristics. Its main defects are difficult peeling and an excess of rind oil, which may modify its organoleptic qualities.

Others. Genotypes of interest in the Mediterranean *Region* are: *Basol*, *Beatriz*, *Capola*, *Clemenpons*, *Cultiford*, *Esbal*, *Orogros* y *Prenules*, *Primosole*, *Simeto*, *Spinoso* as early maturing varieties; other *mid-late season* varieties are *Mandalate*, *Mandared*, *Alkantara*, *Rubino*, *Guillermina*, *Orogrande*, *Caffin*, *Clementard*, *Tomatera*, *Larache*, *Cadoux*, *Muska*.

2.4. Other Species

Species cultivated only in some areas of the Mediterranean *Region* and are only of limited economic importance are included in this group.

Grapefruit (*Citrus paradisi* Macf.). This is mainly cultivated in Israel and Cyprus, but Spain (Valencia province) and Turkey also produce grapefruits. The principal varieties are *Duncan*, *Marsh*, *Red Blush*, *Star Ruby* (Jaffa Sunrise) and *Jaffa Sweetie*. Jaffa is the commercial brand used in Israel for citrus fruit, in particular for grapefruit.

Citrus medica Linnaeus. This is cultivated locally in Southern Italy (Calabria), where the most popular variety is *Diamante*. Bergamot (*C. bergamia* Risso) is also grown there for the production of essential oils (the cultivars *Castagnaro*, *Femminello* and *Fantastico*).

3. PRINCIPAL ROOTSTOCKS AND THE CHOICE OF GENOTYPES

The use of rootstocks is widespread in modern citriculture, because of the specific benefits accruing from each rootstock. Due to the fact that more than twenty separate traits can be affected by the genotype of the rootstock, its choice is basic for the performance of a citrus orchard. Rootstock genotype can modify tolerance, resistance or susceptibility to biotic and abiotic constrains, and also affects scion characteristics like vigor, productivity, fruit quality, plant longevity and others.

A major aspect to be considered in rootstock choice is its affinity with the various varieties; some scion-rootstock combinations are incompatible, which results in reduced crops. Scions are normally grafted on seedlings, which makes apomixis a very important trait for citrus rootstock genotypes. Apomixis enables practitioners to obtain many plants that are genetically identical to the seed source tree, thus yielding uniform rootstocks for propagation and planting. To improve rootstock propagation, it is also advantageous that the rootstock genotypes produce a large number of seeds. The resulting seedlings should grow quickly, be erect and lack ramifications, in order to facilitate variety graft.

Several diseases have strongly influenced the choice of rootstocks in the Mediterranean Region during last century. In 1884 a sudden outbreak of the disease Gomosis (caused by the Oomycetes fungus *Phytophthora*) infected many of the citrus trees in the Azores Islands. The disease quickly spread in Portugal and Spain, where it destroyed all citrus plants, except those grafted on sour orange (*Citrus aurantium*). From that time all citrus trees cultivated in the Mediterranean Region were grafted onto sour orange, because of its resistance to *Phytopthora*, better tolerance to moderate salinity, and excellent agronomic qualities. Between 1910 and 1930 a great number of new varieties were introduced into Europe from California, with the principal objective being to adapt citrus production to international demands. The limited sanitary inspection of the introduced varieties brought about the invasion of new pathogens and viruses, of which, first and foremost, the Citrus Tristeza Virus (CTV).

Although this virus was found to be widespread in the Mediterranean Region, it caused serious economic damage only in Spain and in Israel, where the sour orange rootstock had to been replaced by the citrange '*Carrizo'*. In other countries, like Greece, France, Italy, Egypt, Morocco, Tunisia and Algeria, CTV had been detected but sour orange continues to be the most common rootstock in citrus orchards. *Citrange Troyer* is only now becoming the prevalent rootstock for citrus in Italy, due to the fast and devastating spread of CTV in Southern Italy. Eradication strategies were implemented in that country to overcome the CTV problem. Monitoring for virus presence has now constantly to be conducted in Italy in order to prevent serious damage and a rootstock genotype change will be undertaken in the next years.

It is not possible to identify genotypes that will be usable and adaptable to all conditions. It follows that rootstock choice depends on the principal limiting factors in each area of cultivation: climate, soil, variety, diseases. Monoculture is always a risk for plant production, because infections by some pathogens, like CTV, could unexpectedly cause serious damage. The principal rootstock genotypes are described below:

Carrizo citrange. All citrange genotypes are hybrids between *Poncirus trifoliata* and *C. sinensis*. It is probably one of the most common rootstocks in the world and commonly used in Spanish citriculture, where it is used in 60 % of the new plantations. It is tolerant to CTV and to *Xyloporosis*, and induces very high productivity and good fruit quality. The negative traits of this genotype are sensitivity to the citrus Exocortis viroid, to calcareous soils and to salinity. *Citrange Troyer* has very similar characteristics.

***Sour orange* (*Citrus aurantium*).** A rootstock with excellent agronomic traits, which is resistance to *Phytophthora sp.*, *Armillaria mellea*, *Exocortis*, *Psoriasis* and to *Xyloporosis*. It is adapted for cultivation in saline soils and drought conditions. However, the rootstock's high sensitivity to CTV much reduces its utilization.

***Wester* (*Citrus macrophylla*).** A rootstock largely employed in lemon cultivation, but susceptibility to CTV limits its utilization with others *Citrus* species. On the other hand, *Wester* is particularly suited for early ripening varieties, a trait that allows its utilization with some oranges, mandarins and hybrids. It is very vigorous and productive.

Wester is resistant to *Phytophthora* and to salinity. Hindering its utilization are sensitivity to low temperature and reduced fruit quality.

Poncirus trifoliata. Trees grafted on this rootstock are resistant to low temperature and to wet and acidic soils. Its negative traits are sensitivity to calcareous soils and to salinity. *Poncirus trifoliata* is resistant to CTV, *Psoriasis, Xyloporosis, Phytopthora* and A*rmillaria mellea*. Budwood grafted on this rootstock has to be free from *Exocortis*, which can cause serious damage. This rootstock genotype induces good fruit quality and productivity.

Cleopatra mandarin (Citrus reshni). A rootstock that grows slowly in the nursery and delays the first production, and the tree size is not homogeneous. It is tolerant to CTV, *Exocortis*, *Psoriasis* and *Xyloporosis*. Tolerance to *Phytophthora* is unclear and inconstant. It is especially adapted to saline conditions and shows resistance to chlorosis.

Citrus volkameriana. This genotype has good tolerance to CTV and to *Exocortis*, but is susceptible to "woody gall", X*yloporosis, Phytophthora*, low temperatures and 'Mal secco'. It can be used in calcareous soils. It is a vigorous and very productive genotype, whose fruit quality is lower than that of other rootstocks.

Swingle citrumelo. This is a hybrid between *C. paradisi* x *P. trifoliata*. It is tolerant to CTV, *Exocortis*, *Psoriasis*, *Xyloporosis* and *Phytophthora* and is highly resistant to drought. However, it is sensitive to calcareous soils and salinity. This rootstock is vigorous and induces late ripening in the grafted varieties.

Others rootstocks used in the Mediterranean Region include hybrids *Forner-Alcaide n° 418* and *n° 5*, citrange *C35*, *Rough lemon* (*Citrus jambhiri*), *Sunki mandarin*, *Flying dragon*.

4. PLANT PROPAGATION AND SALINITY CONTROL

Plant propagation is critical when striving to obtain a profitable plantation. Plant homogeneity, vigor and sanility are direct consequences of nursery management and propagation strategies, strongly affecting crop longevity and fruit quality.

Commercial rootstocks are normally apomictic and can directly be obtained from seed cultivation. Nucellar embryony, although an obstacle in citrus breeding, provides a simple and inexpensive method to obtain clonal rootstocks. Furthermore, seedlings are free from pathogens, representing a basic condition for plant production.

Scion propagation is normally realized by grafting, which offers various advantages. Firstly, not all genotypes are apomictic and seedlings could result in highly heterozygotic plants. Seedling juvenility traits also delay fruit production and reduce profits. On the other hand, grafting represents the risk of disease propagation, so preventive measures have to be applied. Use of tolerant or resistant genotypes, exclusion of disease from citrus areas and plant certification are the only means to prevent epidemic infections. This is accomplished by the application of a specific program that is applied in citrus nurseries.

The first step in the implementation of a plant certification program involves the use of healthy plants to be used as principal sources of propagation. Different techniques are available in order to obtain pathogen-free plants, the most common being shoot tip grafting [8]; but thermotherapy and obtaining nucellar plants have also been applied in citriculture. Not all plants obtained with these techniques are completely virus-free, which necessitates systematic plant health control tests.

Pathogen-free plants constitute protected foundation blocks (or germplasm banks), where plants grow in containers within insect-proof screen houses. One or two plants per accession are usually maintained and they are usually under the responsibility of public agencies.

Buds from protected foundation blocks are used in private nurseries (or private associations) to make their own foundation blocks, normally from two to six mother trees per accession. These trees must periodically be inspected in order to ensure genetic trueness and plant health.

Private nurseries are in charge of making budwood multiplication blocks; buds from mother trees are grafted on vigorous rootstocks that allow an exponential increase in the amount of high quality budwood. Modern multiplication blocks are normally maintained in greenhouses but can also be established in the field; the former environment ensures better growth and reduces the risk of infections. Every three/five years these blocks must be completely renewed, in order to avoid bud sport mutations and diseases.

The buds thus obtained are grafted, creating blocks of certified nursery trees. Grafting is normally during August-September or in spring, using seedling rootstocks 30-40 cm high and 0.8-1.5 cm diameter. Trees are inspected to guarantee that they meet the horticultural quality required in the certification regulations and the plants can be sold by November.

Certification programs are usually operated by a provincial or state agency with legal authority to impose restrictions and to inspect all steps of the propagation process in private nurseries. The correct implementation of this program avoids the spread of diseases, guarantees genetic trueness and finally allows the maintenance of high productivity levels.

Sanitary control is not however efficiently realized in all Mediterranean countries. Spain can be considered as a model because the application of a certification program has enabled the total renewal of its citrus industry.

5. CULTURE MANAGEMENT

5.1. Planting Systems

The growing systems used by the citrus industry in the Mediterranean Region are quite different, and many typical, historical groves still exist for each species. In Italy, for instance, there are old lemon orchards, characteristic for the Peninsula of Amalfi and Sorrento. The lemon industry by-products, like the well known "*Limoncello*", depend on the extraordinary trees growing in the terraces facing the Gulf of Amalfi. Very old citrus orchards, 60-80 years old and with closely - spaced trees of *Avana* and *Tardivo di Ciaculli*, Mediterranean mandarins, surround the city of Palermo in Sicily. Old-fashioned along with young orange orchards, consisting of different *Tarocco* clones or '*Navel'* cultivars, grow at the foothills of the Etna volcano. Citriculture thus differs with the different species, in terms of tree spacing, training system, rootstocks and in orchard management. Old orchards came from trees grafted at the height of 0.80-1.00 cm, with densities of 700-900 trees ha-1; orchard layout nowadays depends on rootstock and scion vigor, both of which vary with the species. Vigorous cultivars can be placed 6-7 x 4-5 m apart, whereas smaller distances are often adopted for less vigorous, early ripening oranges or mandarins (6 x 3-4 m). Closer spacing can be used with dwarf rootstocks, such as 'Flying Dragon', though their commercial utilization is rare. Also, orchard layout depends on environmental conditions, which may or may not affect tree vigor, as well as on the personal choices of the farmers (for instance, a dynamic system with close spacing within the rows for the first 5-6 years after planting).

The vegetative growth threshold is 12°C for most species, whereas the optimum range is between 25°C and 30°C. It is well known that mandarins are the most tolerant to cold temperatures, whereas lemons are not. The fruit quality of grapefruits depends on the temperatures during ripening: cooler the climate, more acidic and sour the fruit. This explains why the best grapefruits in the Mediterranean Region can be found in its very mild winter areas. Citrus trees perform best on loamy soils, with a low active carbonate content (<7%), and pH values between 6.5 and 7.5 [9].

5.2. Nutrition and Fertilization

The ultimate goal of soil fertilization is to increase fertility and, in turn, plant production and fruit quality. However, if other limiting factors exist, *i.e.* water stress, the effect of an adequate nutritional plan may be rather low.

Data on the annual nutrient requirements needed to meet vegetative (new and secondary growth), root, and reproductive growth is critical to assess the timing and extent of any fertilization strategy, as well as the phenological stages of tree requirements.

Moreover, we should take into account: a) the efficacy of the fertilization methods; b) nutrient leaching and losses in the soil; c) the presence of natural vegetation in the orchard, which could play a negative or positive role in the annual balance; d) within-tree turnover of nutrients.

We should also know the composition of the irrigation water, which could considerably contribute to fertilization; the nutrient status in the soil and in the tree (foliar analysis) should also be measured periodically. Annual vegetative growth includes leaf turn-over, new sprouts and root growth.

As with many other fruit trees, the root system alternatively serves as a source during spring burst, flowering and fruit growth, and as a sink for nutrients if the root system, at the end of the annual cycle, does not recover its loss of nutrients. In such cases the plant may show different symptoms of starvation. Old leaves usually return part of their nutrient content to the tree before dropping, so that in the long run, their nutrient content is recycled into the soil.

Tables **1-3** present data concerning trees with optimum nutritional levels and with nitrate concentrations in the water not higher than 50 mg/l. The nutritional status for different citrus species is shown in Table **4**.

Table 1. N supply, through drip irrigation, to different citrus species in Spain [11].

		Maximum dosage (g tree-1)						
Year after planting/		1	3	5	7	9	11	Species
Tree spacing	Number of trees ha^{-1}							
4.0 x 2.0	1250	20	60	100	140	190	190	Clausellina and Okitsu
5.0 x 3.5	570	30	90	160	240	360	420	Marisol
5.5 x 4.0	454	40	120	210	310	450	530	Mandarins
6.0 x 4.0	416	45	135	250	380	530	580	Oranges
7.0 x 5.0	285	50	150	280	450	660	840	Lemon and Pomelo

Table 2. P_2O_5 supply, through drip irrigation, to different citrus species in Spain [11].

		Maximum dosage (g tree-1)						
Year after planting/		1	3	5	7	9	11	Species
Tree spacing	Number of trees ha^{-1}							
4.0 x 2.0	1250	5	15	25	40	55	55	Clausellina and Okitsu
5.0 x 3.5	570	10	25	45	70	105	125	Marisol
5.5 x 4.0	454	15	35	60	90	130	155	Mandarins
6.0 x 4.0	416	15	40	70	110	155	170	Oranges
7.0 x 5.0	285	15	45	80	130	190	245	Lemon and Pomelo

The distribution of nutrients in the trees proceeds from April to October, with differences related to the fruit ripening times of each variety. Half the N is delivered from May to July. In orchards with no drip irrigation, trees are N fertilized twice: at the end of winter and at the beginning of summer.

Table 3. K_2O supply, through drip irrigation, to different citrus species in Spain [11].

Maximum dosage (g tree-1)								
Year after planting/		1	3	5	7	9	11	Species
Tree spacing	Number of trees ha^{-1}							
4.0 x 2.0	1250	10	30	50	80	110	110	Clausellina and Okitsu
.0 x 3.5	570	20	50	90	140	210	245	Marisol
5.5 x 4.0	454	30	70	120	180	260	310	Mandarins
6.0 x 4.0	416	30	80	140	220	310	335	Oranges
7.0 x 5.0	285	30	90	160	260	380	490	Lemon and Pomelo

Table 4. Nutrient levels by foliar analysis of different citrus species^{Z}.

		% Dry weight				
		Very low	**Low**	**Normal**	**High**	**Very high**
Orange	N	< 2.30	2.30-2.50	2.51-2.80	2.81-3.00	> 3.00
	P	< 0.10	0.10-0.12	0.13-0.16	0.17-0.20	> 0.20
	K	< 0.50	0.50-0.70	0.71-1.00	1.01-1.30	> 1.30
Clementine	N	< 2.20	2.20-2.40	2.41-2.70	2.71-2.90	> 2.90
	P	< 0.09	0.09-0.11	0.12-0.15	0.16-0.19	> 0.19
	K	< 0.50	0.50-0.70	0.71-1.00	1.01-1.30	> 1.30
Satsuma	N	< 2.40	2.40-2.60	2.61-2.90	2.91-3.10	> 3.10
	P	< 0.10	0.10-0.12	0.13-0.16	0.17-0.20	> 0.20
	K	< 0.40	0.40-0.60	0.61-0.90	0.91-1.15	> 1.15

z: measured on 7-9 month-old leaves in twigs with no fruit.

5.3. Irrigation Systems and Water Management

All cultivated *Citrus* spp. had originated in the humid tropical and subtropical areas of eastern Asia, where water shortage is occasional. In the Mediterranean Region an extended period of water shortage is associated with high temperatures, creating a water deficit that may last, in the warmest areas, from March through October. It should be borne in mind that citrus trees are evergreens; in other words, they need water all the year round.

Moreover, citrus trees have a very long fruit development period. Orange trees bear their fruit from 6 months (early ripening cultivars) to 12 months (late cultivars); lemon trees bear fruit at different growth stages all the year around, depending on the bloom period from which they had originated; and mandarins and their hybrids bear their fruit during 6-10 months.

The root systems of most rootstocks are rather superficial, being mostly 10-70 cm below ground. These roots need relatively high soil water potentials, and cannot afford water potentials in the soil >80 kPa [10].

The seasonal requirement of irrigation water changes with environmental conditions, depending on soil characteristics and on the specific evapotranspiration demands. In the warmest areas of North Africa and the Negev, with prevailing sandy soils, it may last for 12 months, with a seasonal water requirement of 60-80 mm, whereas in

Southern Italy, the irrigation season usually begins in April-May and ends in October, with a seasonal requirement of 30-40 mm irrigation water [10].

Water deficit, particularly during the early stages of fruit set and development, reduces fruit growth and increases its drop, delays peel pigmentation, and affects fruit quality in term of size, juiciness, increase of peel thickness, ripening time, and may even contribute to alternate bearing. An appropriate management of water deficit in lemons, during early summer, was utilized in Sicily to get a July summer bloom, resulting in the *Femminello* fruit, known as 'verdelli' (green, small fruit with a very acid juice), ripening 12 months later. On the other hand, excess of water may delay fruit ripening, reduce fruit growth in terms of sugar/organic acid ratio, and cause root rot.

Citrus may be cultivated in a wide range of soils [9], provided they are not too heavy and salty, and the rootstocks differ in their suitability to water logging. For example the major key to the success of *C. aurantium* as the main rootstock in Italy was its suitability to the submersion method of irrigation, particularly in clay soils, where other rootstocks could be affected by *Phytophthora.*

5.4. Management of Soil and Ground Cover

Management of the orchard floor and its ground cover has been changing, along with planting systems and irrigation methods. In intensive orchards, which are irrigated by the submersion method, floor management included the preparation of a basin around each tree and the continuous and complete removal of any spontaneous vegetation. The reduction of the wetted area, resulting from modern localized irrigation methods, led to a different strategy of floor management. This includes: a) reduction of mechanical weeding, thus reducing soil erosion; b) allowing a winter soil cover and c) maintaining or increasing soil organic matter content. In traditional citrus farming the soil was annually mechanically tilled 3-6 times, and the concept of keeping the soil permanently free from natural vegetation was neglected. In most cases, the presence of a winter cover, such as *Oxalis cernua* Thumb, is beneficial. This plant reduces soil erosion during the rainy period, brings about better nutrient circulation within the orchard during the winter harvest (of oranges and mandarins), soundly contributing to the C balance in the orchard ecosystem.

Spontaneous vegetation that develops during early spring must be removed during late spring and left on the soil as mulching. During summer, the absence of rains and the localized irrigation do not allow any further growth of a spontaneous ground cover, and thus there is no need for any soil management. In mature groves, the shady environment provided by large tree canopies further reduces the growth of a spontaneous flora, even in wet areas.

In organic orchards, the spontaneous flora can be replaced by an artificial cover made of legumes such as *Trifolium* spp. or *Vicia* spp., which account for a large contribution to soil fertility. In young orchards, chemical control of the spontaneous flora is achieved with the herbicide Glyphosate, which is relatively common in many areas, and rapidly decays into the soil. Chemical control of the spontaneous flora reduces the cost of floor management but may increase erosion and decrease soil fertility.

5.5. Harvest and Fruit Quality

The citrus fruit is a berry called hesperidium. The pericarp (flavedo) must be smooth, uniform in color (mostly due to carotenoids), relatively thin and easy to separate from the segments. The peel is rich in pigments, essential oils and vitamins; the spongy, inner part of the peel (albedo) is rich in pectins, cellulose, carotenoids and flavonoids (hesperidin in oranges and naringin in grapefruits) and should not be thick and linked to the segments. The edible part of the fruit, the segments, develops from carpellar leaves; they are composed of multicellular vesicles filled with water and rich in organic acids (mostly citric acid), reduced sugars, pigments and vitamins as well as aromatic components. The stalk, which is surrounded by the segments, is present in fresh fruit and disappears after their senescence [12]. Navel fruits contain a secondary fruit at their stylar end, which is covered by the epicarp of the main fruit. Most commercial citrus fruits are seedless due to parthenocarpy, but many mandarins, particularly in the Mediterranean Region, and lemons, contain seeds. Citrus fruits are not climacteric and have low respiration rates. This means they must be tree-ripe when harvested, in terms of juice components, because the epicarp can be changed by an exogenous ethylene supply. The most important index for the harvest is the ratio between total soluble solids and organic acids content (TSS/TTA), which must be attained on the tree. Harvest and quality parameters are specific for each species and cultivar. Many orange, mandarin and grapefruit cultivars incur substantial fruit drop before harvest. Overripe fruit show a sharp decrease in peel brightness, consistence and essential oil content, accompanied by a decrease in TTA in the juice, whilst TSS decreases at a slower rate. Fruit consistence and juiciness also decline due to transpiration. Commercial parameters for fruit quality include size, shape, minimal TSS/TTA and juice content (%), and an acceptable color. These traits change with species, cultivar and country.

One of the most important problems in maintaining citrus fruit quality is its variability within the tree, within the orchard, and between seasons. This is particularly true for mandarins that may show alternate or biannual bearing, with small fruits in the 'on' years. Prevailing light regimes and temperatures also affect fruit quality. Red orange cultivars, like *'Tarocco'*, show an environmental effect in the extent of their pigmentation, because the development of the pigments (anthocyanins) that are dissolved in the cell sap is greatly affected by the prevailing temperatures, being higher in cool but not cold climates. Citrus fruits are rich in ascorbic acid, both in the epicarp and the juice (40-60 mg per 100 cc^{-1}). Navel oranges may greatly differ in peel thickness and smoothness, being affected by poor irrigation and fertilization management.

5.6. Training Systems and Pruning

Traditional training systems were used in high density orchards, with trees trained to the form of a globe, consisting of a main trunk with 3-5 primary scaffold branches placed at a height of 1.2-1.5 m from the soil. The canopy was pruned to avoid any proximity with the soil. Nowadays trees are also trained to the form of a globe, but with the growth almost reaching the soil and the trees left unpruned for 3-4 year after planting, till they reach a bushy shape. Vertical branches and vigorous suckers growing in the inner parts of the canopy are shortened. Trees are pruned every second or third year, except mandarins that usually require annual pruning. Pruning is carried out at the end of winter, in April-March, after harvest or during summer for late ripening orange cultivars. Mechanical pruning is scarcely used in the Mediterranean area.

6. FUTURE PERSPECTIVES AND GENETIC IMPROVEMENT FOR THE MEDITERRANEAN CITRICULTURE

The increase of citrus world production was relatively constant during the last decades of the 20th century. Citrus fruits are consumed mainly in the developed countries, although the consumption *per capita* is also increasing in developing countries. Most of the citrus fruit (62% in 2003) is imported into the Northern hemisphere.

Citrus fruit production in the Mediterranean Region is primarily intended at the fresh market. The Mediterranean area is the major exporter of fresh citrus, with 60% of the volume. The European countries are the principal destinations of fresh citrus exports from the Mediterranean Region. Southern hemisphere countries, such as Argentina, Australia and South Africa, are increasing their exports by providing out-of-season citrus fruit to the Northern market.

The European Union is also the largest orange juice importer, with more than 80% of world imports. Transformed citrus is a very limited part of the total production of the Mediterranean countries and has not much increased during last years, due to competition from the USA and Brazil on the international market. Only blood oranges are considered important in Italian citriculture, because this transformed product is not produced elsewhere while there is a great demand for it.

Latin American countries obtain good production results with low costs; for this reason different objectives must be realized in order to keep the hegemony of Mediterranean countries in the international fresh fruit market.

Reducing production costs along with increasing productivity are considered to be the basic steps for the future in Mediterranean citriculture. Competitiveness could be maintained with the mechanization of handling practices, like harvesting and pruning. Modernization of harvesting, pruning, fertilization, irrigation and other treatments, as well as post-harvest techniques, is a way to improve production and to get better fruit quality. Genetic improvement should also be developed in relation with specific market demands and environmental conditions (biotic and abiotic constraints). Fruit quality and the extension of the harvesting season are the main selection objectives, although such specific goals of breeding vary dramatically between the producing countries.

New genotypes should come into production quickly, and demonstrate high and constant productivity. The international market demands seedless fruit with good juice content, good organoleptic qualities (flavor, taste, sugar, acidity, etc), easy peeling and with an attractive appearance (size and color are very important in this context). Nutritional quality based on vitamin C, carotenoid and polyphenol contents are now considered as breeding criteria in some improvement programs. Finally, disease resistance that is related to scion genotype has to be considered in genetic improvement programs. In the Mediterranean Region "Mal secco" is probably the most important disease affecting lemons, and *Alternaria* is a problem for mandarin cultivars such as *Fortune* in Spain.

Rootstocks are used to improve genotype adaptations to soil and environmental conditions and to increase tree tolerance to pathogens. Thus the most widespread needs in the Mediterranean Region would be tolerance to pathogens like *Phytopthora*, CTV, "Mal secco" and to nematodes (mainly *Tylenchulus semipenetrans* Cobb). Related to these environmental constraints for the Mediterranean citriculture are tolerance to salinity, to iron chlorosis in calcareous soils and to scarce water availability.

Most of selected genotypes originated from spontaneous mutations, indicating the great importance of monitoring existing genotypes in modern citriculture. Mutations can be induced in order to produce limited changes in the original genotypes [13, 14]. This strategy was successfully applied in Israel and in the USA, where new seedless mutants mandarines were recently obtained (*e.g.* cultivars *Mor* and *Tango*).

Sexual breeding programs have a limited success in citrus genetic improvement because of the complex reproductive biology of these species (apomixis, sterility, incompatibility, juvenility, high heterozygosis levels). New rootstock genotypes were recently obtained by sexual intergeneric hybridization between citrus species and *Poncirus* [15] or by somatic hybridization [16], but it is more difficult to obtain new scion genotypes.

The selection of triploid lines is nowadays a breeding strategy of great promise, intended to obtain new seedless cultivars, as triploids are generally female as well as male sterile [17]. Several methods have been developed for triploid citrus creation, the most suitable are 2x X 2x crosses or interploidy crosses (4x X 2x or 2x X 4x). Biotechnology has further facilitated the selection of new triploid genotypes by: 1) flow cytometry facilitates the analysis of great plant progenies; 2) *in vitro* embryo rescue improves the efficiency of the triploid obtained; 3) techniques like somatic hybridization and "colchicine treatment of shoot tips grafted *in vitro* [18, 19] which increase the availability of new tetraploid parents. Promising triploid genotypes were obtained in Italy (*Mandalate* and *Tacle*) [20] and in Spain (*Garbí, Safor y Avasa pri-43*) [21].

REFERENCES

[1] Tolkowsky S. Hesperides: a history of the Culture and Use of Citrus fruit. London: John Bale, Sons and Curnow Ltd 1938.

[2] Tanaka T. Species problem in Citrus (Revisio aurantiacearum, IX). Tokyo: Jpn. Soc. Promo. Sci Veno 1954.

[3] Gmitter Jr FG, Hu X. The possible role of Yunan, China, in the origin of contemporary citrus species (Rutaceae). Econ Bot 1990; 44: 267-77.

[4] Zaragoza S. Aproximación a la historia de los cítricos. Origen, dispersión y evolución de su uso y cultivo. PHD Tesis. Valencia: Universidad Politécnica de Valencia 2007.

[5] Aubert B. Genèse et développement de la culture des agrumes et patrimoine génétique méditerranéens de l'histoire naturelle des orangers. Dans la réédition de l'Histoire Naturelle des Orangers. Paris: Risseau et Poiteau. Tome 2. Connaissance et mémoires éditeur 2001.

[6] Crescimanno FG. Arancio Washington Navel: storia e valore colturale. Giornale di Agricoltura, REDA 1978; 33: 41.

[7] Crescimanno FG. Il mandarino 'Tardivo di Ciaculli'. Rivista di Ortoflorofrutticoltura Italiana 1954; (79) 5-6: 181-8.

[8] Navarro L. In: Bajaj JPS, Ed. Biotechnology in Agriculture and Forestry. Vol. 18. High-Tech and Micropropagation. Berlin: Springer-Verlag 1992: pp. 327-38.

[9] Calabrese F, Dazzi C. In: Costantini EC, Ed. Manual of Methods for Soil and Land Evaluation. Plymouth: Science Publishers 2009; pp. 341-52.

[10] Sardo IV. In: Vacante V, Calabrese F., Eds. Citrus, Trattato di agrumicoltura. Bologna: Edagricole, Il sole 24 Ore Editoria specializzata 2009; pp. 245-58.

[11] Quiñones OA, Martínez-Alcántara B, Primo-Millo E, Legaz F. Fertilización nitrogenada, fosfórica y potásica de los cítricos: cálculo de las dosis fertilizantes de cada elemento y correciones necesarias en riego por goteo. Vida rural, ISSN 1133-8938, N° 237, 2006; 44-8.

[12] Monselise SP. In: Monselise SP, Ed. Handbook of fruit set and development. Boca Raton: CRC Press Inc. 1986; pp. 87-117.

[13] Hearn CJ. Development of seedless orange, *Citrus sinensis*, cultivar Pineapple and grapefruit, *Citrus paradisi*, cultivars through seed irradiation. J Am Soc Hort Sci 1984; 109: 270-3.

[14] Roose ML, Williams E. In: Khan IA, Ed. Citrus genetics, breeding and biotechnology. Wallingford: CAB International 2007; pp. 345-52.

[15] Forner JB, Forner-Giner MA, Alcaide A. Forner-Alcaide 5 and Forner-Alcaide 13: Two new citrus rootstocks released in Spain 2003; 38: 629-30.

[16] Grosser JW, Ollitrault P, Olivares-Fuster O. Somatic hybridization in Citrus : an effective tool to facilitate variety improvement. *In Vitro* Cell. Dev. Biol - Plant 2000; 36: 434-49.

[17] Ollitrault P, Dambier D, Luro F, Froelicher Y. Ploidy manipulation for breeding seedless triploid citrus. Plant Breeding Review 2008; 30: 323-52.

[18] Juárez J, Aleza P, Olivares-Fuster O, Navarro L. Recovery of tetraploid clementine plants (*Ctrus clementina* hort. ex Tan.) by *in vitro* colchicine treatment of shoot tips. Proc. Int. Soc. Citriculture 2004; 1: 151-4.

[19] Navarro L, Juarez J. In: Khan IA, Ed. Citrus genetics, breeding and biotechnology. Wallingford: CAB International 2007; pp. 353-64.

[20] Starrantino A. Tacle, a new triploid clementine x Tarocco hybrid. Rivista di Frutticoltura e di Ortofloricoltura 1999; 61(1): 45-7.

[21] Navarro L, Juarez J, Aleza P, *et al.* Selección de nuevos mandarinos triploides. Agraria Comunitat Valenciana. 2ª época, año 2, 2006; 7: 23–6.

CHAPTER 2

The History of IPM in the Mediterranean Citriculture

Vincenzo Vacante*

Dipartimento Patrimonio Architettonico ed Urbanistico (PAU), Mediterranean University of Reggio Calabria, Via Melissari, Località Feo di Vito, 89060 Reggio Calabria, Italy

Abstract: The origins of biological control, the influence of its early successes, and relevant observations and experiments conducted in the late 19th and early 20th centuries in the Californian citrus industry are briefly presented. These practices, along with the growing disappointment with chemical control, naturally led to the move from strict biological control to the concept of Integrated Pest Management (IPM). The history of IPM in the Mediterranean citriculture throughout that period is presented against this background, and the most important projects of classic and augmentative biological control carried out in the region are discussed. Brief information on the role of international institutions (IOBC, EU, FAO) and the scientific community in the development of IPM is also provided.

Keywords: Citriculture, Mediterranean Region, citrus pests, biological control, IPM.

1. INTRODUCTION

According to Rosen [1] "biological control is not just another tactic of pest control and the utilization of natural enemies should be regarded as the backbone of any IPM program in citriculture". From this point of view the history of Integrated Pest Management (IPM) largely coincides with that of biological control and illustrates the steps that marked the development of this method in the citrus industry of the Mediterranean Region. It shows the impact of the major experiences of biological control conducted in the Region and in other countries since the end of the 19th century.

The use of natural enemies against agricultural pests, documented since AD 300, began to be interpreted with scientific rigor in 1700 and more fully since the 19th century [2]. The history of human populations in the latter period was accompanied with a gradual improvement of agricultural production, associated with technological innovation that resulted from the industrial revolution. However, the advantages of increased production are associated with an increase of the ecoresistance of agroecosystems. This lead to huge increases in the populations of harmful organisms (pests and pathogens) as well as of competitors (weeds), against which we did not have adequate means of control. The increased trading opportunities between continents increased the risk of the accidental introduction of animals and plants, and therefore of crop and forest pests, and substantially aggravated the problem [3]. To cope with this ever-increasing problem, the entomological community of that time encouraged the recognition of the irreplaceable role played by natural enemies (parasitoids, predators, pathogens), and encouraged their use [4].

With these premises the American entomologist C.V. Riley, between 1888 and 1889, realized, in the citrus orchards of California, the first application of classic biological control in the modern era, by introducing the vedalia beetle (or cardinal ladybird), *Rodolia cardinalis* (Mulsant) from Australia. The beetle is a natural enemy of the cottony cushion scale, *Icerya purchasi* Maskell, and its use brought about, within two years, an almost total reduction of scale infestations in Californian citrus orchards. The entire project cost approximately $5,000, whereas the gain to the citrus industry was annually estimated at millions of dollars. Success with *R. cardinalis* in controlling the pest was repeated in over 50 countries [5].

The first demonstration of augmentative biological control was also conducted in California, where around 1917 an insectary was set up for mass-producing the predator, *Cryptolaemus montrouzieri* (Mulsant), a natural enemy of the citrophilous mealybug, *Pseudococcus calceolariae* Maskell.

In 1919 the American H.S. Smith first coined the method of using living organisms to control pests with the term "biological control", within the context of phytosanitary strategy. The success of this approach encouraged its

***Address correspondence to Vincenzo Vacante:** Dipartimento Patrimonio Architettonico ed Urbanistico (PAU), Mediterranean University of Reggio Calabria, Via Melissari, Località Feo di Vito, 89060 Reggio Calabria, Italy; Tel: + 39-0965-385201; Fax: +39-0965-385219; E-mail: vvacante@unirc.it

Vincenzo Vacante and Uri Gerson (Eds)

application, so that by 1930 there were 16 insectaries capable of annually producing about 20 million individual natural enemies [6-8], and approximately 40 million by 1946 [9]. In the following years a general reduction in the number of insectaries took place; by 1963 only three insectaries were present in California, annually producing about 30 million individuals [10]. Augmentative biological control and the mass production of natural enemies became integral parts in the control strategy of citrus pests in California at the time. The insectaries, representing an operative tool, were transferred to the management of grower cooperatives [11] or commercial companies [12]. Many insectaries were initially built by Regional governmental organizations (*e.g.* county agricultural commissions, *etc.*) that offered the land, supervised the building, organized the facilities, produced and distributed the natural enemies. From the economic point of view, the costs of the facilities and of the natural enemies were supported by the citrus industry of a given area, through a system of taxes that was substantially related to production [7]. Currently the Californian producers are private or grower-owned insectaries [11], associated with licensed pest-control advisers who help the growers in implementing IPM programs [13]. For a detailed discussion of the issue see Luck and Forster [14].

Since then the Californian experience had a resounding and rapid echo in the international scientific community and has stimulated several programs of biological control of citrus pests (and other crops) in the Mediterranean Region. The fundamental steps in the history of the method and the contribution of the international institutions are presented for this Region, focusing on the fundamental experiences of classical and augmentative biological control that has allowed the creation of the most important IPM programs.

2. CLASSIC BIOLOGICAL CONTROL PROJECTS

The main classic biological control projects implemented in the Mediterranean Region during a century or so are listed in Tables **1** and **2** [15]. At least half of the 26 natural enemies introduced during approximately one century are still present today. Many projects were successfully concluded in a relatively short time due to the assistance provided by the Experimental Station of Citrus, (later the Division of Biological Control) of the Department of Entomology University of California, Riverside (USA). Their personnel oversaw the introduction and acclimatization of several natural enemies in Californian citrus orchards and readily made available different species for introduction into several European countries. These opportunities were not overlooked by F. Silvestri [16, 17], a world-renowned entomologist working in Portici (Naples, Italy) from 1903 to 1949. During half a century he introduced more than 50 entomophagous species from the USA and other countries around the world into the Mediterranean Region, where they became acclimatized [18]. As shown in Tables **1** and **2**, the greatest efforts were made in introducing the natural enemies of scale insects. This reflects the significant damage that this group of pests causes to citrus, together with, in different times, natural enemies of other insect pests, such as the citrus leafminer [19] or mites [20, 21], *etc.*

Table 1. Entomophagous insects of various citrus pests introduced into the Mediterranean Region by classical biological control programs [15].

Entomophages	Target Pest	First Use	Countries	Introduction	Origin
Cryptolaemus montrouzieri	*Planococcus citri*	1908	Italy (E; Sardinia P; Sicily E), Israel (P), Portugal, Greece (F), Cyprus (T), France (P), Spain (P), former USSR (Georgia) (F?) (Mediterranean distribution)	Single/Multiple	Australia
Rhyzobius lophantae	Diaspididae	1908	Italy, Spain	Single	Australia
Rhyzobius forestieri	*Saissetia oleae*	1981	Italy (S), France (S), Greece (E), Cyprus (N), Israel (E)	Single	Australia
Rodolia cardinalis	*Icerya purchasi*	1897	Portugal (C), former USSR (Georgia) (C), Italy (S), former Yugoslavia (N), Israel (C*), France (C), Spain (C), Switzerland (F), Greece (S), Malta (C), Cyprus (S) (Mediterranean distribution, CIS)	Single/Multiple	Australia
Scymnus reunioni	*Planococcus citri*	1967	Israel (N), Italy (Sardegna) (P), former USSR (Georgia) (F?)	Single	India
Serangium parcesetosum	*Dialeurodes citri*	1973	former USSR (Georgia, C*; Azerbaijan, C*; Uzbekistan, F), France (Corsica) (E), Israel, Turkey (S?)	Single	India, former USSR

Table 1. cont….

Cryptochetum iceryae	*Icerya purchasi*	1987	Israel (S)	Single	Australia
Amitus spiniferus	*Aleurothrixus floccosus*	1971	France (C*), Italy (S*)	Single	Central America
Aphytis holoxanthus	*Chrysomphalus aonidum*	1956	Israel (C*) (Mediterranean distribution)	Single	Hong Kong
Aphytis lepidosaphes	*Cornuaspis beckii*	1956	Israel (C), Cyprus (N), France (N), Greece (S), Spain (S), Italy (Sicily) (P)	Single	China
Aphytis lingnanensis	*Aonidiella aurantii*	1960	Cyprus (P), Israel (E), Italy (Sicily), Morocco (E), Spain (P) (Mediterranean distribution)	Single	China
Aphytis melinus	*Chrysomphalus dictyospermi*	1962	Greece (S; Kriti, C), Italy (C), Morocco (C), France (Corse) (P), Spain (P) (commercially available for inundative releases since 2008), former USSR (Georgia) (E?) (Mediterranean distribution)	?	India, Pakistan
	Aonidiella aurantii	1961	Cyprus (P), Israel (P), Italy (Sicily) (P), Morocco (E) (Mediterranean distribution)	Single/Multiple	India, Pakistan
Cales noacki	*Aleurothrixus floccosus*	1970	Spain (S), France (C), Italy (S), Morocco (C), Portugal (E), Tunisia (C), Malta, Greece (Mediterranean distribution)	Single/Multiple	Chile
Encarsia herndoni	*Lepidosaphes gloverii*	1979	Italy, Spain, France (Corsica)	Single	East Asia
Encarsia lahorensis	*Dialeurodes citri*	1973	Italy (S; Sardinia, P; Sicily, C), former USSR (Georgia) (C*), France (E), Turkey, Greece (S, E), Israel (C*)	Single/Multiple	India, Pakistan
Eretmocerus debachi	*Parabemisia myricae*	1992	Israel (C), Turkey (C), Italy (S) (Mediterranean distribution?)	Single	Japan, North America
Pteroptrix smithi	*Chrysomphalus aonidum*	1956	Israel (C*)	Single	Hong Kong

Legend: C, complete; S, substantial; P, partial; E, established but not contributing to control or status unknown; F, failed to become established; N, no information on outcome; T, established but believed to have died out; *, cases where more than one organism contributed to the result.

Table 2. Entomophagous insects introduced into the Mediterranean Region by classical biological control programs of various citrus pests [15].

Entomophages	Target Pest	First Use	Countries	Introduction	Origin
Ageniaspis citricola	*Phyllocnistis citrella*	1994	Israel, Morocco, Algeria, Tunisia, France, Greece, Cyprus, Spain (C), Italy (Sicily)	Single/Multiple	Thailand, Florida
Clausenia purpurea	*Pseudococcus citriculus*	1940	Israel (C)	Single	Japan
Comperiella bifasciata	*Aonidiella aurantii*	1924	Israel (P*), Italy, France (N), Spain (P) (Mediterranean distribution)	Single	South China, California ?
Metaphycus anneckei	*Saissetia oleae*	2000	Greece (C), Israel (E), France (E), Italy (F), Egypt, Cyprus	Single	South Africa
Metaphycus flavus	*Coccus hesperidum*	1959	Italy (P), former USSR (Ukraine) (C) (Mediterranean distribution)	Single	Morocco
Metaphycus helvolus	*Saissetia oleae*	1968	Israel (E), France (Corsica) (P*), Greece (C; Kriti, S), Italy (P), Spain (S), Cyprus (E), former USSR (F)	Single	South Africa
Metaphycus lounsburyi	*Saissetia oleae*	1971	France (P), Israel (C), Greece (Kriti) (P), Italy (E), Cyprus (S) (Mediterranean distribution)	Single	California, Australia, Hawaii, South Africa
Metaphycus swirskii	*Saissetia oleae*	1973	Israel (E), France (E), Greece (Kriti) (P), Italy (P)	Single	Kenya
Neodryinus typhlocybae	*Metcalfa pruinosa*	1989	Italy, France, Slovenia, Switzerland	Single	USA

Legend: C, complete; S, substantial; P, partial; E, established but not contributing to control or status unknown; F, failed to become established; N, no information on outcome; T, established but believed to have died out; *, cases where more than one organism contributed to the result.

In addition to the species reported, there are several parasitoids and predators associated with naturally-occurring biological control (the so-called "natural biological control"), which are essential for the maintenance of equilibrium conditions. Relevant details can be found for the northern Mediterranean countries in specialist contributions [22, 23] or in practical handbooks [24].

3. AUGMENTATIVE BIOLOGICAL CONTROL EXPERIENCES

Table **3** [15] lists mass-produced insects intended for the citrus industry in Italy, Spain and Israel (in addition to insectaries that were built for scientific research in several other countries).

Table 3. Entomophagous insects commercialized in the Mediterranean Region and employed in augmentative biological control programs of various citrus pests [15].

Species	Target Pest	First Use	Distribution	Introduction	Origin
Chilocorus bipustulatus	Diaspididae, Coccidae (*Saissetia oleae*)	1959	Widespread in the EPPO Region	Single	South Palaearctic
Coccinella septempunctata	Aphididae	1980	Widespread in the EPPO Region	Single	Palaeartic
Cryptolaemus montrouzieri	*Planococcus citri*	1985	Mediterranean	Single/Multiple	Australia
Rodolia cardinalis	*Icerya purchasi*	1980?	Mediterranean, CIS	Single/Multiple	Australia
Anagyrus pseudococci	Pseudococcidae	1995	Mediterranean	Single/Multiple	Mediterranean
Comperiella bifasciata	Diaspididae (*Chrysomphalus aonidum, Aonidiella aurantii*)	1985	Mediterranean	Single/Multiple	South China, California ?
Leptomastix dactylopii	*Planococcus citri*	1992	Mediterranean	Single/Multiple	Neotropic
Metaphycus helvolus	Coccidae (*Saissetia oleae, Coccus hesperidum*)	1992	Greece, Italy	Single	South Africa
Metaphycus lounsburyi	Coccidae (*Saissetia oleae*)	1997	Mediterranean	Single	California, Australia, Hawaii, South Africa
Microterys nietneri	Coccidae (*Saissetia oleae*)	1987	Italy	Single	California, Pakistan
Tetracnemoidea peregrina	Pseudococcidae	1992	Israel	Single	North America
Scutellista caerulea	Coccidae (*Saissetia oleae, S. coffeae, Ceroplastes rusci*)	1990	Mediterranean, CIS	Single	Africa
Aphytis holoxanthus	Diaspididae	1996	Mediterranean	Single/Multiple	Asia
Aphytis lingnanensis	*Aonidiella aurantii, Chrysomphalus dictyospermi*	1960	Mediterranean	Single/Multiple	China
Aphytis melinus	*Aonidiella aurantii*	1985	Mediterranean	Single/Multiple	India, Pakistan
Cales noacki	*Aleurothrixus floccosus*	1970	Mediterranean	Single/Multiple	Chile

In Italy a representative insectary is located at Ramacca (Catania, Sicily), built in the '90s with a financial contribution from the Sicily Region. This insectary is currently managed by the private sector and produces the aphelinid *Aphytis melinus* DeBach, the encyrtid *Leptomastix dactylopii* Howard and the coccinellid *Cryptolaemus montrouzieri*.

Approximately 60,000,000 adult *A. melinus*, considered to suffice for about 1,500 ha citrus orchards (coming to 40,000 individuals/ha) were produced in 2007 (personal communication). Furthermore, 1,000,000 *L. dactylopii* adults, which were released at a rate of 1,200 individuals/ha on about 800 ha of cultivated citrus area, along with 150,000 *C. montrouzieri* adults, released at a rate of 200 individuals/ha on about 750 ha of citrus groves were produced in the same insectary. To the latter should be added another 100,000 *C. montrouzieri* adults that were not produced in Sicily and applied, at a rate of 400/ ha on about 250 ha of cultivated citrus. The Italian National Agency for New Technologies, Energy and Sustainable Economic Development (ENEA) is planning to build an insectary for the mass production of the same entomophages in southern Italy (Lamezia Terme, Calabria) but at this date the structure has not yet been started. Other small insectaries operate in Pantanello (Matera, Basilicata) and Cagliari (Sardinia). The Company Bioplanet (Cesena, Emilia Romagna), which also operates in Italy, each year sells about 150,000 *C. mountrouzieri* individuals. A certain number of predators are occasionally imported from Australia. For a more detailed discussion of the Italian situation see Kapranas *et al.* [25].

From the available Italian data it appears that the production of entomophages is probably insufficient, and for other Mediterranean countries the mass production of natural enemies seems to be even more limited than the Italian.

Spanish insectaries are small and active insectaries, such as SPV (Servicio de Protección de los Vegetales) of Almassora in Castellón initiated by J.L. Ripolles in 1976 [26]; several commercial companies, including Biobest, which mass-produces entomophages, are present in the same region. The Israeli company Bio-Bee specializes in the mass production of different beneficials, including the encyrtid *Anagyrus pseudococci* (Girault), used against mealybugs, and *C. montrouzieri*.

In general, all companies engaged in the production and/or marketing of natural enemies (Koppert, Biobest, Syngenta Bioline, BCP Bioplanet) sell *C. montrouzieri* [25].

4. IPM EXPERIENCES

Following the first successful use of *R. cardinalis*, about 65 cases of complete control and 83 of substantial biological control of insect pests by the introduction of different entomophagous arthropods were reported from different area of the world during the next 70 years. Of these introductions, about 61% of the former and 22% of the latter were aimed at citrus pests [27]. However, natural enemies applied according to classical or augmentative biological control practices sometimes fail to maintain constant and adequate pest control. Thus the need arose to integrate the activities of natural enemies with other control means, including insecticides, in order to provide the necessary guarantees of eco-toxicological nature. This requirement was realized from the very first use of natural enemies and resulted in the first practical application of IPM in the late 19th and early 20th centuries.

The first practical experience of IPM in citrus was in 1922 in California, by founding the "Fillmore Citrus Protective District" by a group of citrus growers who intended to reduce the cost of defense against pests [11]. But IPM was formally defined in the '50s [28], concomitant with increasing attention from researchers and growers. Only in the '80s was the first protocol for implementing IPM in citrus published by the University of California [29]. The Texas Cooperative Extension Service [30], the University of California [31] and others. The University of Florida [32] subsequently published guidelines for the implementation of IPM in citrus orchards.

In the Mediterranean Region the lack of registered chemicals, until the end of World War II, led to the empirical use of control strategies similar to IPM, which was based on native and/or introduced natural enemies, on the use of petroleum oils, on the HCN fumigant and on other means of control [33]. The first true practical application of IPM in citrus orchards was carried out in 1954-57 in Israel by the Citrus Marketing Board of Israel, which encouraged by successful biological control of *Chrysomphalus aonidum* (Linnaeus), founded a biological control laboratory, now called as the Israel Cohen Institute for Biological Control. This laboratory was charged with the task of promoting biological control in citriculture and in the introduction of natural enemies for the most important citrus pests in Israel. Until 1966 over 45 species of natural enemies (predators and parasitoids) were introduced by this Institute [34]. A large project involving several countries in the Mediterranean Region was undertaken in the 1980's by the Commission of the European Communities (CEC), which during 1979-1981 financed a first program of integrated and biological control of citrus pests. The first protocols on sampling methods and intervention thresholds of the principal citrus pests of the Mediterranean Region were published at the same period [35]. The strategy was operational in Israel by 1986 [36] and in 2006 it was applied to the entire cultivated Region [37].

In the 1990's the EU regulated biological control with specific legislation (Annex II B, EEC Regulation 2092/91, repealed in June 2007 by Reg. No 834/2007) and more recently it defined IPM (Directive 2009/128/CE, 21/10/2009). The northern Mediterranean countries that belong to the EU (comprising about 60% of the entire Mediterranean citrus area) are encouraged to apply IPM by economic incentives to member states that implement the relevant legislation, which supports the extension and management of the method. The Spanish Ministry of "Agricultura, Pesca y Alimentación" thus indicated in the 1990's a protocol intended to implement IPM in citrus [38], which followed directions in the district of Tarragona [39]. In Greece Katsoyannos [23] indicated the lines of "Integrated Insect Pest Management for Citrus" for northern Mediterranean countries. In Italy, Viggiani [22] gives the same information and in addition to EU legislation, Regional protocols [40] are known along with those from private organizations [41]. As regards non-EU countries, in Israel the Ministry of Agriculture and Rural Development published "A Guide to the Control of Citrus Pests" [42]. In other Mediterranean countries the development of these operational tools is being launched or is in progress [37].

Data on the areas under IPM in the Mediterranean Region were not available until 1995 [43], and only in the last decade was a reversal of this trend seen. The last relevant census conducted in the Mediterranean Region [37] indicated that except Israel, where 100% of the cultivated citrus areas are under IPM programs, and Greece, where most of the cultivated area is under IPM, the remaining countries show rates ranging from 30% (Turkey) to less than 1% (France). In Italy 19,549 ha of organic citrus orchards were surveyed in 2006, representing (mainly in Sicily and Calabria) 11.3% of the total citrus cultivated area.

Of 789,433 ha of citrus area harvested in the Mediterranean during 2006, only 150,000 ha (18-19%) were under IPM programs. However, the use of IPM in citrus orchards is constantly increasing. In Spain, in the Community of Valencia in 1999, the citrus surface with IPM programs was about 2.4% of the total [44], increasing gradually over the last decade to 35% in 2009, with a forecast for 2009 of an integrated orange production of 750,000 tons, representing an increase of 54% over the volume achieved in 2008 [45].

Regardless of the value of the data provided, it seems that in the Mediterranean Region a progressive and innovative process of transforming the concepts of pest control and production of citrus (and other crops) is underway, due to the changed socio-cultural conditions and to a different political management.

5. THE ROLE OF THE INTERNATIONAL INSTITUTIONS AND OF THE SCIENTIFIC COMMUNITY

Among various international institutions, the IOBC (International Organization for Biological and Integrated Control of Noxious Animals and Plants) has an important role in developing world-wide IPM. This organization was founded in 1948 by eminent entomologists, including the Italian F. Silvestri of Portici University (Naples) and the French A. S. Balachowsky of the Pasteur Institute of Paris. The organization aims to promote the spread of biological control and IPM in the world. The world-wide dimensions of the problem required organizing the IOBC into 6 Regional Sections, corresponding to geography, including the West Paleartic Regional Section (WPRS) which encompasses the Mediterranean Region. Each section consists of various working groups, which include the group "Integrated Control in Citrus Fruit Crops". This group currently has over 40 researchers, and although their approach is multi-disciplinary, encompassing issues of plant pathology and weed science, the majority are entomologists. Reports from meetings convened in various countries are published in Bulletins of the various working groups, in order to encourage and facilitate the exchange of ideas amongst the different specialists. Since its foundation the IOBC has developed a privileged relationship with the Mediterranean citriculture. The implications and successes of IPM in the citrus industry of the Mediterranean Region were sadly almost completely ignored in the IOBC Internet Book of Biological Control [46]. This deprived the Mediterranean Community of knowledge about the history IPM and of the considerable benefits associated with its application. This omission did not, however, affect to any appreciable degree the role of IOBC in the development of citrus IPM in the Region.

The European Union (EU) has several times encouraged the application of biological control and IPM. The legislation for biological control, cited above, provides almost exclusively, only data on the use of substances of natural extraction, omitting any references to natural enemies of pests, whether indigenous and/or artificially introduced. The absence of restrictions on the use of natural enemies and the request to preserve and/or increase the indigenous biodiversity, do not however preclude their application on citrus (and on other crops). The rules on the application of IPM that have recently been issued (Annex III Directive 2009/128/CE, 21/10/2009) were inspired by a concrete practical vision. In the citrus industry, however, it is difficult to separate between biological control within IPM and the use of such important substances like petroleum, narrow range oils, natural pyrethrins and other "soft" pesticides. The two strategies are consistent with the above-noted EU legislation and with "agricultural production methods compatible with the requirements of environmental protection and the maintenance of the countryside" (Reg. 2078/92 and beyond). Currently there are no similar regulations for non-EU Mediterranean countries, but as they share in the technical-scientific knowledge, it is to be hoped that relevant policies will be advocated there for a common design on a Regional scale. Innovation would primarily be technical and should result in economic benefits, along with solutions to serious phytopathological problems that result from coordinated actions at the Regional level.

Several IPM programs have been, and continue to be, implemented in developing countries of the Near East Region, in collaboration with international and Regional organizations, such as the Food and Agriculture Organization of the United Nations (FAO), the United Nations Development Program (UNDP) and the International Centre for Advanced Mediterranean Agronomic Studies (CIHEAM), or directly through the national institutions in member countries. The different programs have been very successful, although there is still a long way to go. Examples include IPM programs for Cyprus, Egypt, and Syria on citrus and other fruit crops [47].

The scientific community, consisting of many researchers, research centers and universities, has a central role in the development of IPM in the Region. Among the positive aspects that characterize their work should be mentioned a predominant Mediterranean Regional vision, that overcame political and geographical constraints, and the refined development of IPM along with vigorous scientific analyses. As has taken place in other areas of applied research, IPM studies were supported by and enhanced the increasing use of modern research tools. These include biochemical and genetic analyses for the discrimination of species (beneficials and pests), of sophisticated statistical methods intended to study the bio-ecology of various species, of chemical and physical analyses in ethological and monitoring research and others. At the same time, many problems remain unsolved and the further development of IPM requires increased research. Relevant areas include adequate solutions to taxonomic problems, more bio-ecological data, better biotechnical means to combat pathogens, monitoring pest resistance, the mass production of natural enemies, establishing economic injury levels and economic thresholds for may pests and preventing the establishment of exotic new pests [25].

6. CONCLUSIONS

The data presented above show that the development of IPM in Mediterranean citrus orchards was in many ways, especially during the first half of last century, a direct consequence of the North American experience. The development of IPM in the Mediterranean Region has now continued for over a century, but had resulted only in a few production studies, probably due to the different ages of citrus plants. The political fragmentation of the areas involved and the different chronology of pest control activities point to the need for a common set of recommendations that would better harmonize Region-wide procedures. This indicates the need for joint action, including a common strategy intended to prevent the introduction of exotic pests into the Region and the realization of joint research and training, leading towards a set of uniform protocols for citrus pest IPM. This optimization requires a broader approach that has not yet been used.

REFERENCES

[1] Rosen D. Methodologies and strategies of pest control in citriculture. In: Cavalloro R, Di Martino E, Eds. Proceedings of the Experts' Meeting "Integrated pest control in citrus-groves"; 1985: Acireale: Balkema, Rotterdam 1986a; pp. 521-30.

[2] Vacante V. Il contributo delle scuole entomologiche italiane alle applicazioni di lotta biologica contro gli insetti dannosi all'agricoltura e alle foreste nel mezzogiorno. In: Scarascia Mugnozza GT, Ed. Atti Convegno "La scienza nel mezzogiorno dall'unità d'Italia ad oggi", Reggio Calabria 2009: Roma, 2010, pp. 171-84.

[3] Flint ML, van den Bosch R. Introduction to Integrated Pest Management. New York: Plenum Press 1981.

[4] Wilson F, Huffaker CB. The philosophy, scope and importance of biological control. In: Huffaker CB, Messenger PS, Eds. Theory and practice of biological control New York: Academic Press 1976: pp. 3-15.

[5] DeBach P, Rosen D. Biological Control by Natural Enemies, New York: Cambridge University Press 1991.

[6] Smith HS, Armitage HM. Biological control of mealybugs in California. California Department of Agriculture 1920; Monthly Bulletin 9: pp. 104-58.

[7] Smith HS, Armitage HM. The Biological Control of Mealybugs Attacking Citrus. California Agricultural Experiment Station 1931; Bull 509.

[8] Quayle HJ. Insects of Citrus and Other Subtropical Fruits. Ithaca, New York: Comstock Publishing 1938.

[9] Flanders SE. Culture of entomophagous insects. Can Entomol 1949; 81: 257-74.

[10] Fisher TW. Mass culture of *Cryptolaemus* and *Leptomastix*, Natural Enemies of Citrus Mealibug. California Agricultural Experiment Station, Bulletin 797, 1963.

[11] Graebner L, Moreno DS, Baritelle LL. The Fillmore Protective District: a success story in integrated pest management. Bull Entomol Soc Am 1984; 30: 27-33.

[12] Hunter CD. Suppliers of Beneficial Organisms in North America. PM 94-03. Sacramento: Department of Pesticide Regulation, California Environmental Protection Agency 1997.

[13] Dietrick EJ. Commercial production of entomophagous insects and their successful use in agriculture. In: Papavizas GC, Ed. Beltsville Symposia in Agriculture Research, 5, Biological Control in Crop Protection; 1981: Allenheld, Osmun, London; pp. 151-60.

[14] Luck RF, Forster LD. In: van Lenteren JC, Ed. Quality Control and Production of Biological Control Agents: Theory and Testing Procedures. Wallingford: CAB International 2003: pp. 231-46.

[15] EPPO 2009 http://archives.eppo.org/EPPO Standards/ biocontrol_web/biolist.htm#classic

[16] Silvestri F. A survey of the actual state of agricultural entomology in the United States of North America. The Hawaian Forester and Agriculturist 1909a; 6: 297-336.

[17] Silvestri F. Sguardo allo stato attuale dell'entomologia agraria negli Stati Uniti del Nord America e ammaestramenti che possono derivarne per l'agricoltura italiana. Bollettino Società Agricoltori Italiani, Roma 1909b; 14: 305-68.

[18] Silvestri F. Compendio di entomologia applicata. Vol. 1-2. Portici and Naples: Tipografia Bellavista 1934-51.

[19] Argov Y, Rössler Y. Introduction, release and recovery of several exotic natural enemies for biological control of the citrus leafminer, *Phyllocnistis citrella*, in Israel. Phytoparasitica 1996; 24(1): 33-8.

[20] Argov Y. Rearing and colonization of predatory mites from Australia to control the citrus rust mite, *Phyllocoptruta oleivora*. Hassadeh 1993; 73(8): 893-6 (in Hebrew).

[21] Argov Y, Amitai S, Beattie GAC, Gerson U. Rearing, release and establishment of imported predatory mites to control citrus rust mite in Israel. BioControl 2002; 47(4): 399-409.

[22] Viggiani G. Lotta biologica e integrata nella difesa fitosanitaria. Vol. 1-2. Naples: Liguori 1994-97.

[23] Katsoyannos P. Integrated insect pest management for citrus in Northern Mediterranean countries. Athens: Benaki Phytopathologique Institute 1996.

[24] Garcia-Marì F Plagas de cítricos y sus enemigos naturals. Valencia: Phytoma España 2009.

[25] Kapranas A, Vacante V, Colazza S. Il controllo biologico degli insetti fitofagi degli agrumi. In: Vacante V., Calabrese F. Eds. Citrus, Trattato di agrumicoltura. Bologna: Il sole 24 Ore Editoria specializzata 2009; pp. 357-71.

[26] Llorens Climent JM. Homoptera I, Cocchinillas de los cítricos y su control biológico. Alicante: Pisa Ediciones 1990.

[27] DeBach P. Biological Control of insect Pests and Weeds. London: Chapman and Hall Ltd 1964.

[28] Stern VM, Smith RF, van den Bosch R, Hagen KS. The integrated control concept. Hilgardia 1959; 29: 81-101.

[29] UC IPM. Statewide Integrated Pest management Project. Publication 3303. Division of Agriculture and Natural Resources 1984.

[30] Anciso RJ, French JV, Skaria M, Sauls JW, Holloway R. IPM in Texas Citrus. Texas Cooperative Extension, B-6121, 6-02, The Texas A & M University System 2002.

[31] UC IPM. Pest Management Guidelines: Citrus. Publication 3441, 2006.

[32] McCoy CW, Nigg HN, Timmer LW, Futch SH, Rogers ME. Florida Citrus Pest Management Guide: Use of Pesticides in Citrus IPM. University of Florida, IFAS Extension 2010.

[33] Vacante V. Generalità sui mezzi e sulle strategie di lotta contro i fitofagi degli agrumi. In: Vacante V., Calabrese F. Eds. Citrus, Trattato di agrumicoltura. Bologna: Il sole 24 Ore Editoria specializzata 2009; pp. 289-311.

[34] Rössler, Y, Rosen D. A case history: IPM on citrus in Israel. In: Rosen D, Ed. Armored Scale Insects. Vol. 4B. Amsterdam: Elsevier 1990; 519-26.

[35] Cavalloro R, Prota R. Integrated control in citrus orchards: sampling methodology and threshold for intervention against the principal phytophagous pests. In: Cavalloro R, Prota R, Eds. Proceedings E.C. Experts' Meeting "Integrated control in citrus: comparison of results achieved by applying a standardized methodology"; 1982: Siniscola-Muravera: Commission of the European Communities, 1983; pp. 44-51.

[36] Rosen D. IPM on Citrus in Israel. Rehovot, Israel: The Hebrew University, Faculty of Agriculture 1986b.

[37] Franco JC, Garcia-Marí F, Ramos AP, Besri M. Survey on the situation of citrus pest management in Mediterranean countries. Bull IOBC/WPRS 2006; 29(3): 335-46.

[38] Garrido A, Ventura JJ. Plagas de los Citricos: Bases Para el Manejo Integrado. Madrid: Ministerio de Agricultura, Pesca y Alimentacion 1993.

[39] Ripolles JL, Marsa M, Martinez M. Desarrollo de an prograrna de control integrado de las plagas de los citrico en las omarcas del Baix Ebre i Montsih (Tarragona). Amposta: IRTA/Estacion Experimental de l'Ebre 1994.

[40] Areddia R, Conti F, Frittitta C *et al.* Difesa integrata agrumi - Guida per l'applicazione dei programmi agro-ambientali. 2[a] ed. Palermo: Regione Siciliana, Assessorato agricoltura e foreste 2000.

[41] Catara A, Longo S, Polizzi G, Siscaro G. La difesa integrata degli agrumi (parassiti epigei ed ipogei). Catania: Agrobioservice s.r.l., Tipografia LFD 1996.

[42] Drishpoun Y. A Guide to the Control of Citrus Pests. Bet Dagan: Ministry of Agriculture and Rural Development, Publ. No. 17003, 2000.

[43] Vacante V, Tropea Garzia G. Lutte biologique et intégrèe en vergers d'agrumes: activité de la recherche et succès dans la région ouest-paléartique. Bull IOBC/WPRS 1995; 18, 5: 113 - 17.

[44] Peris Moll EM, Juliá Igual JF. Effects of Regulation (EEC) 2078/92 on citrus growing in Calabria (Italy) and the Region of Valencia (Spain). Span J Agric Res 2005; 3, 1: 34-42.

[45] Gigli R. Spagna: cresce a Valencia la produzione integrata di agrumi 2009. www.freshplaza.it.

[46] Van Lenteren JC. IOBC Internet Book of Biological Control 2006.

[47] FAO. Twenty-Eighth FAO regional Conference for the Near East. The Role of IPM Policies in the Promotion of Crop Quality Improvement in the Near East Region; 2006: Sana'a, Republic of Yemen 2006.

CHAPTER 3

Implementation of IPM in Citriculture

Vincenzo Vacante* **and Carmelo Peter Bonsignore**

Dipartimento Patrimonio Architettonico ed Urbanistico (PAU), Mediterranean University of Reggio Calabria, Via Melissari, Località Feo di Vito, 89060 Reggio Calabria, Italy

Abstract: In this chapter the goals of Integrated Pest Management (IPM) are discussed, along with the steps necessary for their implementation in citriculture. Pest prevention is treated first, and then available choices and horticultural practices, climate monitoring, the identification of pests and their natural enemies, the sampling and monitoring of pest populations, the prediction by Degree-Days (°D) and related decision tools and finally calculation of the Economic Injury Level (EIL) and Economic Threshold (ET) and a general approach to the Action Threshold (AT). A discussion of the means of pest control completes the chapter.

Keywords: Citriculture, Mediterranean Region, citrus pests, means of control, IPM.

1. INTRODUCTION

All ecosystems (including agroecosystems) are affected by flows of energy and matter that lead from the lowest level of the producers (*e.g.* plants) to the higher level of the consumers. These flows affect the integrity of the food webs that involve different trophic levels from producers to consumers, and any disturbance, whether climatic or due to pests, diseases and their control can alter their structure, interfering with ecosystem (or agroecosystem) stability. This stability may therefore be considered to be an index of the sensitivity of the ecosystem to a disturbance. Ecosystem stability can be estimated through the properties of the "resistance", *i.e.* the ability of the ecosystem to avoid deviations from this stability, and "resilience", the speed in which the ecosystem returns to its stable state after a disturbance [1]. The structure of a food web may affect the resilience of an ecosystem in response to changes in energy and nutrient availability [2, 3]. The flow of energy through a system can affect its resilience and higher resilience values remove disturbance effects faster from the system. Stability can thus vary with the status of the community and the nature of the disturbance [1].

The citrus orchard may be considered an agroecosystem with an elevated stability, with a modest degree of resistance and with much resilience, resulting in a significant capacity for self-regulation. This simplifies pest management and includes various technical choices affecting the mechanisms underlying the ecological stability of the citrus orchard [4]. Integrated Pest Management (IPM) provides an adequate response to this need, without altering the stability of the orchard and ensuring long-lasting pest containment, along with significant eco-toxicological and economical benefits.

2. INTEGRATED PEST MANAGEMENT

Different definitions of IPM can be found in the literature; Rosen's definition [5] is closest to the requirements of citriculture, as follows: "Integrated pest management (IPM) provides a reasonable compromise, taking into account both the desirability of biological control and the need for some form of chemical control. Like diplomacy, IPM is the art of the possible. It represents a holistic approach, recognizing the unit of the ecosystem and harmonizing all available measures in an attempt to optimize pest control and crop production. Effective IPM may be achieved through the development of a vigorous program of applied biological control, in combination with a relatively judicious use of selective pesticides, only when absolutely necessary in the least disruptive and modes of application. Other selective tactics should be incorporated into the program whenever applicable ".

The implementation of an IPM program thus encompasses the use of different means of control (biological, chemical, physical, biotechnical and others) along with horticultural practices, in order to preserve populations of natural enemies, indigenous and/or introduced. The strategy requires adequate technical support (*e.g.* the mass

***Address correspondence to Vincenzo Vacante:** Dipartimento Patrimonio Architettonico ed Urbanistico (PAU), Mediterranean University of Reggio Calabria, Via Melissari, Località Feo di Vito, 89060 Reggio Calabria, Italy; Tel: + 39-0965-385201; Fax:+39-0965-385219; E-mail: vvacante@unirc.it

Vincenzo Vacante and Uri Gerson (Eds)

production of natural enemies, mainly used in programs of augmentative biological control [6], and relates the total cost of control with the effectiveness of intervention, thus becoming cost-effective over large areas [7]. IPM is not the final step of a technical process; rather it is the prerequisite for a holistic approach with production systems aimed at the realization of "integrated production". In this sense, the farm represents a basic unit and the agroecosystem is a key point, of which the conservation and improvement of soil fertility and environmental diversity are essential components. The use of biological, technical and chemical means must be balanced and be respectful of the environment, income and social needs [8].

3. IMPLEMENTATION PROTOCOL

From a practical point of view the implementation of an IPM protocol can consist of the following steps:

1. Prevention.
2. Choices and cultural practices.
3. Monitoring climate conditions.
4. Identification of pests and their natural enemies.
5. Sampling and monitoring of pests, natural enemies and diseases.
6. Determining injurious economic levels, economic thresholds and action thresholds.
7. Means of control.

3.1. Prevention

The implementation of IPM demands the prevention of pest outbreaks along with long-term and economical control. The accidental introductions of pests can be a severe problem, due to the increasing ease of transport, displacement of people and goods and other factors, including the illegal introduction of infected and/or infested plant material. Over the last decade various pests have been introduced into the Mediterranean citriculture [the mites *Eutetranychus orientalis* (Klein) and *E. banksi* (McGregor), the insects *Trioza erytreae* (Del Guercio), *Toxoptera citricida* (Kirkaldy), *Aleurocanthus spiniferus* (Quaintance) and others, and pathogens (*e.g.* Citrus Tristeza Virus)], some of which cause huge economic losses. Prevention measures aimed at limiting the risk of accidental introductions of pests and pathogens from different regions of the world are formulated in official documents (EPPO Alert List, Directive 2000/29/EC, *etc.*) and their implementation requires strict planning, thorough expertise as well as considerable human and material resources. In principle, it is possible to limit the risk of introductions by formulating plant quarantine measure [9] (see also Chapter 19). Prevention also includes the removal of sources of new infestation, the correct choices of rootstocks and scion cultivars that are less susceptible to pests and diseases, appropriate irrigation systems, and enhancing the activities of natural enemies.

3.2. Choices and Cultural Practices

Choice of soils and their preparation, of rootstocks and scion cultivars and appropriate horticultural management methods require careful planning and should be made during the planning stage. These topics are examined below, in the chapters dedicated to agronomic aspects. In this section we discuss the contribution of common horticultural practices in limiting pest development. Other information on the direct and indirect effects of the various citrus hosts, their age, the in-between cover crops and the surrounding vegetation on the pests and on their natural enemies are in Chapter 19.

Fertilization. The net primary productivity of citrus, its fertilization and the development of pest populations are often interrelated. As known from other ecosystems, where nitrogen, potassium and phosphorus-potassium fertilizers affect the development of pest insects [10] or of pest mites [11, 12], fertilizers also stimulate the development of citrus pest numbers. Such adaptive responses differ from each other. For example, excess nitrogen, phosphorus and potassium induced an increase of fertility in citrus whitefly, *Dialeurodes citri* (Ashmead), on clementines [13]; nitrogen and potassium fertilizers encouraged the development of purple scale, *Cornuaspis beckii* (Newman), but did not have the same effect on populations of the California red scale, *Aonidiella aurantii* (Maskell) [14]. Increasing rates of fertilization with urea, applied to lemon trees, increased the numbers of live larvae, pupae and parasitized larvae of the citrus leafminer, *Phyllocnistis citrella* Stainton [15]. On the other hand, moderate doses of fertilizers and other horticultural practices limited the development of the latter pest [16] and in general also that of other pests.

Tillage and soil management. The choice of a system of soil management depends on the agronomic and phytosanitary benefits that are expected in result. For example, mechanical soil tillage impedes the development of ants, voles and molluscs, by mechanically damaging their eggs, juveniles and adults as well as their nests and galleries. It also exposes them to adverse physical and weather factors and to predators. However, in some environments soil tillage may be incompatible with the proper economic management of the citrus orchard. Conversely, in other areas and soil types, non-tillage results may be beneficial, but may lead to an increase in the development of specific pests such as ants (in citrus orchards of all ages) or soil insects (*Gryllotalpa* spp.) after planting. This example shows that all choices must be carefully related to the environment. In general, one should prefer, when and where possible, mechanical soil tillage. It improves soil fertility, further contributing to some pest control due to mechanical weeding.

Maintaining orchards (and surrounding areas) fairly free of weeds has ecological and phytosanitary aspects that are not yet fully understood. This is due to contrasting results: on the one hand weeds facilitate the development of pests that live commonly (molluscs and voles) or partly (ants) in the ground, but on the other it hinders the development of other pests (like plant bugs or the leafhopper *Asymmetrasca decedens* Paoli, which develops on weeds and/or horticultural plants associated with citrus). Weeds, or intercropping with *Ageratum conyzoides* Linnaeus and *Eupatorium pauciflorum* Kunth [17, 18] or *Neonotonia wightii* (Wight and Arnott) [19] help the development and/or maintenance of predatory mites or other species. In general, monoculture provides fewer resources (prey or alternative hosts, nectar, pollen and refuges) to natural enemies [20]. Thus preserving the biodiversity of citrus orchards, as a uniform ground cover or with herbicides, may have repercussions that are seldom assessed before the intervention. Thus we need to consider choices that allow for the maintenance of plant diversity. In the case of aphids, this condition ensures the presence of other species that serve as alternative prey for their predators and parasitoids.

Intercropping with other plants and windbreaks. Intercropping of horticultural plants within citrus orchards encourages infestations by various pests, like the two spotted spider mite, *Tetranychus urticae* Koch, the broad mite, *Polyphagotarsonemus latus* (Banks), or leafhoppers. Intercropping with fruit trees (apricot, fig, loquat, peach, persimmon plum and prickly pear), while without positive effects, enables the Mediterranean fruit fly, *Ceratitis capitata* (Wiedemann), to continue its development throughout the year [21]. Intercrops also encourage infestations of scale insects on fig and olive trees. The use of trees or shrubs as hedges and windbreaks tends to make the system more complex and heterogeneous, facilitating the development of pests and consequently also that of their natural enemies [22, 23]. The choice of botanical species as windbreaks thus requires careful evaluation. For example, the association of *Pittosporum tobira* (Thunb.) Ait. with citrus increases the risk of infestations by scale insects, thrips and aphids, especially of the black citrus aphid, *Toxoptera aurantii* Boyer de Fonscolombe. The replacement of Pittosporum with *Nerium oleander* Linnaeus promotes the development of oleander aphid, *Aphis nerii* Boyer de Fonscolombe, which provides alternative hosts for the parasitoids of T. aurantii without becoming a problem for citrus [24]. Hedges of *Viburnus* spp. may facilitate the development of greenhouse thrips, *Heliothrips haemorrhoidalis* (Bouché), or of the citrus planthopper, the flatid *Metcalfa pruinosa* (Say). Equally important for the maintenance and high fecundity of pollen-requiring adults of syrphids [25] and chrysopids is the presence of flowering plants [26].

Pruning. Pruning may differentially affect the bio-ecology of pest populations, facilitating or hindering their development. Pruning encourages ventilation of the internal canopy parts and directly exposes the juvenile stages of scale insects and mites to solar radiation and rain, which results in their high mortality rates. In this manner vigorous pruning hinders the development of the Mediterranean black scale, *Saissetia oleae* (Olivier), or the chaff scale, *Parlatoria pergandei* Comstock. Pruning orange and grapefruit trees is useful for the control of the citrus mealybug, *Planococcus citri* (Risso) [27]. Summer pruning, when the pruned branches are placed under citrus trees, result in fewer live larvae and pupae of the citrus leafminer, without affecting the numbers of parasitized larvae [15]. Horticultural practices that discourage new growth by the moderate use of fertilizers and irrigation during the period of major spring and summer flushes reduced the numbers of the citrus leafminer [16], whereas repeated pruning can facilitate its population increase. On the other hand, vigorous pruning of red orange trees (CV *Moro*), which produce much flowering every two years in Sicily, coincident with a minimal flowering season, intensified the damage of the plant bug, *Closterotomus trivialis* (Costa). In environments not subjected to winter or spring frosts, early pruning may enable the host-tree to escape attack by some aphids. In conclusion, there are no general rules to guide the growers (in the context of plant protection) about pruning, and each case should be decided according to the specific environmental conditions and pest populations.

Irrigation. Changes in the amount of water in the soil depend on various factors, including soil texture and seasonal climatic conditions, and can lead to changes in the osmotic pressure and in sap concentration. During arid seasons this may result in a form of resistance to sucking pests (aphids and scale insects), caused by an increasing concentration of the sap [28]. The rational application of irrigation could thus affect the development of pests. Examples include the mealybug *Pseudoccoccus affinis* (Maskell) on lemon and sweet orange, where levels of attack were reduced in comparison to pest numbers in citrus orchards that had been irrigated at ground level than by overhead sprinklers [29]. Irrigation management, the degree of soil water and its drainage help in limiting the development of certain root diseases and optimize cultural management.

In Sicily, the advance of irrigation in lemon orchards, due to the practice of water stress in July (in order to produce out-of-season green or "Verdelli" lemons), results in early flowering that largely escapes attacks by the citrus flower moth, *Prays citri* Millière [30].

Harvest. The harvest of yellow oranges (cv Valencia, Navel and others) during the prescribed period limits attacks by the Mediterranean fruit fly. Conversely, keeping mature fruits in spring, pending the harvest, or (as in some southern Italy localities) during summer, in order to sell the fruit as a typical summer production, increases the numbers of this pest, thereby creating conditions for serious fruit fly injury in the following productive seasons.

3.3. Monitoring Climate Conditions

Available data on regional temperatures and humidity trends, recorded by agro-meteorological stations, help cultural management and pest control. At times such information is integrated with that provided by other instruments, like evaporimeters, anemometers and others. Monitoring climate conditions can rationalize the management of cultural choices, such as irrigation, tillage, and aspects of pest control, including chemical treatments and the employment of natural enemies. In the absence of an available agro-meteorological station, some of the required information can be obtained by using data from a few miles away correcting them for attitude, integrating them into outputs of meteorological models as input data, with threshold and maximum temperatures for Degree-Days (°D) models. These can predict the occurrence of the developmental stages of pests with an accuracy of between one day and one week. The data needed for preparing pest predictive models can be re-constructed at the spatial resolution of a few miles with a suitable meteorological model. This could suffice in order to use its provisional outputs as input data, in order to calculate a predictive Degree-Days (°D) model, and might be used as a tool for reasonable, season-long predictions about the appearance and duration of the susceptible life stages of the monitored pests with remarkable accuracy.

3.4. Identification of Pests and Their Natural Enemies

The correct identification of pests and their natural enemies is an absolute prerequisite for the implementation of a biological control program [31]. The example of *Aonidiella aurantii* in California and the prolonged confusion between that and similar pests, and between their parasitoids, is very instructive. The encyrtid *Comperiella bifasciata* Howard and the aphelinids *Aphytis lignanensis* Compère and *A. melinus* DeBach were introduced quite late, due to their misidentification with *A. chrysomphali* (Mercet), already present in California and considered to be of secondary importance. A taxonomic revision of these insects showed that the entity considered to be *A. chrysomphali*, a parasitoid of the California red scale in the East and in other regions of the world, which had accidentally been introduced into California, actually consisted of at least seven species. They were characterized by different biological adaptations, and include *A. lignanensis* and *A. melinus*, currently considered the main parasitoids of the pest in California and other Mediterranean countries. Another example is *A. holoxanthus* DeBach, whose misidentification for 80 years prevented the remarkable biological control of the Florida red scale, *Chrysomphalus aonidum* (Linnaeus), in Florida, Israel and other regions of the world [32, 33].

The California red scale populations that infested citrus in Sicily in the 1960's and their current populations probably belong to different strains. In the 1960's *A. aurantii* infested lemons almost exclusively and did not develop on other citrus species, and occurred only in coastal lemon groves. Currently the pest is injurious to all citrus species, particularly to oranges, and develops at high density in the drier inland areas. It is not known whether these different adaptations are due to the activities of the most common parasitoids, but it is certain that natural enemy action was not decisive and indicates the need of a re-examination of the taxonomic and bio-ecological aspects of this pest.

3.5. Sampling and Population Monitoring

Sampling and monitoring of populations may be based on "conventional" and "innovative" methods, such as the use of Degree – Days and related decision tools.

Sampling and monitoring. The sampling and monitoring of the populations of pests and their natural enemies are intended to determine the risk arising from the former and the control provided by the latter, helping to adopt decision-tools for control. Distribution, aggregative responses and the use of food patches by pests and beneficials may complicate the applicative protocols.

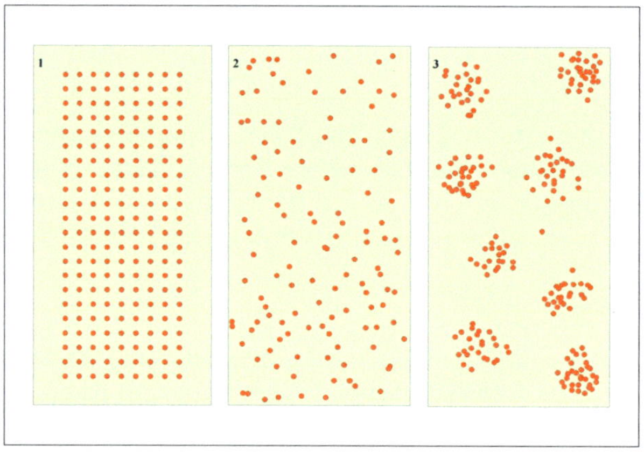

Fig. (1). Regular (1), random (2) and aggregate (3) distribution of pest populations.

The distribution of pest populations (and other organisms) depends on the spatial scale used to examine them and may be "regular", "random" or "aggregate" (Fig. 1) [34]. In regular distribution the individuals making up the population are evenly separated from each other in the area. A common example can be seen during heavy attacks by various pests. In random distribution the presence of an individual is independent of that of the others. Aggregate distribution is characterized by irregular aggregations of many individuals and is characteristic of the population dynamics of certain pests, like aphids or whiteflies.

The distribution of pests and their natural enemies on citrus (and other cultivated plants) is related to the heterogeneity of the environment and to the resulting distribution of food as "patches." A single leaf, a tree or the whole orchard may be food patches for pests, which chose the most profitable site and thus define an "aggregate response" [35, 36]. This behavior is seen in aphids, whiteflies or eriophyid mites during spring-summer, when they avoid old foliage and develop on new growth, which is the richest food resource. The persistence of pests on a patch is proportional to their risk of being parasitized and/or preyed upon; patches of low density allow individuals to escape from their natural enemies and may be identified as "partial refuges" [37]. The aggregation of some pests can be independent of the density of their host plants, thus creating an aggregated distribution of adverse effects that confers stability to the dynamics of interacting populations [38]. Examples are broad mite, *Polyphagorasonemus latus*, or the European leafroller, *Archips rosanus* (Linnaeus), injurious to new growth but harmless to the old.

In general, pests tend to create optimal foraging patterns that result in a continuous redistribution process which produces equal profitability of all food patches, leaving, in the case of different profitabilities, the less profitable patches. An "ideal free distribution" state [39] is thus reached, that should simplify the monitoring and sampling of populations. This occurs during the attack of certain pests. On the other hand, heterogeneity confers stability in the interactions between host plant and pests [40, 41]. A citrus orchard consists of a set of patches and the population dynamics of its pests relates to the availability of these different food patches. That does not however facilitate the estimation of pest populations. For instance, it was difficult to estimate the numbers of *Dialeurodes citri* and

Saissetia oleae in Corsica in order to determine their economic threshold, because the analysis of variance of the population densities of both pests showed much heterogeneity in their distribution among the trees [42]. Natural enemies show similar patterns of behavior and distribution.

Thus there are serious difficulties in deciding on reliable methods for population sampling. The validity of a sampling method depends on the number of organisms detected per sample and on the estimated density of whole populations in the field. The most common method of quantitative estimation is based on values obtained by direct counting under a stereoscopic microscope or by other analytic tools. The complexity of the protocols and the time required to achieve them suggest the use of alternative methods. For several pests it is preferable to assess their density directly in the field, without destructive samplings [43], by examining their distribution on their host plants [44]. This avoids laborious laboratory work and reduces logistical and economic difficulties (*e.g.* fruit harvesting), although not much time may thus be saved. Field examinations that are based on a "visual class system", usually consisting of less than 10 units, can describe trends in the density of species examined and simplify the sampling. This method requires that the personnel conducting the survey recognizes the different stages of the monitored organism, so that all stages are included, in order to determine the density values of organisms per sampling unit in the observation period and in different conditions. Application of these methods, when compared to accurate counts, may save about 90% of the necessary time [45]. For example, a scale of 1 to 7, in which 1 is the lightest and 7 is the heaviest citrus pest infestation has been suggested in California [46].

Monitoring of pest populations faces several difficulties, including the aggregate behavior of many species, which requires increasing the number of sampling units in order to make reliable estimates. This becomes particularly arduous when the practitioner must supply an early estimated level of attack. Exploiting the relationship between variance and the mean density of the attack in sequential sampling, taking into account the spatial heterogeneity of the population, is not reflected in the field [42]. In Spain binomial sampling of the woolly whitefly, *Aleurothrixus floccosus* (Maskell), the citrus whitefly, *Dialeurodes citri*, and the Japanese bayberry whitefly, *Parabemisia myricae* (Kuwana), was rejected due to the large size of the necessary sample. The enumerative method was the most suitable for the second and third pests, whereas for *A. floccosus* a visual sampling of 150 leaves with an index of occupation (from 0 to 10) linearly was reliable and time-saving [47]. A similar conclusion was reached in regard to the citrus mealybug, when enumerative counts were less labor-intensive than binomial sampling [48]. These difficulties highlight the complexity of the topic and indicate the need to use alternate sampling solutions, even if only temporarily, without excluding future innovations. For this purpose we usually rely on stratified random sampling of a percentage (between 2 and 10%) of the trees and/or infested organs, avoiding absolute counts of the pest population as far as possible. Visual monitoring, along with data from pheromone traps of some pests and with other data simplifies operations.

Taking into account the current difficulties in developing innovative populations sampling and monitoring protocols, it is necessary to adopt a visual methodology, derived from that proposed for the Mediterranean region by experts of the Commission of European Communities [49], and integrated with personal suggestions and with those of other authors (Tables **1** and **2**). Data collected in the field should be transferred in the form of a summary record sheet for monitoring of pests and natural enemies (Table **3**). Visual sampling and monitoring of pest populations (and their natural enemies) simplifies the process and lowers control costs. This method must be integrated with the phenology of the host plant and the basic aspects of the pests' bio-ecology. Garcia Marì recently formulated a sampling method and thresholds for the key citrus pests of Spain [50], but this protocol appears to be more suitable to small areas or to a few trees.

Table 1. Visual sampling and monitoring for major insect pests of citrus based on suggestions of different experts.

Pests	**Sampling and monitoring**
Aphis spiraecola *A. gossypii* *Myzus persicae* *Toxoptera aurantii* *T. citricida*	• On 10% of plants at random observe every week all infested shoots within a frame of ¼ m^2 from March to June taking into account that higher densities of afid populations normally occur in April and May [49]. • Observe 100 new shoots from March to June taking into account that higher densities of afid populations normally occur in April and May [50]. • *A. spiraecola* and *A. gossypii*: see [58, 60].

Table 1. cont….

Dialeurodes citri *Aleurothrixus floccosus* *Parabemisia myricae*	• *D. citri*: on 10% of plants observe at random every month all last infested vegetative growths [49]. The development of pest populations occur from April to November and presents 3 peaks of density generally in May, July and October. • *D. citri* and *P. myricae*: enumerative method sampling 40 leaves and counting all biological stages [47]. The development of populations of *P. myricae* occurs from March to October, with higher density in March and April. • *A. floccosus*: sampling 250 leaves and counting all biological stages [47]. The development of pest populations occurs from April to November and higher densities occur from May to October. • *A. floccosus*: visual sampling of 150 leaves with an index of occupation from 0 to 10 [47]. • *D. citri* and *P. myricae*: counting larvae on 40 leaves of last vegetation. • *A. floccosus*: observe 100 new growts [50].
Planococcus citri	• On 20% of plants at random, observe every 3 weeks 10 randomly fruits from plant [49]. It is possible use 1 sex pheromone trap/ha to count males captures every week. The populations occur on fruits and leaves from May to October, with higher densities from June to September. In this last period the sampling must conducted every 1 or two weeks. • Enumerative sampling and required sample sizes for the range of population densities in the calyx, fruit, and trunk equal to 50, 110–330, and 30 respectively [48]. • Observe 200 fruits [50].
Aonidiella aurantii *Aspidiotus nerii* *Chrysomphalus aonidum* *Lepidosaphes beckii* *L. gloverii* *Parlatoria pergandei*	• *A. aurantii*: on 10% of plants at random observe at random 4 shoots 10 cm long and 20 fruits per plant to count adult females [49]. Use also 2-3 sex pheromone traps/ha to count males captures every week. The same protocol can be applied to the remaining pests, taking in to account that they are present on citrus during the year and the monitoring can be conducted every two weeks from July to February, while for *A. aurantii* can begin before in the June. • All recorded pests: observe 100 ripe fruits, 100 fruits during development [50].
Ceroplastes floridensis *C. rusci* *C. sinensis* *Saissetia oleae*	• *S. oleae*: on 10% of plants observe at random 4 shoots 10 cm long and count adult females [49]. Monitoring must be conducted with the higher densities of young larvae in August and September and in December-January with females and various larval stages. • The same protocol of *S. oleae* can be applied to remaining pests, taking into account that the populations of *C. floridensis* present 2 peaks in spring-summer (May-June) and autumn-winter (November-December) respectively, *C. rusci* presents higher development from June to November, and *C. sinensis* from July to November [50].
Ceratitis capitata	• Use 10 traps/ha to trimedlure or other kinds and count adults captured every week [49]. Monitoring depends on citrus species and varieties. On Valencia and other varieties harvested in spring can be necessary monitoring the populations from April to June. In other species and varieties from August to November. • Traps to trimedlure; observe 200 fruits [50].
Phyllocnistis citrella	• On 10% of plants at random observe every month all last infested vegetative growth from May to October, taking particular account into nurseries and young plants. • Observe 100 growts [50].
Prays citri	• On 10 % of plants observe 50 flowers per tree at random every week both in spring flowering and were occurs in summer flowering [49]. • 200 bottoms or flowers [50].
Pezothrips kellianus *Scirtothrips inermis*	• *P. kellianus*: counting average presence of larvae in 25x25 cm quadrate portions of the canopy circumference [109]. • Both pests: 100 young fruits [50]. Attacks occur from March to June.
Empoasca spp., *Asymmetrasca* sp.	• On 10% of plants at random, observe every one or two weeks 10 randomly fruits from plant from August to November. It possible use trap monitoring. • Observe fruits [50].

Table 2. Visual sampling and monitoring for major mite pests of citrus based on suggestions of different experts.

Pests	Sampling and Monitoring
Aceria sheldoni	• On 10% of plants at random observe 4 shoots of last vegetative growth [49]. Sampling and monitoring can be conducted from last August to September in lemon subjected to the technique of water stress in July for the production of out-of-season green lemons; in other lemon groves commonly in winter from November to January. • Observe buds of 100 shoots [50].
Aculops pelekassi *Phyllocoptruta oleivora*	• On 10% of plants at random observe all fruits and shoots of last vegetative growth from June to October every one or two week.
Tetranychus urticae	• On 10% of plants at random observe 4 shoots of last vegetative growth and 20 fruits per plant [49]. Sampling and monitoring every two weeks from June to October. • Observe 100 leaves of newest vegetation, 100 fruits [50].
Panonychus citri	• On 10% of plants at random observe 4 shoots of last vegetative growth [49]. Sampling and monitoring every month and from summer (August - September) to winter (December and later) when the pest develops also on fruits every 1 or two weeks. • 100 leaves of last vegetation (4/tree) [50].
Eutetranychus orientalis *E. banksi*	• On 10% of plants at random observe 4 shoots of last vegetative growth from July to November, taking into account that high densities occur commonly from August to October. • 100 leaves of last vegetation [50].
Polyphagotarsonemus latus	• On 10% of plants at random observe all fruits and shoots of last vegetative growth from August to November.

The constraints associated with the ecology of populations (*e.g.* distribution) suggest the need to integrate visual sampling and monitoring - when and where possible - with an inspection of the entire harvested surface. Experiments conducted in the Italian citriculture indicate that, depending on the environment, a technical personnel unit can handle the sampling and monitoring of a cultivated area of between 200 and 300 ha.

Degree – Days and decision tools. In addition to the visual sampling and monitoring of populations, for some pests it is possible to use the Degree-Days (°D) model. This exploits the relationship between the environment and the rate of development of various insects, and especially the influence of diverse factors (nature and quantity of the trophic resource, RH, temperature, photoperiod, *etc.*) on the growth of various species. Temperature especially affects all biological parameters and through the calculation of °D [51] constitutes a useful tool for rationalizing pest control. Moreover, °D is a useful forecasting tool for estimating the development rates of host plants and their fruits.

The concept of °D is based on the "poikilothermic" condition of invertebrates, namely that their body temperature changes as a function of environmental temperature. In practice, even with an unlimited availability of nutrients, temperature is the factor limiting the rate of increase and development of poikilothermic organisms, including insects and mites injurious to citrus. Temperatures that rise within the range compatible with life positively stimulate the metabolism of various organisms and their rate of development. Depending on the thermal properties of a given environment, every species (and each of their stages) has a specific rate of development. The development time of a species can best be described as "physiological time", which is the amount of heat needed in time for the completion of its development, and not as a chronological rate. Physiological time is the cumulative product of the total development time (in hours or days) multiplied by the temperature (in degrees) above the threshold for development (or growth), which is the temperature below which no development occurs. Physiological time is indicated in °D or hours-grade (h°). The involvement of many factors (temperature variations, true temperatures registered in different habitats, microhabitats) complicates the application of the method to natural insect populations. This then requires validating the physiological time data obtained in the laboratory with the calculation of degree-hours or degree-days required in nature, using the threshold of development values established in the laboratory [52].

Information on the relationship between temperature and development, and the use of physiological time in order to assess the biological cycle of a species in an agroecosystem, can be used to predict its behavior (presence of harmful stages, generation time) at the temperatures recorded in the field. The use of °D allows us to provide reliable predictions to help rationalize the pests' control [53].

Table 3. Summary record sheet for monitoring of pests in a citrus orchard.

Country____________________Species________________Cultivar__________________________

Age_______________Treatments____

Notes____

	Tetranychus urticae			***Aonidiella aurantii***				
Date	**% infested leaves**	**% infested fruits**	**predators**	**females/cm shoot**	**females/fruit**	**% infested fruits**	**adults/trap**	**parasites**

The calculation of °D is usually undertaken with computer programs, but practitioners may obtain a rough estimate of the °D by adding the daily maximum and minimum temperatures and dividing by 2. This provides the daily average temperatures. Subtracting the lower temperature threshold of the pest from the average daily temperatures offers the accumulated number of °D of that day. For example, in the case of the California red scale, the minimum threshold for development is 11.7°C and 1100 °D are required above this threshold to complete one generation. Thus if 10 °D are daily accumulated, a generation develops in 105 days, or with 20°D accumulated every day, a generation develops in 53 days. The use of a starting point for accumulating the °D totals usually begins in one of two ways. For many pests on a calendar date (like January 1) is commonly used to start keeping track. This is simple but is disadvantageous if the temperatures rise long before any insect development occurs. The second method, used for other pests, starts from a specific biological event called biofix, the date of the first adult capture by a pheromone (or other) trap.

The application of °D enables us to plan pest control over large areas, but requires basic knowledge about the fundamental parameters of the key pests in the environment and a strict management of interventions. In California decision tools to control various citrus pests were developed by adopting °D models with different temperature thresholds and cutoff methods (http://www.ipm.ucdavis.edu/WEATHER/ddretrieve. html). These models predict the stages of development of various pests with a precision ranging from one day to one week. Due to their simplicity they significantly improve control, including the use of pesticides and the intensification of field surveys during periods of maximum pest development. Their use improved the economy of the agricultural environment in general through reduced use of pesticides and by cost optimization of sampling [53]. In Italy a °D model has been calibrated at various locations for the California red scale on citrus and allowed the implementation of some interesting decision tools. In Sicily the processing of data about male catches with pheromone traps showed a positive response of the model (proposed by the University of California). It was based on a model using the sine-wave curve in order to calculate the number of degree-days within a horizontal cutoff with a lower threshold of 11.5°C and above 30°C; this operative tool is available to SIAS (www.sias.regione.sicilia.it). In Basilicata (South Italy) monitoring of *A. aurantii* has however suggested the adoption of the Grout model [54], which employs a double sine-method with a horizontal cutoff and minimum and maximum thresholds of 11.7 and 37.8°C, respectively [53].

3.6. Economic Injury Levels, Economic Threshold and Action Threshold

The assessment tools of injury include the "Economic Injury Level" (EIL) and the "Economic Threshold" (ET). Both concepts were proposed in the 60's [55] and enable us to assess the real injury caused by the pest populations and the tolerance to some levels of attack or even the theoretical role of certain pests. Regrettably relevant data are not available for all citrus pests. The relationship between the EIL and the overall equilibrium position of population dynamics of a pest permits us to define the fundamental pest status. The issue can be examined from both the entomological and the economic points of view [56] and herein the latter is discussed first.

Economic Injury Level. The EIL can be defined as "the lowest population density of a pest capable of producing a loss, *e.g.* the amount of harm that justifies the economic cost of control" [57], and can be calculated by the following equation:

$$EIL = C/VIDK$$

where C is the management costs per production unit (*e.g.* €/ha), V is the market value per production unit (€/kg), I is injury per pests equivalent, D is the damage per unit injury (reduction in kg/ha/injury unit) and K is the proportional reduction in injury obtained by the control. The index I converts the insect populations in "insect damage equivalent" and it corresponds to the potential total damage caused by a pest, by calculating the structure and density of the population and its potential damage. Some EIL values were calculated for citrus; for instance, that of the cotton aphid, *Aphis gossypii* Glover, on clementines (271 aphids/m^2 of canopy) [58]. Another example is the citrus leafminer on *Citrus aurantifolia* Swing, where the damaged leaf area percentages were 16 and 23%, and between 18 and 85% caused a significant reduction in citrus production in orchards of 15 and 5 years old, respectively [59]. Yet another case is that of spiraea aphid, *Aphis spiraecola* Patch, on clementines (370 aphids/m^2 of canopy) [60] or of broad mite (9-15 mites/ day) [61].

The practical importance of EIL lies in its ability to indicate when control measures are needed, by pesticides or by other means. The use of EIL is unprofitable if the pest always causes economic injury, if the estimate of damage is difficult and if control measures are difficult or impossible. An example of the latter case is the Mediterranean fruit fly infestation in various species and varieties of citrus, *e.g.* Satsuma Miyagawa, Tangelo Mapo and others. In this case the damage and the commercial value of the fruit result in a low EIL value, whose adoption is not recommended. The same situation pertains to the pink citrus rust mite, *Aculops pelekassi* (Keifer), or the broad mite, against which control measures are undertaken as soon as infestation is first seen on the fruits.

The estimated market value of production (V) can be problematic due to the varying degrees of cosmetic damage that accompany biological control or IPM. The solution to this problem involves technical, economic and socio-cultural aspects.

The impact of pesticides involves the relationship between EIL and the environment. Changes in calculating EIL are related to operations intended to improve environmental sustainability. If we include environmental costs in the cost of managing, EIL can be described as "Economic Injury Level Environmental" (EILE) [57]; in the case of *Aphis spiraecola* it is equal to 614 aphids/m^2 of canopy [60].

Economic Threshold. The ET can be defined as the "density of the pest against which control measures should be undertaken in order to avoid reaching the EIL". This means that pest populations have become large enough to cause economic level damage and they must have a minimum density in order to justify control. Pest density would thus lead to an increase of damage that is equal to the cost of prevention, and population density should decline after the use of a pesticide (or other means) in order to maximize profit. Knowledge of pest density, supported by appropriate sampling and monitoring methods, assists the practitioner in taking control measures before the population reaches the EIL. Pesticide use should include marginal costs equal to marginal benefits and the marginal costs that arise with the dose, and profits should be non-negative [62].

These conditions have to be assessed in any IPM program, as control may require the use of more chemicals. This defines the ET as "the lowest level of the population that justifies the use of pesticides when their amount is calculated to maximize profits". Each specific ET involves the choice of a control strategy and relates to its optimal use and the timelines of action. If during the growing season more interventions are required, the use of pesticides is identified with a dynamic decision-making process and the calculation of ET must take into consideration host plant phenology and its different susceptibility to attack. For example, for *Aphis spiraecola* the ET value is equal to 322 aphids/m^2 of canopy [60].

In calculating the ET, using only the direct benefits of pesticides will bias the estimate towards lower values. Usually we tend to identify the mean density of the pest with ET, assigning importance only to the private marginal cost, while considering the social cost (the sum of costs of external effects of pesticides and of market cost) profits becomes positive only in the presence of high pest density. This indicates that the social costs are moving towards a higher ET and to a lower consumption of pesticides [62].

Action Threshold. Several difficulties, such as the availability of a model pest-crop system, procedures to assess optimal dosage each time a pest-control decision must be made, evaluation of infestation levels, tables of dosages maximizing profit, *etc.*, do not allow easy implementation of ET, whereas the application of Action Threshold (AT) is simple, using only the population threshold and the dosage rates of chemicals. This conclusion is based on the fact that the ET can be also seen as "the pest level population subsequent to application of a computed, maximizing dosage-profit rate" [63]. It also represents the residual population following treatments with optimal dosage; noting that the dosage varies continuously with the initial density of the population and that this limit determines the lack of acceptance of ET as an applied management tool. The AT corresponds to "pest density prior to taking an action " [63], according to the concept of ET of Edwards and Heath [64] and now commonly referred to as the AT and interpreted in empirical studies as "the minimum population level for which it is profitable to apply a pre-specified, fixed amount of pesticides, ordinarily at label or recommended dosage rate [63]. The AT is based on a discrete choice decision problem, where the choice consists of either not treating or applying a fixed dosage. Pest management is thus to apply a certain dosage if the pest density exceeds a certain level and not to treat with at lowest level. The method is simple and its application requires a farmer only to calculate if the pest population exceeds a specific level and if it does, to treat with a specific dosage. However, one cannot underestimate the risk of arbitrariness of AT and the dosages used, and that we must report any decisions with a serious economic calculation [63]. In this context the fixed dosage related to threshold equates the expected value of a marginal product when a treatment equals the cost of chemicals. Moreover, the dosage might differ from that associated with a change in the average population size.

An application example of AT on citrus is from California's San Joacquin Valley, where citrus growers are advised to control attacks of the California red scale found at packing houses by spraying before the number of second-generation males caught in pheromone traps in ha^{-1} reaches 13.000 [46].

The AT may fall short of profit maximization. This risk has resulted in the formulation of pest-management rules intended to achieve economic profitability. Several researchers defined a decision rule used as AT with more satisfactory results [63].

In conclusion, an AT integrated with economic-efficiency provides a practical decision rule for management of pest problems. The method represents an improvement over the ET concept, being simpler for farmers than ET and more productive. The situation in the Mediterranean region indicates that currently we do not have sufficient information on ET values of all pests. This deficiency requires compromises about the use of empirical values of thresholds, as suggested by experts of the Commission of European Community [49] and other authors (Table **4**). These indications seem closer to the concept and objectives of the AT. The topic has not been sufficiently studied and requires responses that only joint efforts from different research fields and different countries can supply.

Status of pests. In the Mediterranean Region 108 insect and 10 mite species were tallied in total on citrus; of which 26 insects and 4 mites are considered to be key pests in at least one country [65]. This suggests that key and secondary pests may be confused, which could lead to incorrect control decisions, from both the economical and eco-toxicological points of view.

Table 4. Thresholds for different citrus pests based on suggestions of different experts.

Aphis spiraecola *A. gossypii* *Myzus persicae* *Toxoptera aurantii* *T. citricida*	• *T. aurantii*: 25% of infested shoots; *A. spiraecola* and *M. persicae*: 10% of infested shoots on orange and 5% on clementine [49]. • EIL *Aphis gossypii* Glover on clementines: 271 afids/m^2 of canopy [58]; ET *Aphis spiraecola*: 322 afids/m^2 of canopy [60]. • *A spiraecola* and *M. persicae*: 30% of infested shoots; *A. gossypii*, *T. aurantii* and *T. citricida*: 50% of infested shoots [50].
Aleurothrixus floccosus *Dialeurodes citri* *Parabemisia myricae*	• D. citri: 5-10 2nd instar or older nymphs/leaf on clementine and 20-30 on orange and lemon [49]. • *A. floccosus*: 25% of infested shoots without parasitism [50]. • *P. myricae*: ?
Planococcus citri	• 5-10% infested fruits in summer and 15% in autumn by young larvae in little colonies [49]. • 20% infested fruits in summer and autumn [50].
Aonidiella aurantii *Aspidiotus nerii* *Chrysomphalus aonidum* *Lepidosaphes beckii* *L. gloverii* *Parlatoria pergandei*	• *A aurantii*: 1 female/cm of twig and/or 4 females/fruit [49]. • All reported pests: 2% of infested fruits with more than 10 individuals [50].
Ceroplastes floridensis *C. rusci* *C. sinensis* *Saissetia oleae*	• *S. oleae*: 3-4 females/40 cm of twig and or 3-5 nymphs/leaf [49]. • All reported pests: 4 females/40 cm of twig and or 4 nymphs/leaf [50].
Ceratitis capitata	• Clementine (approaching ripeness): 20 adults/trap/week in autumn; orange (preripe and ripe): 40-50 adults/trap/week in autumn; orange (late ripening varieties): 10 adults/trap/week in spring [49]. • 1 adult/trap/day or 1% of infested fruits [50].
Phyllocnistis citrella	• On adult groves no treatment; presence of attack on young plants, newly transplanted seedlings or grafting. • Presence of attack on young plants, newly transplanted seedlings or grafting [50].
Prays citri	• 50% of infested flowers and 2-3% of fruits attacked [49]. • 65-70% infested flowers with rich flowering [110]. • 10% of presence of bottoms and/or flowers with larval entry holes or 5% larval damage [50].
Pezothrips kellianus *Scirtothrips inermis*	• Both pests: 5-10% fruits with larvae [50]. • P. kellianus: 10 larvae in 25x25 cm quadrate portions of the canopy circumference [109].
Empoasca spp. *Asymmetrasca* sp.	• Presence of damage on fruits in September-October [50]. • 1 patch of 1 cm diameter per 250 fruits [114]
Aceria sheldoni	• On adult plants more than 70% of infested buds in August-September and/or in winter; on young groves less than 70% of infested buds [72]. • 20% of infested shoots from March to June [50].
Aculops pelekassi *Phyllocoptruta oleivora*	• *A. pelekassi*: 2-3% of infested fruits [49]. • *Ph. oleivora*: 2-3% of infested fruits.
Tetranychus urticae	• 8-10% of infested leaves and 2% infested fruits [49]. • 7% of infested leaves and 2% infested fruits [50].
Panonychus citri	• 20-40% of infested leaves and 3 mites/leaf [49]. • 50% of infested leaves from August to October and 80% in the other months [50].
Eutetranychus orientalis *E. banksi*	• 80% of infested leaves [50].
Polyphagotarsonemus latus	• 2-3% infested fruits.

The analysis of EIL and of the general equilibrium position of the population dynamics can place the population dynamics of pests and their related risk in a more rational context. The relationship between EIL and the general equilibrium position of the pests' population dynamics indicate 4 pest status, respectively called "non-economic", "occasional", "perennial" and "key" [55, 66].

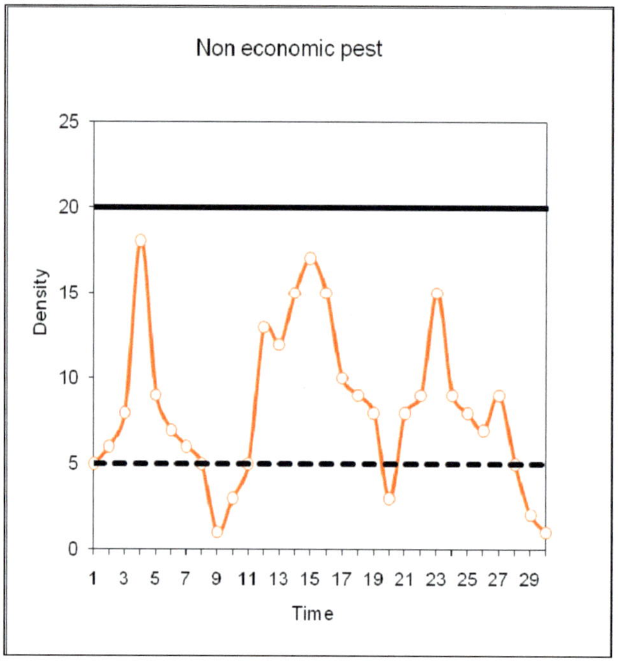

Fig. (2). Population dynamics of a non-economic pest. Economic Injury Level (EIL) (-); general position of equilibrium of population (--); density of pest (○).

Non-economic pests show a general position of population equilibrium located far below the EIL and their population density does not reach the EIL (Fig. **2**). Many citrus pests are in this category, such as the citrus bud mite, *Aceria sheldoni* (Ewing), some thrips and many leafhoppers.

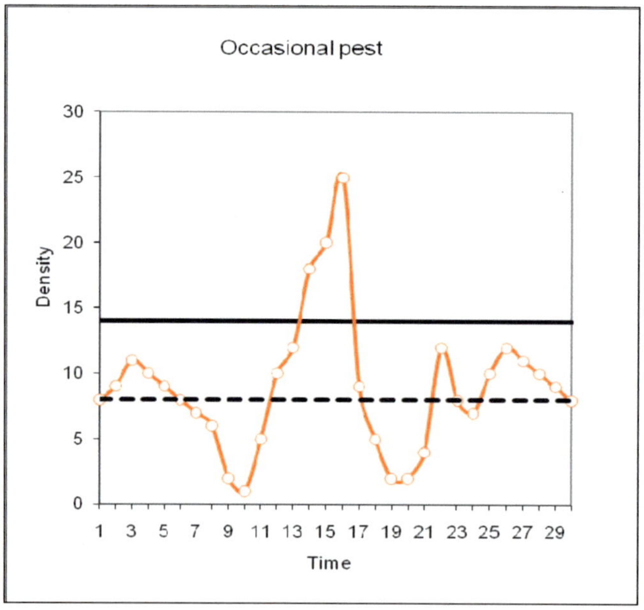

Fig. (3). Population dynamics of an occasional pest. Economic Injury Level (EIL) (-); general position of equilibrium of population (--); density of pest (○).

The occasional pests show a general equilibrium population position below the EIL, but they may sometimes reach the level of EIL (Fig. **3**). The cottony cushion scale, *Icerya purchasi* Maskell and plant bugs are occasional pests, but under some environmental conditions and/or the irrational use of pesticides, they can become pests. The general equilibrium position of perennial pests is usually under, but close, to the EIL and their population density frequently attains and exceeds the EIL (Fig. **4**). This is characteristic of many pests (aphids, white flies, scale insects, spider mites and others). The general population equilibrium position of these key pests falls above the EIL, representing severe control problems (Fig. **5**). The Mediterranean fruit fly, the California red scale and the citrus mealybug are key pests.

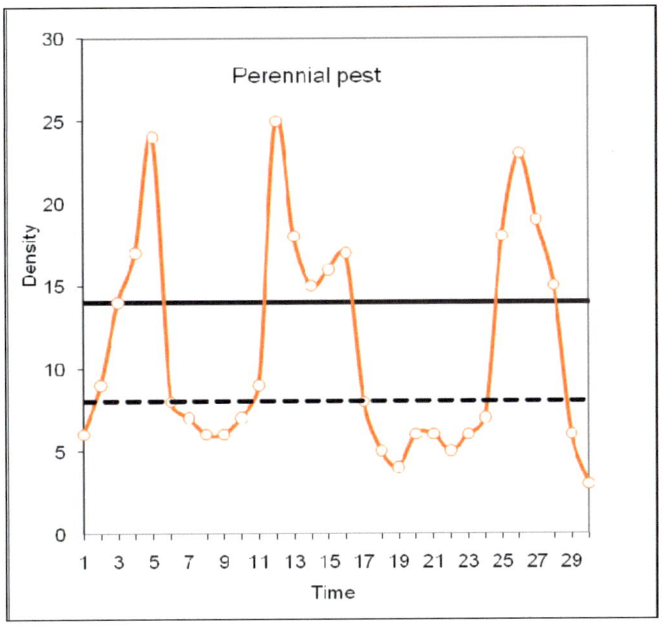

Fig. (4). Population dynamics of a perennial pest. Economic Injury Level (EIL) (-); general position of equilibrium of population (--); density of pest (○).

Non-economic or occasional pests may sometime represent the secondary pests.

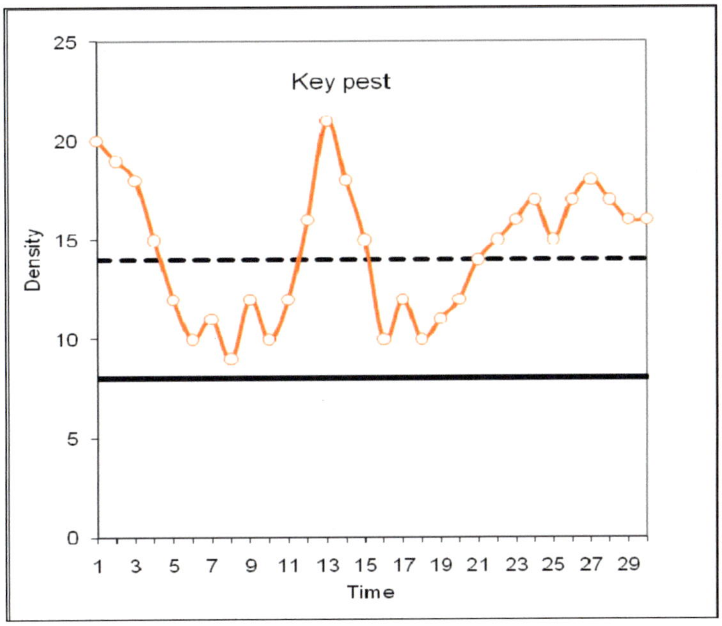

Fig. (5). Population dynamics of a key pest. Economic Injury Level (EIL) (-); general position of equilibrium of population (--); density of pest (○).

Secondary pests. There are many secondary pest species and their populations can develop to high densities in consequence of different factors, including the failed use of insecticides. From a practical point of view, reductions in the efficacy and/or the disappearance of chemicals with broad-spectrum activity have clear eco-toxicological benefits (*e.g.* preservation of natural enemies). However, this might also bring about the appearance and/or development of populations of secondary pests, which are no longer constrained by the use of these substances. For example, the limited use of chlorpyriphos methyl, highly effective against ants and mealybugs in IPM programs, is partly responsible for the resurgence of ants in citrus groves.

The greater occurrence of the carob moth, *Ectomyelois ceratoniae* (Zeller), or of the honeydew moth, *Cryptoblabes gnidiella* (Milliere), both of which are attracted to honey dew excreted by citrus mealybugs infesting citrus fruits that are not chemically treated, may be explained by the same mechanism. Factors such as horticultural practices, like tillage of terrain and pruning further contribute to population increases of secondary species. At this time there is no overall solution to this problem and each species must be treated individually.

3.7. Means of Control

The implementation of an IPM program requires the use of various technical means (biological, chemical, biotechnical, physical and mechanical) in order to lower pest density below the levels of economic damage and/or thresholds, as well as to conserve natural enemies. Among available means the biological options are the most important, and under appropriate conditions their use can assure a long and sustainable pests control program, with economic and eco-toxicological advantages.

Biological means. The biological means, *sensu stricto,* include different organisms belonging to various systematic categories (viruses, bacteria, fungi, nematodes, spiders, mites, insects, reptiles, birds, mammals). A less restrictive conception includes among the biological means also "non-living means", consisting of extracts of various natural substrates (animal, vegetable, *etc.*). In the 1990's the European Union (EU) regulated biological control with specific legislation (Annex II B, EEC Regulation 2092/91, repealed in June 2007 by Reg. No 834/2007) and also included various biological substances of plant origin (*e.g.* azadirachtin, natural pyrethrins), commonly known as biopesticides, as well as certain chemicals (petroleum oils, narrow range oils), which are distillates from petroleum (Table **5**). For a detailed treatment of biopesticides see Grant *et al.* [67].

Table 5. Acaricides and insecticides authorized in the biological control of citrus pests in EU (EEC Reg. 2092/91, Annex II B, and Reg. No 834/2007).

Active ingredient	**Orange**	**Mandarin**	**Clementine**	**Lemon**	**Grapefruit**	**Bergamot**	**Ppm (mg/kg)**
Azadirachtin	3	3	-	3	3	-	0.5
Deltamethrin	3	3	3	3	3	3	0.05
Petroleum oils	20	20	20	20	20	20	
Paraffinic oils o narrow range	20	20	20	20	20	20	
Pyrethrins	2	2	2	2	2	2	1
Lime sulphur	30	30	30	30	30	30	50
Spinosad	3	3	3	3	3	3	0.30
Sulphur	5	5	5	5	5	5	50

For each active ingredient are reported in order security related intervals (in days) and maximum residue limits allowed (mg/kg). Moreover also allowed to use some microorganisms [*Bacillus thuringiensis* Berliner (sub. *aizawai* de Barjac & Bonnefoi), *B. thuringiensis* (sub. Dulmage *kurstaki*), *Beauveria bassiana* (Balsamo) Vuillemin], vegetable oils, protein hydrolysates, potassium salts and of fatty acids and copper. These substances are approved for use or known to be used. Deltamethrin is allowed into Mediterranean fruit fly traps.

Among natural enemies the insects play a major role and have long been used for the control of citrus pests [68]. They include many species ascribed to different orders (the most important being the Hymenoptera and Coleoptera) and representatives of dozens of families. They are briefly discussed in Chapter 5 and readers interested in acquiring more information on this topic are invited to consult specialized texts [69, 70] or, for practical needs, also other contributions [50]. For more specific questions the IOBC Bulletins of the Commission "Identification Service of entomophagous insects" are likewise available. Several species of mites (Acari) also play an important ecological role. Although they are fewer in numbers, mites sometimes play a decisive role in maintaining the biological equilibrium of the system [71, 72]. Spiders are a third group of arthropods equally important to the ecological stability of citrus groves [73-75].

Limiting the population of a pest depends on the joint action of any natural enemies already present in the environment, more than on artificially-introduced species and/or other biological means used for this purpose. This indicates the necessity of knowing the endemic natural enemies, in order to preserve their populations, along with any other required interventions and programs of augmentative biological control. Conservation of natural enemies is fundamental and provides the protection and preservation of the ecosystems in order to keep them safe, providing assurances of biodiversity. The stability of ecosystems is best measured in terms of species diversity [76] (see also Chapter **19**).

If needed, programs of importation and establishment of exotic natural enemies may be initiated. According to Rose and DeBach "effective natural enemies usually exist, but they must be discovered to be utilized" [77]. All available species should be introduced and colonized and any competition between the natural enemies selects the best species, being a fundamentally dynamic phenomenon [77].

Selection of other means of control (*e.g.* chemicals) suitable for integration with the natural enemies is an absolute prerequisite.

Biological control. Biological control is based on the manipulation of interspecific relationships within agroecosystems and on antagonism between living beings. The method emphasizes the role of natural antagonists (parasites, predators and pathogens) [68, 78] without precluding intraspecific activities and the use of biotechnical means, *e.g.* pheromones, autocidals, lethal genes, *etc.* A less restrictive definition includes Behavior-Modifying Chemicals (BMCs), such as food attractants [79] and the "Sterile Insect Technique" (SIT), by chemicals or sterilizing radiation [80]. Another definition suggests the genetic improvement of natural enemies [81].

Regardless of these technical possibilities, the realization of an IPM program is primarily based on the activity of natural enemies. Rosen [5] stated that we cannot deny that when successful, biological control is an ideal form of pest management permanent, inexpensive and non-hazardous. It is by far the most promising, and most desirable alternative to chemical control available wholesale in our arsenal". Natural enemies must have some fundamental attributes consisting of "high searching ability", "adaptability and fitness", "prey specificity and discrimination" and "power of increase" [82].

Similarly to what happens with the pests, the natural enemies are also influenced by environmental factors. These include the effects of the abiotic and biotic factors. The former include temperature, humidity, precipitation and light shading. The effects of pesticides and phenomena of diapause and synchronization are also related to physical factors. Major biotic factors are host plant influences, host pest influences, subsidiary foods and alternate hosts, interspecific competition and hyperparasitism [83].

In general, better (quantitative) estimates of the impact of the natural enemies (or any other IPM program) on pest numbers and (even more important) on yield quantity and quality are necessary (see also Chapter **19**).

Biological control and IPM activities can be complementary, as seen in the European Union (EU) laws on "Methods of Organic Production" [EEC Regulation 2092/91 (Annex II B), repealed in June 2007 by Regulation No 834/2007] and "Agricultural Production Methods Compatible with the Requirements of Environmental Protection and the Maintenance of the Countryside" (Reg. 2078/92 and beyond). The EU has recently set out the basic principles of IPM (Directive 2009/128/CE, 21/10/2009), assigning a central role to biological means and horticultural management. From a practical point of view, biological control can be implemented as one or more of three strategies, respectively termed "classical biological control", "augmentative biological control" and "biological control intensification" [68].

Classical biological control. Classical biological control is based on the introduction of one or more natural enemies from the areas of the pest's origin, or from other regions, with similar climates [68, 32, 84]. Over 26 species of entomophagous insects have been successfully introduced into the Mediterranean citriculture over the last century, of which at least half are now distributed in the Mediterranean region (see chapter **2**). Most of the citrus pests in this region are introduced species, which for various reasons are more amenable to biological control by the importation of their natural enemies [85]. Among the three control strategies, classical biological control is the most advantageous, as it can guarantee, when bolstered by a concrete success, sustainable pest control at low economic costs. Classical biological control may be based on the introduction of the largest possible number of natural antagonists or on that of the best natural antagonist [86]. The advantages of one or more of the other strategies are not always clear. Application of classical biological control is usually entrusted to universities, research centers and/or technical extension service structures.

Augmentative biological control. Augmentative biological control is based on the mass release of laboratory-bred natural enemies, mainly insects and mites, and under some circumstances also bacteria and fungi, in order to obtain rapid reductions in the pest populations. It requires the mass production of the natural enemies, whose theoretical and practical implications were examined by various specialists [87, 6].

The implementation of this method is justified by three main arguments. The first relates to the control of certain developmental stages of a pest at a specific phenologic time. The second reason consists of the enhancement of efficacy of known and established natural enemies until more effective natural enemies can be discovered, imported and established. The third is related to the negative impact of chemicals and farming practices on natural enemies [88].

The microorganisms [*Bacillus thuringiensis* Berliner, *Beauveria bassiana* (Balsamo) Vuillemin and others] are packaged as pesticides (Fig. **6**) and distributed with ordinary sprayers. There are several methods for the introduction and release of entomophagous insects and mites, usually based on the use of adults and/or juvenile stages that are sold and kept in plastic containers (Fig. **7**).

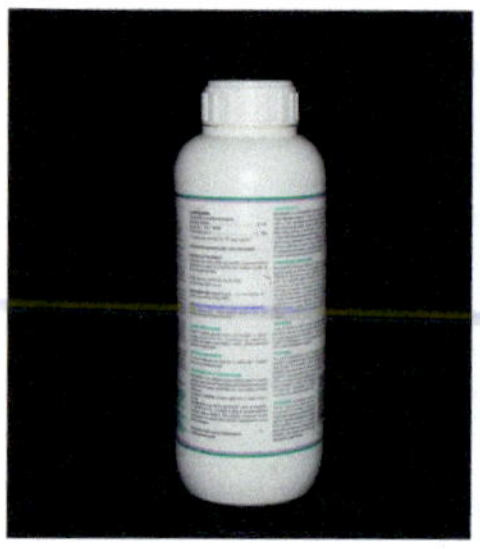
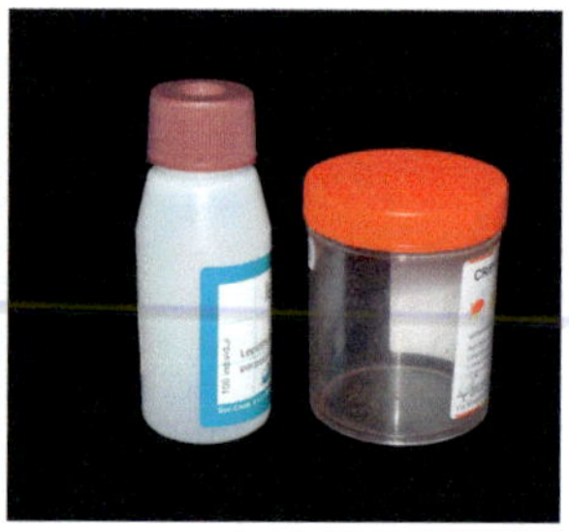

Figs (6 and **7).** Commercial packaging of an entomopathogenic fungus (at left) and commercial packaging of *Cryptolaemus montrouzieri* (Mulsant) and of *Leptomastix dactylopii* Howard (at right).

In augmentative biological control the distribution of natural enemies is considered as the common technique of control, intended to have a rapid, direct impact on the pest populations. It may consist of seasonal inoculative interventions, meant to integrate with the activities of any present species or with those introduced earlier. Augmentative biological control aims to prevent, or to inhibit, the development of pest populations throughout the growing season; it does not however preclude any positive effects in subsequent years. The technique is more expensive than classical biological control and requires more than the mass production of natural enemies; of special importance is the close link between the mass production and the use of the beneficials in the field (storage, transport, employ, technical assistance). In North American it has been calculated that the method is beneficial if the total cost of control is distributed amongst the greatest possible number of farmers [7]. The principal natural enemies used in augmentative biological control programs on citrus are the aphelinid *Aphytis melinus*, the encyrtid *Leptomastix dactylopii* Howard and the coccinellid *Cryptolaemus montrouzieri* (Mulsant). In EU countries such natural enemies have more or less the same cost, whereas variable costs were recorded from other Mediterranean regions. In Italy *A. melinus* is marketed in packs of 5,000 individuals for 35 € plus 4% tax (about € 0.007 for each individual) and 50,000 individuals/ha are commonly used, costing the biological control of the California red scales about 350 €/ha. *Leptomastix dactylopii* is marketed in packs of 100 individuals, costing 10 € plus 4% tax (approximately € 0.1 per individual) and is used in quantities of 1,200 individuals/ha, costing 120 €/ha. *Cryptolaemus montrouzieri* is marketed in packs of 100 individuals at a price of € 30 plus 4% tax (approximately € 0.3 for each individual). This predator is used in quantities of 200-400 individuals/ha, with a total price of between 60 and 120 €/ha. The costs of technical assistance have not been reported, as they are now supported by the producers of the natural enemies or by the commercial organization, but in future these costs will also be taken into account [89].

Biological control intensification. This consists of a series of activities that enhance the numbers of natural enemies in the citrus orchard. The most important are the periodical introduction of natural enemies (the inoculative method), undertaking measures to preserve their existence (protective method) and the containment of the levels of pest attacks. Horticultural practices and the rational use of selective pesticides are of considerable practical importance. The negative impact of non-selective or partially selective chemicals can be mitigated by not treating some plants/ha in order to preserve therein populations of important natural enemies [*Cales noacki* Howard, *Rodolia cardinalis* (Mulsant) and others].

Chemical means. The use of pesticides is a fundamental strategic choice for the implementation of an IPM program, at times being the only option for reducing those pest populations that are not adequately controlled by their natural enemies. The rapid development of technical means and the geo-political situation in the Mediterranean Region does not allow us to present a single framework of the chemical means that are permitted on citrus. This suggests the need to link data on the use of chemicals with the laws of each country and the commercial destination of the treated products. An example of the active ingredients allowed by law in EU countries (Directive 91/414/CEE, Annex 1), as shown in Tables **6** and **7**. For a detailed treatment of insectides, see [90, 91]. As regards the northern countries of the Mediterranean Region, it is possible for example to consult official banca data for Greece [92], Spain [93] and Italy [94]. However, it must be borne in mind that these data may change from year to year. In an IPM program the use of various pesticides must depend on the threshold values of the various pests, on the characteristics of the chemicals used (dosage, action, miscibility, toxicity to natural enemies, *etc.*), on the equipment used for distribution and on the chosen coverage. Moreover, the treatments must be compatible with the economy of the production system. This means that the pests should not be dealt separately - when and where possible – but to apply together the different treatments. In fact, by spreading the cost of control on more acreage the practitioner will have more savings. Regardless of the choices taken, in citrus there are basically two main periods of intervention, in summer and in winter, intended to control scale insects and some mite species. Other treatments can be applied in the spring or summer against other insects and mites.

Table 6. Acaricides authorized in Italy for citrus pests control according to Directive 91/414/CEE, Annex 1.

Active ingredient	**Orange**	**Mandarin**	**Clementine**	**Lemon**	**Grapefruit**	**Bergamot**	**Ppm (mg/kg)**
Abamectin	10	10	10	10	-	-	0.01
Acrinathrin	7	-	7	7	7	-	0.20
Bifenthrin	21	21	-	31	-	-	0.10
Clofentezine	30	30	30	30	30	30	0.50
Etoxazole	14	14	14	14	14	14	0.10
Hexythiazox	14	14	14	14	14	14	1.00
Fenazaquin	21	21	21	21	21	21	0.50
Fenbutatin oxide	60	60	60	60	60	60	5.00
Fenprioximate	30	-	30	30	-	-	0.50 orange 0.30 clementine / lemon
Flufenoxuron	75	75	75	-	-	-	0.30
Propargite	15	15	15	15	15	15	3.00
Pyridaben	14	14	14	14	14	14	0.50
Spirodiclofen	14	14	14	14	14	14	0.50 orange, lemon, grapefruit, bergamot 0.10 mandarin / clementine
Tebufenpyrad	14	14	14	14	14	14	0.50
Sulphur	5	5	5	5	5	5	50.0

For each active ingredient are reported in order security related intervals (in days) and maximum residue limits allowed (mg/kg).

Table 7. Insecticides authorized in Italy for citrus pests control according to Directive 91/414/CEE, Annex 1.

Active ingredient	Orange	Mandarin	Clementine	Lemon	Grapefruit	Bergamot	Ppm (mg/kg)
Abamectin	10	10	10	10	-	-	0.01
Acetamiprid	14	14	14	14	14	14	1.0
Acrinathrin	7	-	7	7	7	-	0.20
Alpha-cypermethrin	30	30	30	30	30	30	2.0
Azadirachtin	3	3	-	3	3	-	0.5
Bifenthrin	21	21	-	31	-	-	0.10
Cypermethrin	14	-	-	14	-	-	2
Chlorpyrifos	60	60	60	60	60	60	0.30 orange, grapefruit, bergamot 0.2 lemon 2. 0 mandarin, clementine
Chlorpyrifos-methyl	15	15	15	15	-	-	0.5 orange 0.3 lemon 1.0 mandarin, clementine
Deltamethrin	3	3	3	3	3	3	0.05
Etofenprox	7	7	7	7	7	7	1
Flufenoxuron	75	75	75	-	-	-	0.30
Fluvalinate	28	28					0.10
Phosmet	30	30	30	30	30	30	0.20
Imidacloprid	14	14	14	14	14	14	1
Lambda-cyhalothrin	7	7	7	7	-	-	0.10 orange, 0.2 lemon, mandarin, clementine
Lufenuron	28	28	28	28	-	-	1.0
Methoxyfenozide	14	14	14	-	-	-	1
Petroleum oils	20	20	20	20	20	20	
Paraffinic oils o narrow range	20	20	20	20	20	20	
Pyrethrins	2	2	2	2	2	2	1
Pirimicarb	14	14	14	14	14	14	3.0
Lime sulphur	30	30	30	30	30	30	50
Pymetrozine	21	21	21	21	-	-	0.3
Pyriproxyfen	28	28	28	28	28	28	0.6
Spinosad	3	3	3	3	3	3	0.30
Tebufenozide	14	14	14	14	14	14	2
Thiamethoxam	28	28	28	28	-	-	0.2
Zeta-cypermethrin	14	-	-	14	-	-	2

For each active ingredient are reported in order security related intervals (in days) and maximum residue limits allowed (mg/kg)

Improper use of pesticides can lead to the emergence of resistant pest strains. This aspect has long been known in citriculture and was first noted by the occurrence of resistant strains of the California red scale to hydrogenocyanide fumigants [46]. Despite their advantages, many pesticides have an ecotoxicological impact that weakens the ecological stability of the citrus orchard, hindering the development and survival of natural enemies of pests, even while stimulating the latter. Much information is therefore required about the selectivity of different active ingredients to natural enemies and to honeybees (and other pollinators), especially in spring during flowering.

Selectivity of chemical means. The selectivity of pesticides to the natural enemies of pests is an essential prerequisite for the success of a citrus IPM program [95]. Information on the selectivity of the main chemicals used on citrus is provided in Table **8**. Tables of selectivity, prepared by commercial producers, based on their own experiments and on the results of studies conducted by specialists of the IOBC working group "Pesticides and Beneficial Organisms", are also available. Failure to follow these guidelines may lead to severe pest outbreaks. Examples are the recent population explosions of *Iceria purchasi* and of *Tetranychus urticae* on citrus in several Mediterranean regions, consequent upon spring applications of imidacloprid or deltamethrin against aphids and other pests. Curiously, only little attention is paid to the problem of pesticide selectivity towards bees and pollinator in general, disregarding the operational need to treat against aphids and/or other pests in the spring. Petroleum oils and "narrow-range" oils are among the chemicals used in citrus and which show greater selectivity [96, 97], due to their action and low persistence on foliage [98]. The use of these substances, applied against scale insects and mites in summer and/or in winter, characterizes citrus IPM programs, when used according on the specific needs and environmental conditions.

Table 8. Active ingredients (acaricides and insecticides) authorized in Italy for citrus pests control (Directive 91/414/CEE, Annex 1) and their selectivity towards some important beneficial organisms living on citrus groves.

Active ingredient	***Lysiphlebus testaceipes***	***Leptomastix dactylopii***	***Cales noacki***	***Euseius stipulatus***	***Cryptolaemus montrouzieri***	***Rodolia cardinalis***
Abamectin	-	-	3-4	2-3	3-4	1
Acetamiprid	-	-	-	-	-	-
Acrinathrin	-	-	-	-	-	-
Alpha-cypermethrin	-	-	-	-	-	-
Azadirachtin	1	3-4	-	1	1	3-4
Bacillus thuringiensis var. *kurstaki*	1	1	1	1	1	1
Bifenthrin	-	1	-	3-4	-	3-4
Chlorpyrifos	3	3	2-3	2	2	1-2
Chlorpyrifos-methyl	3	2-4	1-2	3	1	1
Clofentezine	-	-	1	1-2	2	1
Cypermethrin	1	-	3-4	4	4	4
Deltamethrin	1	4	3	4	3-4	4
Etofenprox	-	-	-	-	-	-
Etoxazole	-	-	-	-	-	-
Fenazaquin	-	-	3	4	2	4
Fenbutatin oxide	1	1	1	2	1	-
Fenprioximate	-	-	-	-	-	-
Flufenoxuron	-	-	1-2	2-3	-	-
Fluvalinate	-	-	-	-	-	-
Hexythiazox	1	-	-	1	-	1
Imidacloprid	1	4	3	2-3	-	4
Lambda-cyhalothrin	-	-	3	-	-	-
Lufenuron	-	-	1	1	1	4

Table 8. cont....

Methoxyfenozide	-	-	-	-	-	-
Petroleum oils	2	1	1-4	1-2	1-2	1
Paraffinic oils o narrow range	-	-	-	-	-	-
Pyrethrins	-	-	-	-	-	-
Pirimicarb	1	1	1	1-2	2	1-2
Propargite	1	-	2-3	4	-	-
Lime sulphur	-	-	-	-	-	-
Phosmet	1	3	4	2-3	4	4
Pymetrozine	-	-	-	-	-	-
Pyridaben	-	-	-	-	-	-
Pyriproxyfen	-	1-2	2-3	1	4	4
Spinosad	4	3-4	-	-	1	1
Spirodiclofen	-	-	-	-	-	-
Tebufenozide	-	-	-	1	-	-
Tebufenpyrad		-	-	-	2	-
Thiamethoxam	-	-	-	-	-	-
Zeta-cypermethrin	1	-	3-4	4	4	4

The classification of various active ingredients adopts the IOBC Working Group "Pesticides and Beneficial Organisms" standards: 1 = harmless; 2 = slightly harmful; 3= moderately harmful; 4 = harmfull). Data presented are taken from works of Jacas and Garcia Marì [111], Pascual-Ruiz and Urbaneja [112], Urbaneja *et al.* [113] and other authors.

Petroleum and narrow range oils. For a detailed treatment of these substances see [95]. Petroleum oils were first used in crop protection around 1870. They were initially based on kerosene emulsions and later on lighter lubricating oils. Their effectiveness as insecticides, including good activity against different scale insects, was first accompanied by phytotoxicity, assuaged in 1914 by reducing the phytotoxic fractions by a process of sulfuration with sulfuric acid. In the 1920's the development of oil-and-water emulsions improved their formulation and resulted in significant practical benefits (coverage of canopy, *etc.*) [99]. The development of synthetic pesticides that occurred after World War II resulted in petroleum oils being relegated to lower importance as pesticides. Only in the 1970s and 1980's, as a result of control failures that resulted from the irrational use of synthetic molecules, was interest in these substances rekindled [98]. Oils of narrow distillation range (paraffinic oils, summer or narrow range) began to appear on the market at the same time. These substances have innovative features and were initially applied in citrus against scale insects, thrips, mites and sooty mold in Florida and California [100 - 102]. The "narrow range oils" are selective [84, 85] and often used in citrus IPM programs [102 - 104].

Advantages associated with the use of narrow range oils depend on their physico-chemical properties and on the technique of distillation. Mineral oils are extracted from a fraction of the fractional distillation of petroleum. The distillation of conventional petroleum oil is between 190 and 280°C at 10 mm Hg pressure and the distillation interval is 80°C, while that of a "summer oil" or "narrow range oil" is between 36 and 50°C. The latter treatment eliminates the outer edges of the range of distillation, limiting the presence of phytotoxic compounds, without altering the insecticidal activity. By this process the percentage of paraffinic compounds exceeds at least 60% of the total hydrocarbons, with a significant reduction in the potentially phytotoxic aromatic fraction and of naphthens devoid of insecticidal activity. For this reason these compounds are called "paraffinic oils".

Narrow range oils are also classified according to the midpoint of oil distillation, as 414 and 440 or 212 oils and 226°C, depending on whether we use the Fahrenheit or Celsius systems. A narrow distillation range (or narrow range) implies a more efficient oil, which due to the uniformity of its molecules evaporates evenly on treated leaf surfaces.

The 414 or 212 oils, when in contact with citrus foliage, quickly evaporate and can thus be used under high temperatures conditions; for example, from September onwards. Unlike other oils, they are not phytotoxic to the fruit. The 440 or 226 oils demonstrate a favorable compromise between speed of evaporation and the risk of phytotoxicity. The insecticidal activity of an oil also stems from its persistence on vegetation [98].

Regardless of their category, all oils act by contact and by asphyxia. They penetrate into the breathing apparatus and occlude the trachea, preventing breathing and causing the suffocation in a few hours. They also act as repellents, reducing oviposition and trophic activity. In general, the use of oils is sometimes treated with mistrust due to the risk of phytotoxicity and from their limited insecticidal efficacy. These are serious drawbacks that preclude generalization. However, it should be noted that the extent of phytotoxicity depends on the oils' physico-chemical properties and on their correct use.

Fig. (8). Sex feromone trap for *Planococcus citri* (Risso).

For instance, when applying oils during winter frosts, or in summer with temperatures above 32-34°C, or in flooded citrus orchards and/or when the trees are exposed to strong southeast winds. Mixing the oils with sulfur, or treating the plants a few days before spraying, increases the danger of phytotoxicity. A similar risk is due to adding organotins to the mixture. The practitioner should also pay attention to the species and varieties of citrus to be treated and to their phenological stages. Regarding effectiveness, it is totally inappropriate to make any comparisons with pesticides that have different modes of action [4]. Moreover, the cost of the oils is considerably lower than that of common insecticides and acaricides.

Biotechnical means. Biotechnical means include sex pheromones [105], food attractants [79] and various types of traps used to monitor adult flights or to mass trap the adults of some pests. Regarding the former, at least 15 sex pheromones of different pests are known, including those of the citrus flower moth, the Mediterranean fruit fly, the California red scale and of the citrus mealybug (Fig. **8**).

The attractants include blue color (for thrips) (Fig. **9**) and yellow (Fig. **10**) (for leafhoppers and the Mediterranean fruit fly).

Fig. (9). Blue trap for thrips.

Fig. (10). Yellow trap for *Ceratitis capitata* (Wiedemann) and leafhoppers.

Sticky traps with glue adhesives can be used, and in the case of the Mediterranean fruit fly, also with trimedlure (Fig. **11**) and/or ammonium salt. The "Sterile Insect Technique" (SIT) applied to control the Mediterranean fruit fly, is among the biotechnical means [80]. The combination of pheromone traps, the °D technique and/or other tools against pests like the California red scale, in order to estimate or forecast the pests' occurrence, strongly streamlines their control.

Physical means. Physical control means include refrigeration technology, used to prevent the development of larvae of the Mediterranean fruit fly in citrus fruits exported to other continents [5]; the method was also proposed against false spider mites infesting lemons [106].

Mechanical means. Mechanical means of pest control have widely been used in the recent past, but today are little used in different regions due to the high cost of labor.

Fig. (11). Adults of *Ceratitis capitata* captured in trimedlure trap.

Fig. (12). Glass wool strips used against weevils (*Otiorrhynchus cribricollis* Gyllenhall).

Fig. (13). Wire mesh to prevent the attack of rabbits to citrus grafting.

These means include hand picking the cetoniid *Tropinota hirta* (Poda) from flowers, rubbing the citrus trunk to remove armored scale insects, the barking of old trunks or branches that are affected by phytophthora gummosis, the collection and/or mechanical destruction of ant nests, the use of glass wool strips against weevils (*Otiorrhynchus cribricollis* Gyllenhall and other species) (Fig. **12**), the layout around the collar of wire mesh or plastic to prevent the attack of rabbits (Fig. **13**) or glues to keep out ants and distributing ashes or lime around the neck of young plants against molluscs.

Fig. (14). Non-woven fabric to protect citrus grafting from attack of *Phyllocnistis citrella* Stainton.

Other mechanical methods include non-woven fabric and insect-proof screens to protect newly transplanted citrus seedlings or grafting against the citrus leafminer [107] (Fig. **14**), or the use of high water pressure to remove scale insects and sooty mold from fruits [5]. Experience in South Africa indicted that "this mechanical method of removing armored scales makes it possible to raise the threshold of economic damage, thus reducing the number of sprays" [108].

4. CONCLUSIONS

The importance of IPM is due to its ability to promote ecological stability in citrus orchards, concurrently ensuring the effective control of its pest populations with economic and eco-toxicology benefits. The correct implementation of IPM enables the practitioner to reduce the costs of control within an acceptable level of economic management. And if the need should arise to apply seasonal chemical treatments, the total cost of control would still usually be lower than a traditional program of strict chemical control. IPM likewise provides convincing answers to environmental and health problems, which are related to the social cost of chemical treatments. Nevertheless, certain aspects remain that require solutions, such as the biological control of several pests, the development of reliable systems for sampling and estimating populations and identifying the relevant EIL and ET values. Of equal important is the study of the impact of pesticides on the citrus ecosystem and the technical training of personnel. Resolving these aspects depends on the preparation and implementation of an active policy of intervention, formulated on a Mediterranean Region scale.

REFERENCES

[1] Begon M, Harper J, Towsend CR. Ecologia, individui, popolazioni, comunità. Bologna: Ed. Zanichelli 1993.

[2] O'Neill RV. Ecosystem persistence and heterotrophic regulation. Ecology 1976; 57: 1244-53.

[3] DeAngelis DL. Energy flow, nutrient cycling and ecosystem resilience. Ecology 1980; 61: 764-71.

[4] Vacante V. In: Vacante V, Calabrese F, Eds. Citrus, Trattato di agrumicoltura. Bologna: Edagricole, Il sole 24 Ore Editoria specializzata 2009; pp. 289-311.

[5] Rosen D. Methodologies and strategies of pest control in citriculture. In: Cavalloro R, Di Martino E, Eds. Proceedings of the Experts' Meeting " Integrated pest control in citrus-orchards"; 1985: Rotterdam: Balkema 1986a; pp. 521-30.

[6] Luck RF, Forster LD. In: van Lenteren JC, Ed. Quality Control and Production of Biological Control Agents: Theory and Testing Procedures. Wallingford: CAB International, 2003; pp. 231-46.

[7] Graebner L, Moreno DS, Baritelle LL. The Fillmore Protective District: a success story in integrated pest management. Bull Entomol Soc Am 1984; 30: 27-33.

[8] El-Titi A, Boller EF, Gendrier JP. Integrated production, Principles and techinical guidelines. Bull IOBC/WPRS 1993; 16 (1): 13- 33.

[9] Burger HC, Ulenberg SA. Quarantine Problems and Procedures. In: Rosen D, Ed. Armored Scale Insects. Amsterdam: Elsevier 1990; pp. 313-27.

[10] McNeill S, Soutwood TRE. In: Harbone JB, Ed. Biochemical aspects of plant and animal coevolution. New York: Academic Press Inc. 1978; pp. 77-98.

[11] Puttaswamy, Channabasavanna GP. Influence of Nitrogen Fertilization of the Host Plant on the Population of *Tetranychus ludeni* (Acari, Tetranychidae). Indian J Acar 1982; 6: 64-71.

[12] Jackson PR, Hunter PE. Effects of Nitrogen Fertilizer Level on Development and Populations of the Pecan Leaf Scorch Mite (Acari: Tetranychidae). J Econ Entomol 1983; 76: 432-5.

[13] Onillon JC, Onillon J, Brun P, Franco E, Decoux G. Effect of fertilizer on the population level of the citrus white-fly *Dialeurodes citri* (Homoptera, Aleyrodidae). In: Cavalloro R, Di Martino E, Eds. Proceedings of the Experts' Meeting " Integrated pest control in citrus-orchards"; 1985: Rotterdam: Balkema 1986; pp. 109-20.

[14] Salama HS., Amin AH, Hawash M. Effects of nutrients supplied to citrus seedlings on their susceptibility to infestation with the scale *Aonidiella aurantii* (Maskell) and *Lepidosaphes beckii* (Newman) (Coccoidea). Z Ang Ent 1972; 71; 395-405.

[15] Ateyyat MA, Mustafa TM. Cultural control of citrus leafminer, *Phyllocnistis citrella* Stainton (Lepidoptera: Gracillariidae) on lemon in Jordan. Int J Pest Manag 2001; 47 (4): 285-88.

[16] Lin YD, Fang JM, Huang SF, Chen W. A study on integrated control measures for *Phyllocnistis citrella* Stainton. Fujian Agricultural Science and Technology 1985; 1: 30-1 (in Chinese).

[17] Gravena S, Coletti A, Yamamoto PT. Influence of green cover with *Ageratum conyzoides* and *Eupatorium pauciflorum* on predatory and phytophagous mites in citrus. Bull IOBC/WPRS 1993; 16 (7): 104-14.

[18] Kong C, Liang W, Hu F, *et al.* Allelochemicals and their transformations in the *Ageratum conyzoides* intercropped citrus orchard soils. Plant and Soil 2004; 264: 149-57.

[19] Mailloux J, Le Bellec F, Kreiter S, Tixier MS, Dubois P. Influence of ground cover management on diversity and density of phytoseiid mites (Acari: Phytoseiidae) in Guadeloupean citrus orchards. Exp Appl Acarol 2010. DOI: 10.1007/s10493-010-9367-7

[20] Root RB. Organization of a plant-arthropod association in simple and diverse habitats: The fauna of collards (*Brassica oleracea*). Ecol Monogr 1973; 43: 95-124.

[21] Bodenheimer FS. Citrus entomology in the Middle East. Gravenhage: Uitgeverij Dr Junk S 1951.

[22] Balachowsky A. Encyclopédie Entomologique, A.15. Paris: P. Lechevalier 1932.

[23] Argyriou LC. On some scales of row-trees and wind-break in Greece. In: Comptes Rendus des 6ème Journée de Phytiatrie et de Phytopharmacie Circum-Méditerranénne, May 1981. Perpignan 1981.

[24] Stary P. The creation of artificial foci of parasites – A new method of aphid parasite release. Acta Entomol Bohemoslov 1968; 65: 76-7.

[25] Proctor M, Yeo P. The Pollination of Flowers. New Naturalist Series, N. 54. London: Collins 1973.

[26] Hagen KS, Tassan RL. In: Rodriguez JG, Ed. Insect and Mites Nutrition. Amsterdam: Elsevier 1972; pp. 323-51.

[27] Alexandrakis VZ. In: Cavalloro R, Piavaux A, Eds. CEC Programme on Integrated and Biological Control, Final Report 1979/1983. Brussells-Luxembourg 1984; pp. 103-28.

[28] Flanders SE. On host plant induced behaviour of scale insects and their endoparasites. Can Entomol 1970; 102: 913-26.

[29] Panis A. Biological features of *Pseudococcus affinis* (Mask.) (Homoptera, Pseudococcidae) as guidelines of its control in water-sprinkled citrus orchards. In: Cavalloro R, Di Martino E, Eds. Proceedings of the Experts' Meeting " Integrated pest control in citrus-orchards"; 1985: Rotterdam: Balkema 1986; pp. 59-65.

[30] Nucifora A, Calabretta C, Caltabiano G. La tecnica del "mass-trapping" e la pratica della rottura anticipata della secca contro *Prays citri* Mill. in limonicoltura. Atti Giornate Fitopatologiche 1984; 2: 325-36.

[31] Rosen D. The role of taxonomy in effective biological control programs. Agricul Ecosyss Environ 1986b; 15: 121-9.

[32] DeBach P. Biological control by natural enemies. Cambridge: Univ Press 1974.

[33] Rosen D, DeBach P. Species of *Aphytis* of the world (Hymenoptera: Aphelinidae). Jerusalem: Israel Universities Press; The Hague: Dr. Junk W 1979.

[34] Horn DJ. Ecological Approach to Pest Management. London: Elsevier Applied Science Publishers 1988.

[35] Hassell MP. Mutual interference between searching insect parasites. J Anim Ecol 1971; 40: 473-86.

[36] Hassell MP, May RM. Aggregation of predators and insect parasites and its effect on stability. J Anim Ecol 1974; 43: 567-94.

[37] Hassell MP. Patterns of parasitism by insect parasitoids in patchy environments. Ecol Entomol 1982; 7: 365-77.

[38] Way MJ, Cammell M. In: Watson A, Ed. Animal Populations in Relation to their Food Resources. Oxford: Blackwell Scientific Publications 1970; pp. 229-47.

[39] Fretwell SD, Lucas HL. On territorial behaviour and other factors influencing habitat distribution in birds. Acta Biotheoretica 1970; 19: 16-36.

[40] Huffaker CB. Experimental studies on predation: dispersion factors and predator-prey oscillations. Hilgardia 1958; 27, 343-83.

[41] Huffaker CB, Shea KP, Herman SG. Experimental studies on predation. Hilgardia 1963; 34: 305-30.

[42] Onillon JC, Panis A, Brun P, Onillon J, Franco E, Marro JP. Recherches préalables à l'estimation du seuil de nuisibilité de *Dialeurodes citri* et de *Saissetia oleae* en vergers de clémentiniers. In: Cavalloro R, Prota R, Eds. Proceedings EEC Meeting; 1980: San Giuliano (Corsica) and Siniscola (Sardinia) 1981; pp. 147-61.

[43] Dewar AM, Dean GJ, Cannon R. Assessment of methods for estimating the number of aphids (Hemiptera: Aphididae) in cereals. Bull Entomol Res 1982; 72: 675-85.

[44] Addicott JF. The population dynamics of aphids on fireweed: A comparison of local populations and metapopulations. Can J Zool 1978; 56: 2554-64.

[45] Lapchin L, Shtienberg GD. In: Albajes R, Gullino ML, van Lenteren JC, Elad Y, Eds. Integrated Pest and Disease Management in Greenhouse Crops. Dordrecht, Boston, London: Kluwer Academic Publishers 1999; pp. 82-96.

[46] IPM Manual Group. Integrated Pest Management for Citrus. University of California, Division of Agriculture and Natural Resources 1984; Publication 3303.

[47] Soto A, Ohlenschlaeger F, Garcia Marì F. Distribution and sampling of the whiteflies *Aleurothrixus floccosus*, *Dialeurodes citri*, and *Parabemisia myricae* (Homopetra: Aleyrodidae) in citrus in Spain. J Econ Entomol 2002; 95 (1): 167-73.

[48] Martinez Ferrer MT, Ripolles JL, Garcia Marì F. Enumerative and binomial sampling plans for citrus mealybug (Homoptera: Pseudococcidae) in citrus orchards. J Econ Entomol 2006; 99 (3): 993-01.

[49] Cavalloro R, Prota R. Integrated control in citrus orchards: sampling methodology and threshold for intervention against the principal phytophagous pests. In: Cavalloro R, Prota R, Eds. Proceedings E.C. Experts' Meeting "Integrated control in citrus: comparison of results achieved by applying a standardized methodology"; 1982: Siniscola - Muravera, 1983.

[50] Garcia Marí F. Plagas de cítricos y sus enemigos naturals. Valencia: Phytoma-España, Valencia 2009.

[51] Daly HV, Doyen JT, Ehrlich PR. Introduction to Insect Biology and Diversity, New York: McGraw-Hill 1978.

[52] Gullan PJ, Cranston PS. Lineamenti di entomologia. Bologna: Zanichelli Editore 2006.

[53] Federico S. In: Vacante V, Calabrese F, Eds. *Citrus* -Trattato di Agrumicoltura. Bologna: Edagricole, Il sole 24 ore 2009; pp. 177- 87.

[54] Grout TG, Dutoit WJ, Hofmeyr JH, Richards GI. California red scale (Homoptera: Diaspididae) phenology on citrus in South Africa. J Econ Entomol 1989; 82: 793-8.

[55] Stern VM, Smith RF, van Den Bosh R, Hagen KS. The integrated control concept. Hilgardia 1959; 29: 81-101.

[56] Tomerlin AT. Economic threshold decision rules as an integrated pest management in Florida citrus production. A dissertation presented to the graduate school of the university of Florida in partial fulfillment of the requirements for the degree of doctor of philosophy, Universiy of Florida 2005.

[57] Peterson KD, Higley LG. In: Pimentel D, Ed. Encyclopedia of Pest Management. New York, Basel: Marcel Dekker 2002; pp. 228-30.

[58] Hermoso de Mendoza A, Belliure B, Carbonell EA, Real V. Economic thresholds for *Aphis gossypii* (Hemiptera: Aphididae) on Citrus Clementina. J Econ Entom 2001; 94 (2): 439–44.

[59] Peña JE, Hunsberger A, Schaffer B. Citrus Leafminer (Lepidoptera: Gracillariidae) Density: Effect on Yield of 'Tahiti' Lime. J Econ Entom 2000; 93 (2): 374–9.

[60] Hermoso de Mendoza A, Arouni R, Belliure B, Carbonell E, Pérez-Panadés J. Umbrales económicos de tratamiento contra *Aphis spiraecola* (Hemipetra, Aphididae) en clementinos. Bol San Veg Plagas 2004; 30: 287-98.

[61] Peña JE. Relationships of broad mite (Acari: Tarsonemidae) density to lime damage. J Econ Entomol 1990; 83: 5, 2008-15.

[62] Regev U. In: Pimentel D, Ed. Encyclopedia of Pest Management. New York, Basel: Marcel Dekker 2002; pp. 831-3.

[63] Moffitt, LJ, Baritelle, JL. Economic Thresholds. In: Rosen D, Ed. Armored Scale Insects. Amsterdam: Elsevier 1990; pp. 357-62.

[64] Edwards CA, Heath, GW. Principles of Agricultural Entomology. Springfield: Charles C. Thomas.

[65] Franco JC, Garcia Marì F, Ramos AP, Besri M. Survey on the situation of citrus pest management in Mediterranean countries. Bull IOBC/WPRS 2006; 29 (3): 335-46.

[66] Funderburk J. In: Pimentel D, Ed. Encyclopedia of Pest Management. New York, Basel: Marcel Dekker 2002; pp. 590-2.

[67] Grant WP, Greaves J, Tatchell M, Chandler D, Prince G, Bailey A. Biopesticides: Pest Management and Regulation. Wallingford: CABI Publishing 2010.

[68] DeBach P. In: DeBach P, Ed. Biological control of insect pests and weeds. London: Chapman and Hall 1964.

[69] Rosen D (Ed.). Armored Scale Insects. Amsterdam: Elsevier 1990.

[70] Katsoyannos P. Integrated insect pest management for citrus in Northern Mediterranean countries. Athens: Benaki Phytopathologique Institute 1996.

[71] Gerson U, Smiley RL, Ochoa R. Mites (Acari) for Pest Control. Oxford: Blackwell Publishing 2003.

[72] Vacante V. Citrus Mites. Wallingford: CABI Publishing 2010.

[73] Benfatto D, Di Franco F, Vacante V. Spiders of Italian Citrus Orchards. Proceedings International Society Citriculture 1992; 3, pp. 938-42.

[74] Ribeiro C, Cardoso P, Franco JC. Abundance and diversity of spiders in lemon orchards with different weed management systems. Bull IOBC/WPRS 2006; 29(3): 167-78.

[75] Monzó C, Mollà O, Montón H *et al.* Ground-dwelling spiders (Araneae) in citrus orchards in Spain. Bull IOBC/WPRS 2008; 38: 267.

[76] Rose M, DeBach P. Conservation of Natural Enemies. In: Rosen D, Ed. Armored Scale Insects. Amsterdam: Elsevier 1990; pp. 461-72.

[77] Rose M, DeBach P. Foreign Exploration and Importantion of Natural Enemeis. In: Rosen D, Ed. Armored Scale Insects. Amsterdam: Elsevier 1990; pp. 417-31.

[78] Van Den Bosch R, Messenger PS, Gutierrez AP. An introduction to biological control. New York: Plenum Press 1982.

[79] Jones OT. The use of behaviour modifying chemicals in the integrated management of selected fruit fly species. In: Economopoulos AP, Ed. Proceedings 2nd Int Symp Fruit Flies; 1986: Crete 1987; pp. 451-8.

[80] Hendrichs J, Franz G, Rendon P. Use of sterile male-only releasees as "biological pesticides" for control of Mediterranean fruit flies in commercial fruit production. Bull IOBC/WPRS 1994; 17(6): 212-24.

[81] Hoy, MA. Genetic Improvement of Natural Enemeies of Armored Scale Insects. In: Rosen D, Ed. Armored Scale Insects. Amsterdam: Elsevier 1990; pp. 441-51.

[82] Huffaker CB, Rosen D. The attributes of Effective Natural Enemies. In: Rosen D, Ed. Armored Scale Insects. Amsterdam: Elsevier 1990; pp. 197- 204.

[83] Huffaker CB. Effects of Environmental Factors on Natural Enemies of Armored Scale Insects. In: Rosen D, Ed. Armored Scale Insects. Amsterdam: Elsevier 1990; pp. 205-20.

[84] DeBach P, Rosen D. Biological control by natural enemies. 2nd ed. New York: Cambridge Univ. Press 1991.

[85] DeBach P, Rosen D, Kennett CE. In: Huffaker CB, Ed. Biological control. New York, London: Plenum Press 1971; pp. 165-94.

[86] Huffaker CB, Simmonds FJ, Laing JE. In: Huffaker CB, Messenger PS, Eds. Theory and practice of biological control. New York: Academic Press 1976; pp. 41-78.

[87] Knipling EF. Principles of Insects Parasitism Analyzed From New Perspectives: Practical Implications for Regulating Insect Populations by Biological Means. U. S. Department of Agriculture 1992; Agriculture handbook 693.

[88] Rose M. Periodic Colonization of natural Enemies. In: Rosen D, Ed. Armored Scale Insects. Amsterdam: Elsevier 1990; pp. 433-40.

[89] Kapranas A, Vacante V, Colazza S. In: Vacante V, Calabrese F, Eds. Citrus, Trattato di agrumicoltura. Bologna: Il sole 24 Ore Editoria specializzata 2009; pp. 357-71.

[90] Riehl LA. Control Chemicals. In: Rosen D, Ed. Armored Scale Insects. Amsterdam: Elsevier 1990; pp. 365-92.

[91] Ware GW, Whitacre DM. The Pesticide Book. 6th Edition. Willoughby: MeisterPro Informations Resources 2004.

[92] Hellenic Ministry of Rural Development and Foof (2008) Authorized plant protection products data base. http://www.minagric.gr/syspest/

[93] MAPA (2008) Registros de productos fitosanitarios.http://www.mapa.es/es/agricultura/pags/fitos/registro/menu.asp

[94] MiPAAF (2008) Banca Dati Fitofarmaci. Centro di Ricerca per la Patologia Vegetale, Roma. http://www.sian.it/fitovis/

[95] Rosen D. Effect of commercial pesticides on the fecundity and survival of *Aphytis holoxanthus* (Hymenopetra: Aphelinidae). Israel J Agr Res 1967; 17: 47-52.

[96] Campbell MM. Duration of toxicity of residues of malathion and spray oil on citrus foliage in South Australia to adults of a California red scale parasite *Aphytis melinus* DeBach (Hymenoptera: Aphelinidae). J Aust Ent Soc 1975; 14: 161-4.

[97] Morse JG, Bellow JR TS, Gaston LK, Iwata Y. Residual toxicity of acaricides to three beneficial species on California citrus. J Econ Entomol 1987; 80: 953-60.

[98] Davidson NA, Dibble JE, Flint ML, Marer PJ, Guye A. Managing Insects and Mites with Spray Oils. IPM Education and Publications, Statewide IPM Project, University of California, Division Agriculture and Natural Resources 1991; Publication 3347.

[99] De Ong ER, Knight H, Chamberlain JC. A preliminary study of petroleum oil as an insecticide for citrus trees. Hilgardia 1927; 2: 351-84.

[100] Trammel K. Properties of petroleum oils in relation to performance as citrus tree sprays in Florida. Univ Fla Citrus Exp Sta Thesis Univ Microfilms 1965; Inc. 65-9621.

[101] Chapman PJ. Petroleum oils for the control of orchard pests. NY State Agr Exp Sta Bull 1967; 814: 1-22.

[102] Rhiel LA. Advances relevant to narrow-range spray oils for citrus pest control. Proceedings 1st Int Symp 1969; 2: pp. 897-07.

[103] Rhiel LA. Relation of properties of spray oil to use on fruit trees for pest control. Proceedings Ann CAPCA Conf 1980; 6: 102-17.

[104] Rhiel LA. Fundamental consideration and current development in the production and use of petroleum oils. Insects and mites pests. Proceedings Int Soc Citricul 1981; 2: 601-7.

[105] Gieselmann MJ, Richard ER. Use of Pheromone Traps. In: Rosen D, Ed. Armored Scale Insects. Amsterdam: Elsevier 1990; pp. 349-52.

[106] JadueJ, Vargas C, Rubio T, Araya TE. Effects of cold storage on the false grape mite, *Brevipalpus chilensis* Baker. Z Pflanzenk Pflanzen 1996; 103: 403-8.

[107] Nucifora A, Nucifora MT. El minador de los brotes de los citricos, *Phyllocnistis citrella* Stainton, en viveros y plantaciones de renjertes en Sicilia (Italia): estrategias de control mecanicas y fisicas. Phytoma Espana 1995; 44: 16-7.

[108] Bedford, ECG. Mechanical Control High-pressure Rinsing of Fruit. In: Rosen D, Ed. Armored Scale Insects. Amsterdam: Elsevier 1990; pp. 507-13.

[109] Perrotta G, Conti F. A threshold hypothesis for an integrated control of thrips infestation on citrus in South Eastern Sicily. Bull IOBC/WPRS 2008; 38: 204-9.

[110] Nucifora A, Vacante V. La "Tignola della zagara" (*Prays citri* Mill.): importanza e valore della soglia d'intervento. I contributo. Atti Giornate Fitopatologiche, Sorrento, Napoli 1984; 2: pp. 315-23.

[111] Jacas JA, Garcia Marí F. Side-effects of pesticides on selected natural enemeis occurring on citrus in Spain. Bull IOBC/WPRS 2001; 24: 103-12.

[112] Pascual Ruiz S, Urbaneja A. Lista de Efectos Secundarios de Plaguicidas sobre Fauna Útil en Citricos. Levante Agrícola 2006; 380: 186-91.

[113] Urbaneja A, Pascual-Ruiz S, Pina T *et al.* Efficacy of five selected acaricides against *Tetranychus urticae* (Acari: Tetranychidae) and their side effects on relevant natural enemies occurring in citrus orchards. Pest Manag Sci 2008; 64: 834-42.

[114] Anonymous. Plant Protection Technical Instructions. Directorate General of Agricultural Researches. Ministry of Agriculture and Rural Affairs of Turkey. Ankara 2008 (in Turkish).

CHAPTER 4

Identification of the Important Groups of Citrus Pests in the Mediterranean Region

Uri Gerson[1*] and Vincenzo Vacante[2]

[1]*Department of Entomology, Faculty of Agricultural, Food and Environmental Quality Sciences, The Hebrew University of Jerusalem, P.O. Box 12, Rehovot, 76-100, Israel and* [2]*Dipartimento Patrimonio Architettonico ed Urbanistico (PAU), Mediterranean University of Reggio Calabria, Via Melissari, Località Feo di Vito, 89060 Reggio Calabria, Italy*

Abstract: The correct identification of pest species is a prerequisite for the implementation of Integrated Pest Management (IPM) programs. Towards this aim a key is presented for the identification of the classes, orders and families of the most important pests infesting citrus in the Mediterranean Region.

Keywords: Citriculture, Mediterranean Region, citrus pests, key, identification.

1. INTRODUCTION

The correct identification of pest species is a prerequisite for the implementation of Integrated Pest Management (IPM) programs. For this purpose we present a key for the identification of the classes, orders and families of the more important citrus pests in the Mediterranean Region. Our system is based on the most updated concepts about the systematics of the major relevant arthropod groups [1, 2]. The key also refers to groups that include only a few citrus pests, such as the Hemipteran Psyllidae, Hymenoptera, Coleoptera and others. The inclusion of these groups is due to the need to increase the understanding of the key and to enable the identification of secondary pests that belong to these groups.

The legends shown in each figure indicate the author of the design, but no comments were added, in order to simplify the presentation. Interested readers may consult specialized texts on the subject.

2. KEY TO THE INSECTA, ACARI AND THEIR ORDERS AND FAMILIES THAT INCLUDE SPECIES INJURIOUS TO CITRUS (ADULT FEMALES)

1. With one pair of antennae; body divided into head, thorax and abdomen; usually with three pairs of legs emanating from the thorax; wings often present (**Insecta**) (Fig. **1**) 2

-. Without antennae; body undivided; with four pairs of legs, without wings (**Acari**) (Fig. **2**)....................... 18

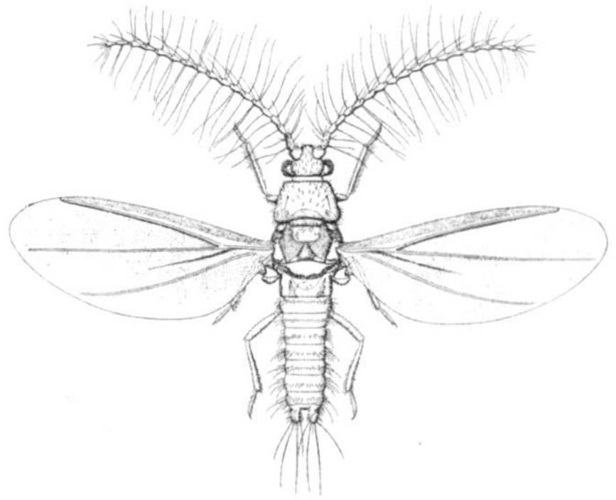

Fig. (1). Male of *Icerya purchasi* Maskell (Insecta) [3].

***Address correspondence to Uri Gerson:** Department of Entomology, Faculty of Agricultural, Food and Environmental Quality Sciences, The Hebrew University of Jerusalem, P.O. Box 12, Rehovot, 76-100, Israel; E-mail: Gerson@agri.huji.ac.il

Vincenzo Vacante and Uri Gerson (Eds)

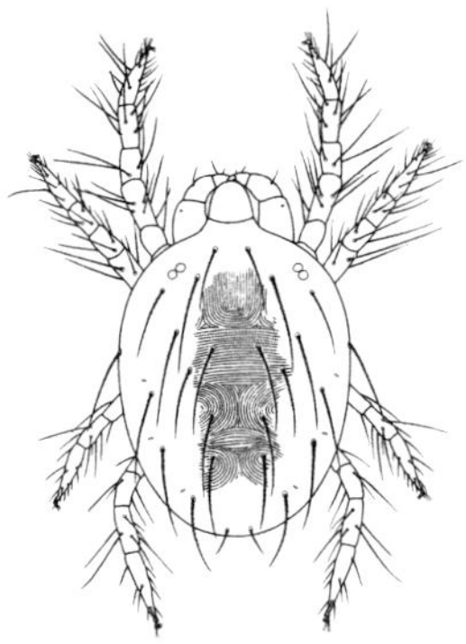

Fig. (2). Female of *Tetranychus urticae* Koch (Acari) [4].

2. Mouthparts adapted for piercing and sucking, located in a beak, without palpi (**Hemiptera**) (Figs. **3** and **4**).. 3

-. Mouthparts otherwise, palpi usually present (Figs. **21, 28** and **29**)............................ 11

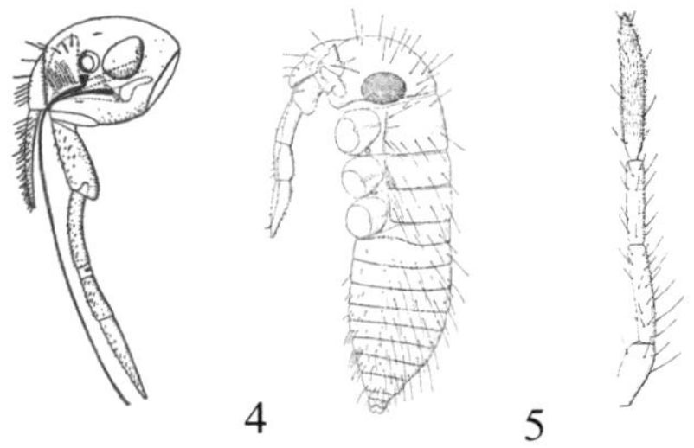

3 4 5

Figs. (3-5). Lateral view of mouthparts of adult (3), lateral view of larva (4) and antenna (5) of *Capsodes lineolatus* (Br.) (Hemiptera, Miridae) [3].

3. Beak located at front of head, antennae with 4-5 segments, forewings with their basal part thickened; their apical portion and the hind wings membranous.. ....**Miridae** (Figs. **3-7**) (suborder **Heteroptera)**

-. Beak located at the back of the head or between coxae I, antennae either with more than 5 segments or bristle-like; forewings, if present, usually similar to hind wings or (in one family) thicker, but not darkly sclerotized 4

4. Antennae short, bristle-like, beak emerging from the back of the head... ....**Cicadellidae** in the suborder **Auchenorrhyncha** (Figs. **8 and 9**)

-. Antennae long, segmented, beak emerging between coxae I (suborder **Sternorrhyncha**) (Figs. **12, 14,** and **19**).. 5

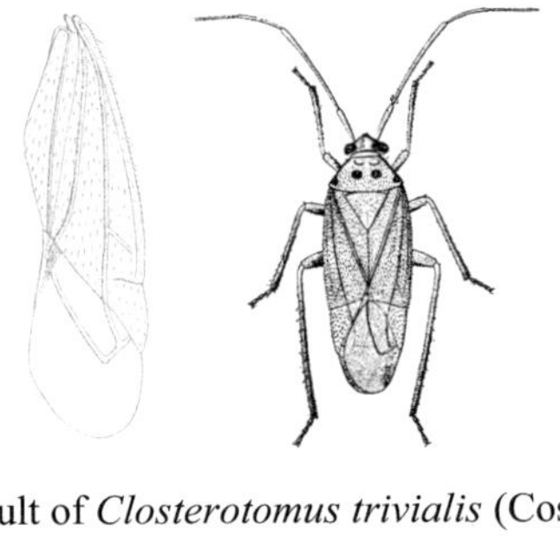

6 7

Figs. (6 and 7). Forewing of Heteroptera (6); adult of *Closterotomus trivialis* (Costa) (7) [3].

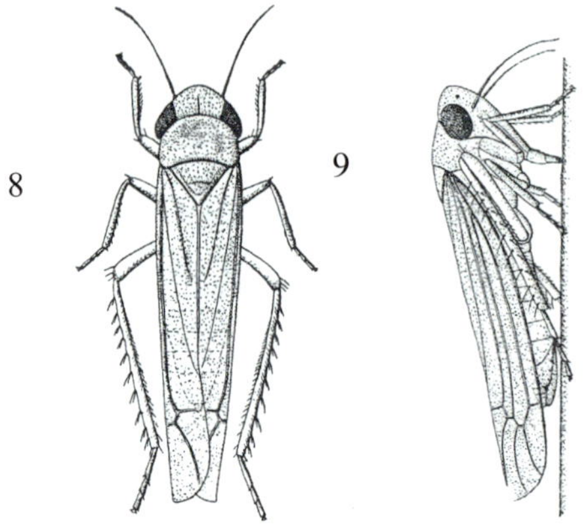

Figs. (8 and 9). Dorsal (8) and lateral view (9) of *Empoasca flavescens* F. (Auchenorrhyncha Cicadellidae) [3].

5. Tarsi 2-segmented, with 2 claws, females with two pairs of well-developed wings 6

- Tarsi with 1 segment bearing a single claw, females always wingless.. 8

6. Antennae with 5-10 segments, forewings may be thickened; jumping insects.......................... **Psyllidae**

The jumping plantlice.

- Antennae with 3-7 segments, not jumping insects.. 7

7. Wings covered with white powder; hind wings and forewings subequal in size; no siphunculi (cornicles)... **Aleyrodidae (**Figs. **10 and 11)**

The whiteflies, whose nymphs are stationary on their host plants. Several species are pests of citrus and vectors of plant viruses.

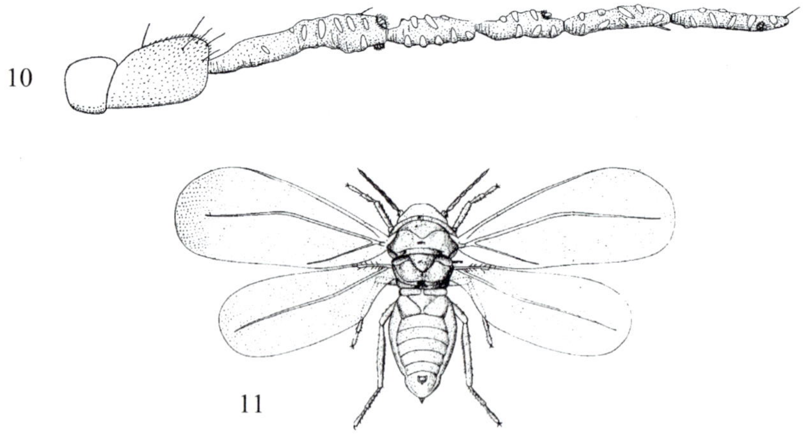

Figs. (10 and 11). Antenna of *Dialeurodes citri* (Ashmead) (10) [3]; adult of *Dialeurodes citri* (11) [5].

- Wings not covered by white powder; hind wings much smaller than forewings; siphunculi (cornicles) present .... **Aphididae (**Figs. **12 and 13)**

Some aphids (*e.g. Toxoptera aurantii, T. citricidus*) are serious citrus pests, due to their feeding, honeydew excretion and transmission of plant viruses.

13

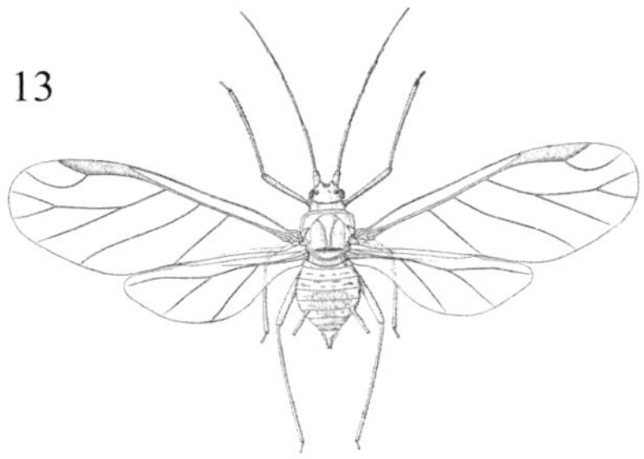

12

Figs. (12 and 13). Female alata (12) and antenna of anphigonic female (13) of *Myzus persicae* (Sulzer) [3].

8. Anterior spiracles present .... **Monophlebidae** (Fig. **14)**

***Icerya purchasi* is a devastating pest of citrus that is usually under complete biological control.**

- Anterior spiracles absent... 9

9. Anal opening covered by two triangular plates, abdomen with a developed anal cleft .... **Coccidae** (Fig. **15)**

The soft scales include *Coccus hesperidum* and *Ceroplastes floridensis*, serious citrus pests.

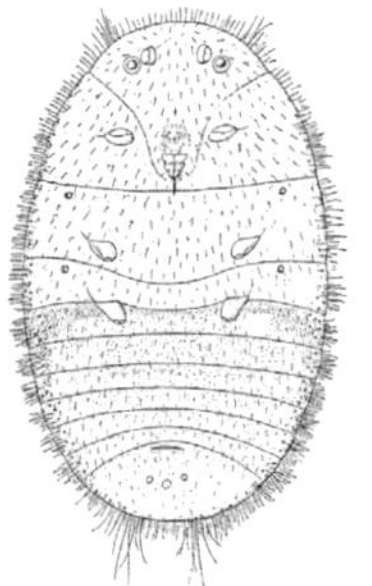

14

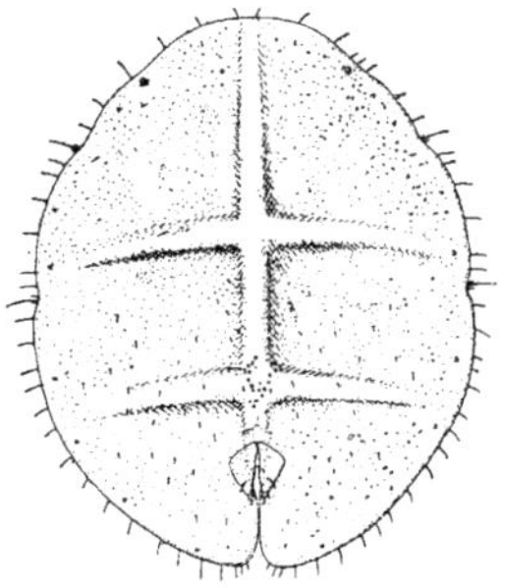

15

Figs. (14 and 15). Ventral view of female of *Icerya purchasi* Maskell (14), with wax removed; dorsal view of female of *Saissetia oleae* (Olivier) (15) [3].

- Without triangular plates covering the anus, anal cleft, if present, not developed 10

10. Posterior abdominal segments fused, forming a pygidium, body covered by a shield (scale), legs absent or vestigial. .... **Diaspididae** (Figs. **16-18)**

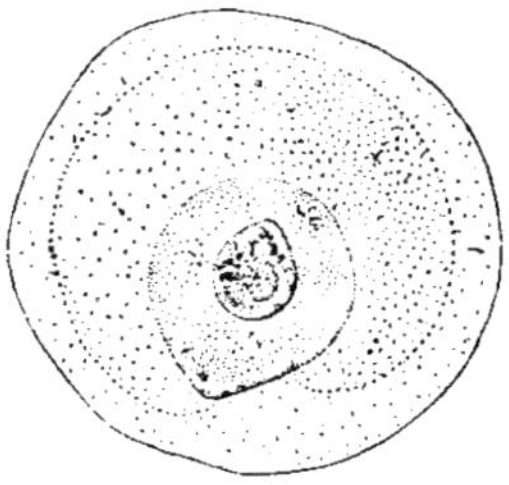

16

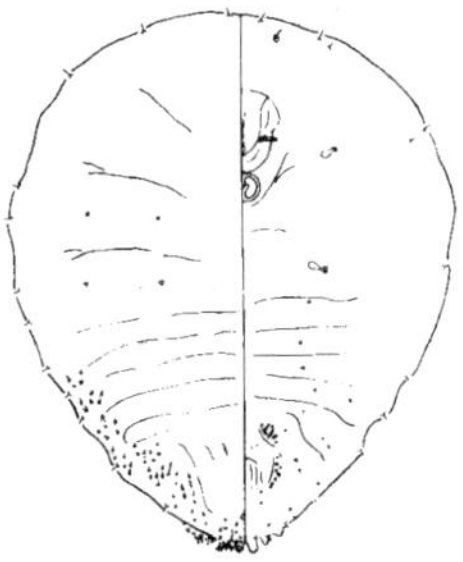

17

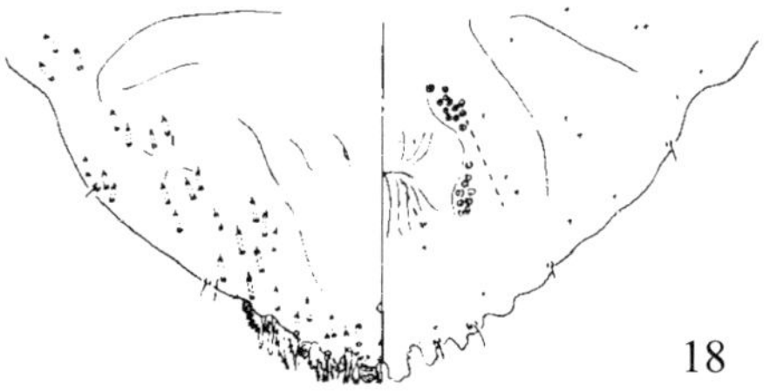

18

Figs. (16-18). Shields of *Aonidiella aurantii* (Maskell) (16) [3]; dorsal (at left) and ventral side (at right) of the body of female of Diaspididae Aspidiotinae (17) [6]; dorsal (at left) and ventral side (at right) of pygidium of female of Diaspididae Aspidiotinae (18) [6].

-. Posterior abdomen clearly segmented, not fused, body covered by a white waxy secretion, legs present .... **Pseudococcidae** (Fig. **19**)

The mealybugs include several serious pests of citrus, including *Planococcus citri*.

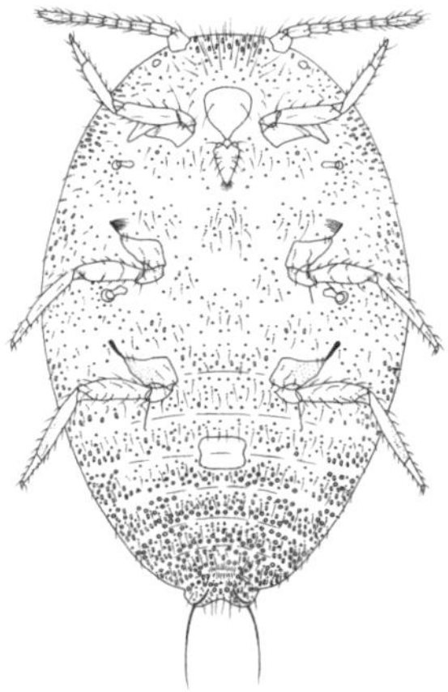

Fig. (19). Ventral view of female of *Planococcus citri* (Risso) (Sternorrhyncha, Pseudococcidae) [7].

11. With a single pair of wings (**Diptera**) (Figs. **20** and **22**) 12

-. With two pairs of well-developed wings 13

12. With long, multi-segmented antennae adorned by whorls of hairs; wings without spots and cross veins .... **Cecidomyiidae** (Fig. **20**)

***Contarinia citri* is the only gall midge known to damage citrus in the Mediterranean Region.**

20

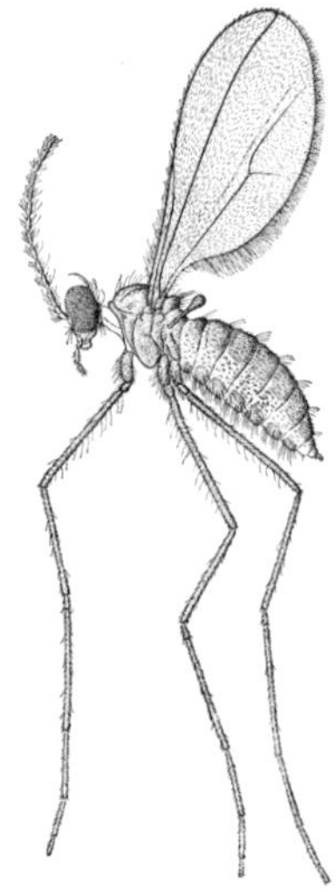

21

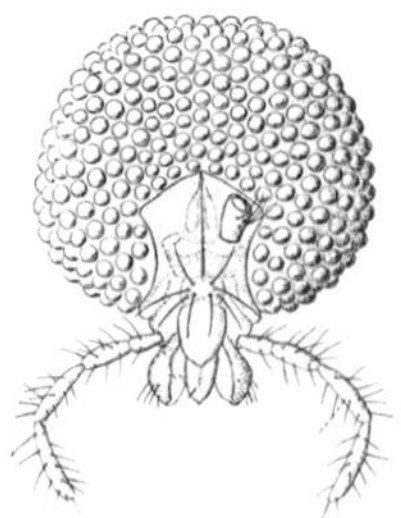

Figs. (20 and 21). Female (20) and frontal view of head with mouthparts of Cecidomyiidae (Diptera) (21). [8]

- With short, bristle-like antennae, devoid of whorls; wings spotted and with several cross veins .... **Tephritidae** (Fig. **22**)

The Mediterranean fruit fly, *Ceratitis capitata*, is a major citrus pest wherever it occurs.

Fig. (22). Female of *Ceratitis capitata* (Wiedemann) [9].

13. Anterior pair of wings completely and often darkly sclerotized... **Coleoptera**

-. Anterior and posterior wings similar (but may differ in size), neither thickened nor sclerotized......... 14

14. Body and wings covered by scales. **Lepidoptera** (Fig. **23**) 15

-. Body and wings without scales 17

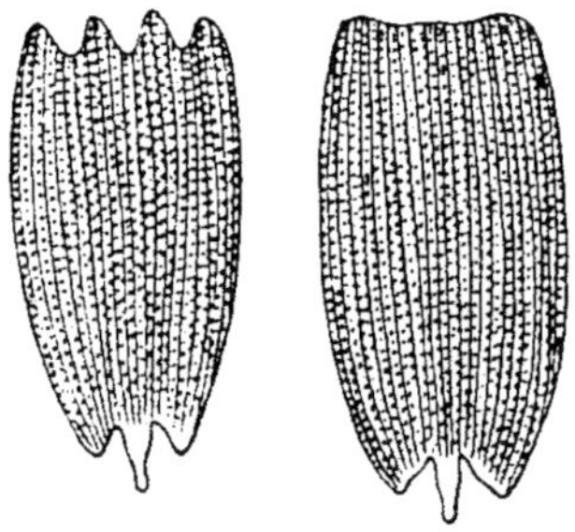

Fig. (23). Scales of wings of Lepidoptera [10].

15. Antennae shorter than forewings, without a fringe of long hairs on the hind wings 16

-. Antennae longer than forewings; with a fringe of long hairs on the hind wings ....
....**Gracillariidae** (Fig. **24**)

The leaf-blotch miners are represented in the Mediterranean Region by one major citrus pest, the citrus leafminer, *Phyllocnistis citrella.*

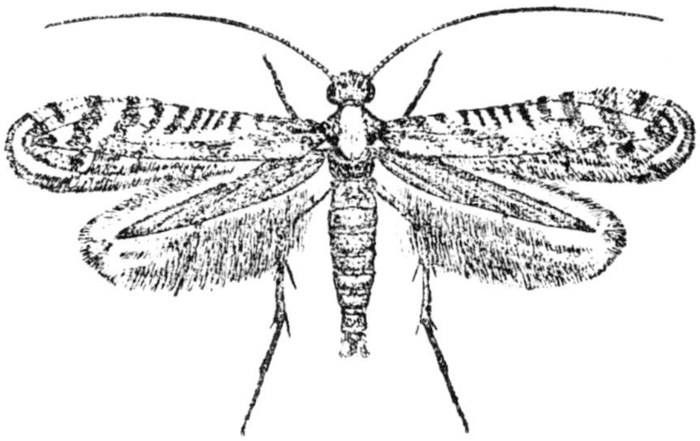

Fig. (24). Lepidoptera Gracillariidae [3].

16. Forewings and hind wings subequal in size, or hind wing narrowing distally; ocelli absent. ....**Yponomeutidae** (Fig. **25**)

The ermine moths are represented in the Mediterranean Region by the citrus flower moth, *Prays citri.*

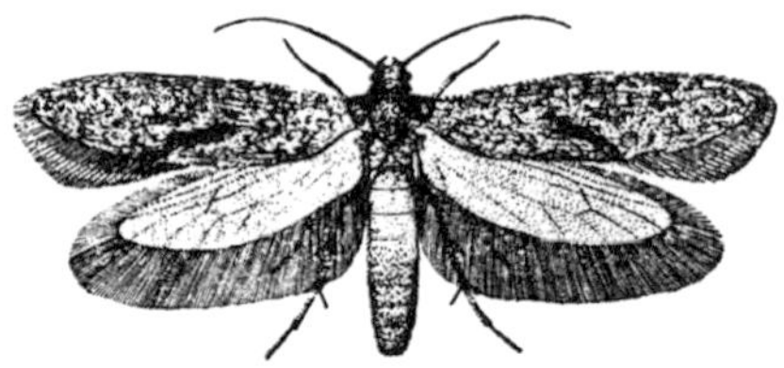

Fig. (25). *Prays citri* Millière [3].

- Forewings much larger than hind wings, which are rounded distally; ocelli present...... **Pyralidae** (Fig. **26**)

Two members of the snout or pyralid moths occur on citrus in the Mediterranean Region: *Cryptoblabes gnidiella* and *Ectomyelois ceratoniae.*

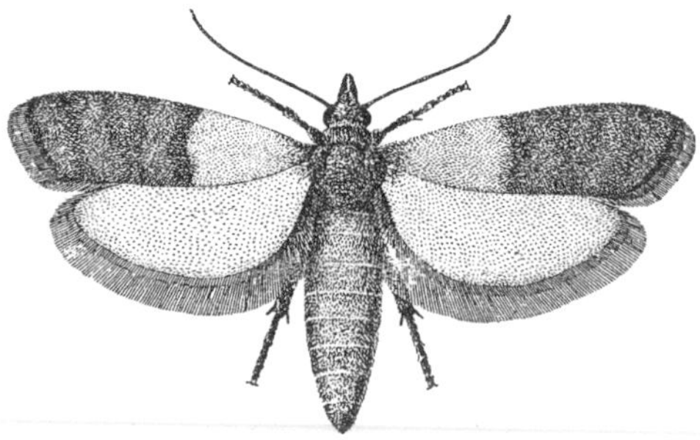

Fig. (26). Lepidoptera Pyralidae [11].

17. Wings with a long posterior fringe of setae; thorax and abdomen not joined by a constriction ....**Thysanoptera** (Fig. **27**)

Several thrips species may occur on citrus in the Mediterranean Region: *Thrips angusticeps, T. major, T. meridionalis, T. tabaci, Frankliniella occidentalis, Heliothrips haemorrhoidalis, Scirtothrips inermis, and Pezothrips kellyanus.* Only the latter is a serious pest.

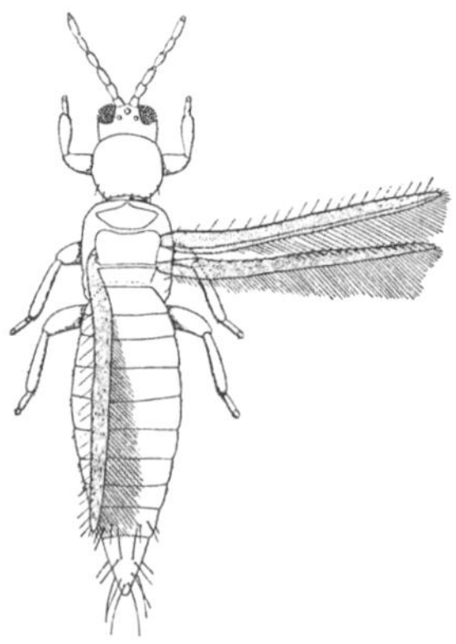

Fig. (27). Adult of Thysanoptera Thripidae [12].

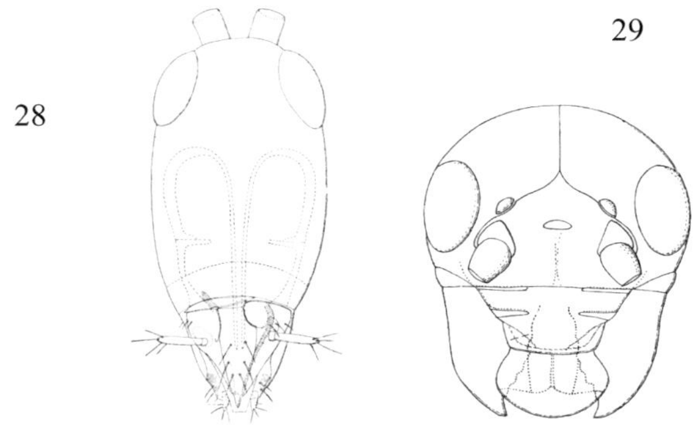

Figs. (28 and 29). Frontal view of head with mouthparts (28) of Thysanoptera Thripidae and Orthoptera Gryllidae (29) [3].

-. Wings without a fringe of long posterior setae; thorax and abdomen joined by a constriction, formed by the anterior abdominal segment .... **Hymenoptera**

The ants, which may disrupt the activities of natural enemies and are sometimes phytophagous, belong to this order (Fig. **30**).

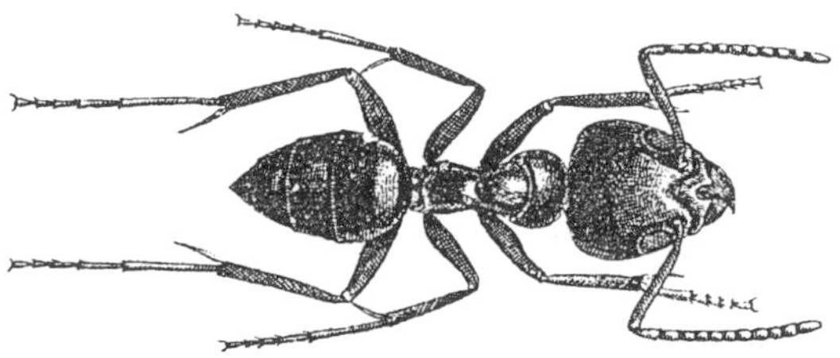

Fig. (30). Adult of Hymenoptera Formicidae [12].

18. Body worm-like, with only 2 pairs of legs ....**Eriophyidae** (Fig. **31**)

Several rust or bud mites, like the pink citrus rust mite, *Aculops pelekassi*, and the citrus rust mite, *Phyllocoptruta oleivora*, are serious pests of citrus.

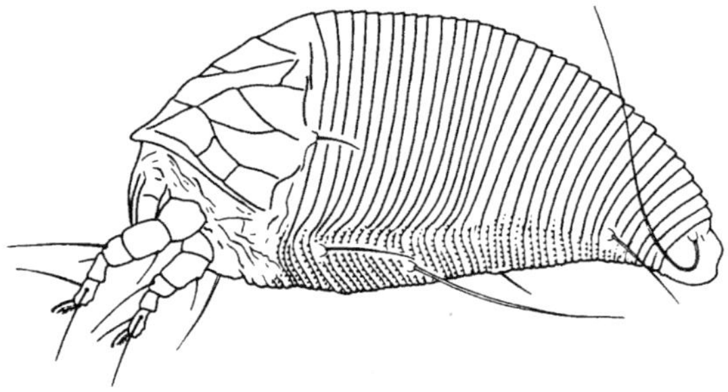

Fig. (31). Adult of *Aculops pelekassi* Keifer [4].

-. Body not worm-like, with 4 pairs of legs 19

19. Tarsus of leg IV different from other tarsi, bearing a terminal whip-like seta but no pretarsus, claw or pulvillus .... **Tarsonemidae** (Fig. **32**)

Only *Polyphagotarsonemus latus*, the broad mite, is a pest of citrus.

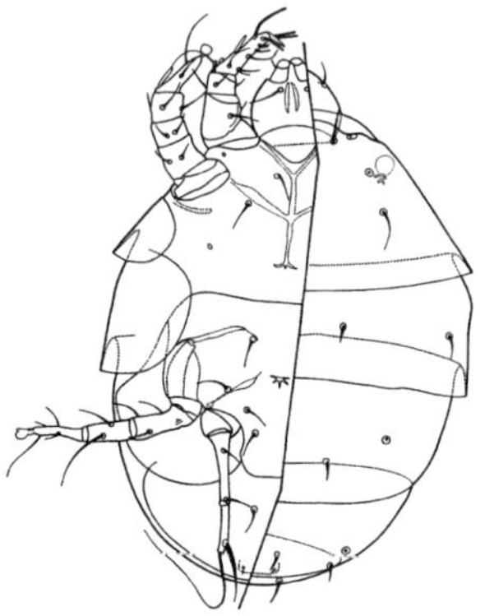

Fig. (32). Ventral (at left) and dorsal side (at right) of the female of *Polyphagotarsonemus latus* (Banks) [13].

-. Tarsus of leg IV similar to other tarsi, without a terminal whip-like seta but with pretarsus, claw or pulvillus 20

20. With a distinct palpal claw or thumb-claw process .... **Tetranychidae** (Figs. **2**, **33 and 34**)

Several spider mites, like the citrus red mite, *Panonychus citri*, are serious, world-wide pests of citrus.

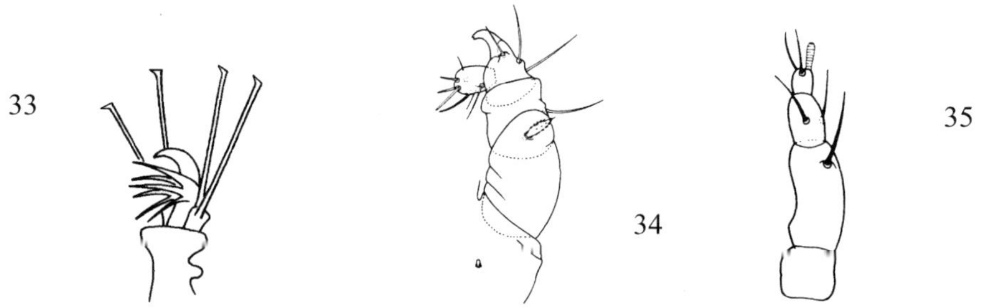

Fig. (33-35). Pretarsus I of female of *Panonychus citri* (McGregor) (33) [14]; palp with thumb-claw process of Tetranychidae (34); palp of Tenuipalpidae (35) [15].

-. Without a distinct palpal claw or thumb-claw process .... **Tenuipalpidae** (Figs. **35 and 36**)

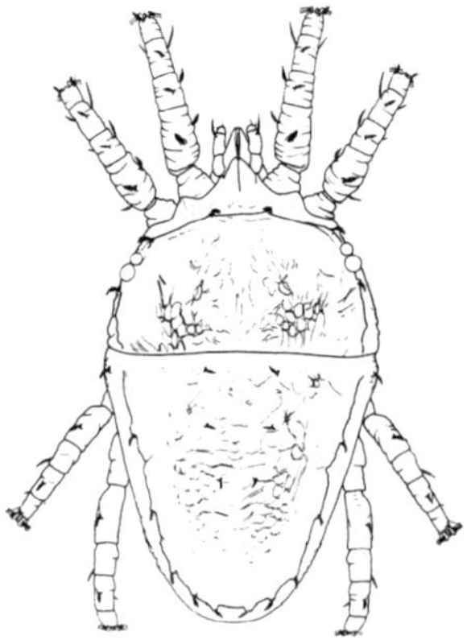

Fig. (36). Female of *Brevipalpus phoenicis* Geijskes (Acari Tenuipalpidae) [4].

Several false spider mites in the genus *Brevipalpus* transmit serious virus diseases of citrus (*e.g.* leprosis).

REFERENCES

[1] Gullan, PJ, Cranston PS. The Insects, an Outline of Entomology. Hoboken, NJ: Wiley-Blackwell 2010.

[2] Lindquist EE, Krantz, GW, Walter, DE. Classification. In: Krantz GW, Walter SE, (Eds). A Manual of Acarology 2009; pp. 97-103.

[3] Silvestri F. Compendio di Entomologia applicata. Vol. I and II. Portici: Tipografia Bellavista 1934-39.

[4] Vacante V, Nucifora A. Gli Acari degli agrumi in Italia. I. Specie rinvenute e chiave per il riconoscimento degli ordini, dei sottordini e delle famiglie. Boll Zool Agr Bachic 1985; II: 115-66.

[5] Priore R. Il *Dialeurodes citri* (Ashmead) in Campania. Boll Lab Ent Agr Portici 1969; 27: 287-316.

[6] Stoetzel MB, Davidson JA. Sexual dimorphism in all stages of the Aspidiotini. Ann Ent Soc Amer 1974; 67: 138-40.

[7] Cox JM. Pseudococcidae (Insecta: Hemiptera). Fauna of New Zealand 1987; 11: 1-228.

[8] Roberti D. La *Phaenobremia aphidimyza* (Rond.) (Diptera Cecidomyiidae) predatrice di Aphis (Doralis) frangulae Koch. Boll Ist Ent Univ Bologna 1946; 15: 233-56.

[9] Costantino G. La Mosca della frutta. Bollettino R. Stazione Sperimentale di Frutticoltura e di Agrumicoltura di Acireale 1941; XX, 78: 1-13.

[10] Berlese A. Gli insetti, loro organizzazione, sviluppo, abitudini e rapporti coll'uomo. Vol. I and II. Milano: Società Editrice Libraria 1909-1925.

[11] Candura GS. La Plodia in Italia. Contributi 1-2. Boll Zool Agr Bachic 1937; 8: 129-55.

[12] Menozzi C. Animali e vegetali dannosi alla barbabietola da zuchero e mezzi per combatterli. Bologna: Soc. Tip. Bolognese 1947.

[13] Nucifora A, Vacante V. Citrus mites in Italy. VII. The family Tarsonemidae. Species collected and notes on ecology. Acarologia 2004; XLIV, 1: 49-65.

[14] Vacante V. Citrus Mites. Identification, Bionomy and Control. Wallingford: Cabi Publishing 2010.

[15] Vacante V, Nucifora A. Gli Acari degli Agrumi in Italia. I. Specie rinvenute e chiave per il riconoscimento degli ordini, dei sottordini e delle famiglie. Boll Zool agr Bachic, ser. II, 1985; 18: 115-66.

CHAPTER 5

Natural Enemies

Carmelo Peter Bonsignore* **and Vincenzo Vacante**

Dipartimento Patrimonio Architettonico ed Urbanistico (PAU), Mediterranean University of Reggio Calabria, Via Melissari, Località Feo di Vito, 89060 Reggio Calabria, Italy

Abstract: The major natural enemies of the citrus pests in the Mediterranean Region are briefly presented. The complexity of the topic has led to a simplification of the treatment. The more common Hemiptera, Neuroptera, Lepidoptera, Diptera, Coleoptera, Hymenoptera and Acari are discussed, along with the general features of each group.

Keywords: Citriculture, Mediterranean Region, citrus pests, natural enemies, IPM.

1. INTRODUCTION

Natural enemies are fundamental for the implementation of IPM (Integrated Pest Management) [1]. Their history, ecological role and technical importance were discussed in chapters **2** and **3**, respectively. The correct identification of the natural enemies represents a fundamental step in the implementation of a successful IPM program. To meet this need, we already presented the basic characters needed in order to identify the major groups of pests. In this chapter we provide information on the major groups of natural enemies (pathogens, spiders, mites and insects). Herein our primary aim is to present the general characters of the common natural enemies that live on citrus in the Mediterranean Region. The treatment has basically taken into account the main natural enemies, particularly insects and mites, and for each group (order, suborder, family) are presented the general morphological features that can guide their identification and the bio-ecological adaptations that underline their importance. Moreover, being a very complex ecosystem, natural enemies that are not treated herein may be found in citrus orchards. In general, lacking any indication of the identification of different species, readers interested in their identification are advised to consult specialist texts [2, 3].

With the exception of the most recent figures, the other illustrations are of historical interest and were chosen for the importance of various authors, their beauty and the fidelity of the details shown. In the legends the authors are indicated in parentheses.

2. MAJOR SYSTEMATIC GROUPS OF NATURAL ENEMIES

Most important natural enemies of citrus pests are pathogens, spiders, mites and insects. Other beneficials, such as protozoa and nematodes, may occur but are not discussed here.

2.1. Key for the Identification of Major Systematic Groups of Natural Enemies (Adults)

1. Four pairs of legs 2
-. Three pairs of legs .... class **Insecta**...7
2. Spinnerets absent .... subclass **Acari**...3
-. Body segments X and XI with 1-4 pairs of ventral spinnerets .... order **Araneae** (Fig. **1**)
3. Without stigmata posterior to coxae II; coxae I-IV fused with the body and the trochanter is the first free leg segment; tarsi intact. supeorder .... **Acariformes**.....4

***Address correspondence to Carmelo Peter Bonsignore:** Dipartimento Patrimonio Architettonico ed Urbanistico (PAU), Mediterranean University of Reggio Calabria, Via Melissari, Località Feo di Vito, 89060 Reggio Calabria, Italy; Tel.: + 39-0965-385201; Fax: +39-0965-385219; E-mail: cbonsignore@unirc.it

-. With 1-4 pairs of stigmata posterior to coxae II; coxae variously articulated to body; tarsi II-IV each with a peripodomeric fissure separating the basitarsus from the telotarsus (superorder **Parasitiformes);** 1 pair of opisthosomal stigmata, lateral to coxae II-IV or posterior to coxae IV, with elongate peritremes oriented anteriorly; subcapitulum with a maximum of 4 pairs of setae; tritosternum with base and 2 laciniae; anal valves nude or at most with a single pair of setae (order **Mesostigmata**); dorsal shield with porelike structures and without an enlarged pair between setal series *J* and *Z*; less than 23 pairs of dorsal setae; less than 4 pairs of marginal setae on female soft integument; fixed and movable cheliceral digits developed; tibia II with 7 setae and tibia IV with 6 setae .... family **Phytoseiidae** (Figs. **2-5**)

Very mobile mites, white hyaline tending to gray, some species brown or orange red; feeding on other mites, small insects and pollen.

4. Chelicerae with bases sometimes fused medially, rarely chelate, with fixed digit often regressed and movable digit variously structured (hook, knife, needle, stylet-like); palpi simple or modified into a thumb-claw process (order **Trombidiformes**); 1 pair of stigmata between bases of chelicerae or on anterior prodorsum .... suborder **Prostigmata**...5

-. Chelicerae with bases always separate, chelate, dentate, and rarely attenuate or styletlike; palpi simple, never with thumb-claw process (order **Sarcoptiformes**); prodorsum without specialized sensory organs; idiosoma weakly sclerotized, with epimeral plates undeveloped or weakly formed between bases of the legs (suborder **Oribatida**, cohort **Astigmatina**); empodial claw absent and condylophores apparently missing; tibiae I-II with 0-1 setae; genital opening between or posterior to coxal fields IV; female ovipore often confluent with anal opening; male with alveoli of setae *4b* fused medially and modified into a sucker anterior to aedeagus .... family **Hemisarcoptidae** (Fig. **9**)

Very slow mites, white hyaline tending to gray and with brown or blackish spots, feeding on armored scale insects.

5. Prodorsum with 1-2 pairs of bothridial setae; peritremes absent; chelicerae free and moving laterally; palpi ending in a stout spine, sometimes with strong spines or knobs; 2 pairs of genital papillae; eugenital setae lacking; tarsi III-IV and tibia I without trichobotria............... family **Cunaxidae** (Fig. **8**)

Large mites, brown to brown-red or brown green, sometimes yellow brown red; feeding on other mites and on little arthropods.

-. Prodorsum without bothridial setae 6

6. Body with one or more dorsal plates separated by striate cuticle, or soft; chelicerae fused medially for at least one third from the base; tibiae and genua with more than 3 setae; palpi 5-segmented, with a thumb-claw process .... family **Stigmaeidae** (Fig. **7**)

Body flattened dorsoventrally, yellow, orange to red; feeding on other mites, little arthropods and various scale insects.

-. Body oval to round, sometimes elongate; stylophore fused with subcapitulum to form a gnathosomatic capsule showing dorsally a complex peritreme; palpi often with fewer than 5 segments; palptibial-claw developed; tibiae and genua with 3 or few setae; genu I with 1 solenidion .... family **Cheyletidae** (Fig. **6**)

Body yellow, orange to red, feeding on little arthropods, pest mites and armored scale insects.

7. Mouthparts consisting of two mandibular and two maxillary stylets lying in the lower labrum (rostrum), adapted for piercing and sucking (order **Hemiptera**); forewings with leathery texture near bases and membranous at apex .... suborder **Heteroptera**....8

-. Mouthparts different 9

8. Head with 2 ocelli; rostrum 3-segmented .... family **Anthocoridae** (Fig. **10**)

Minute pirate bugs, red to dark black; feeding on small arthropods.

-. Head without ocelli; rostrum 4-segmented.. .... family **Miridae** (Fig. **11**)

Minute plant bugs, green to brown in color; feeding on small arthropods.

9. Hindwings reduced to small, club-like structures (halteres) that function as balancing organs during flight .... order **Diptera**...10

-. Two pairs of wings variously structured.. 12

10. Antennae 12-16-segmented and flagellomeres with thin-walled sensoria, long and thread-like loops; forewing with vein R5 unbranched and venation commonly reduced to three main veins.. .... family **Cecidomyiidae** (Figs. **18 and 19**)

-. Antennae with fewer segments.. 11

11. Forewing cell R5 either parallel-sided or narrowed distally; vein 2A short and not reaching wing margin...family **Muscidae**

The hunter fly, *Coenosia attenuata*, preys in flight on aleyrodids and other small arthropods.

-. Wings with spurious veins, located parallel to the fourth longitudinal wing vein. .... family **Syrphidae** (Figs. **20** and **21**)

Flower flies are sometimes very colorful, strikingly swift, and remain in flight at the same point; larvae feeding on aphids.

12. Wings covered with flattened scales (order **Lepidoptera**); larvae feeding on scale insects (*Saissetia oleae, etc.*).......... .... family **Erebiidae** (Figs. **16** and **17**)

Larvae feeding on black scale, on the fig wax scale and on other pests.

-. Wings never covered with flattened scales.. 13

13. Forewings hardened and unsuitable for flight (elytrae); hindwings membranous and adapted to flight (order **Coleoptera**); distal segment of maxillary palpus commonly securiform and often conical or parallel-side; antennae short, club-like, 11-segmented (sometimes 7-segmented); elytrae never striate. ....family **Coccinellidae** (Figs. **22-28**)

Lady beetles are most showy; both adults and larvae feeding primarily on aphids, and also on mites, scale insects and other small insects.

-. Forewing not hardened, suitable for flight.. 14

14. Both wings membranous, transparent, similar to each other, with many veins, sometimes longer than body, at rest forming a roof....order **Neuroptera** 15

-. Hindwings smaller than forewings; both wings held together by small hooks (hamuli)........... ....order **Hymenoptera**...16

15. Body medium to large, green or brown; eyes pronounced with a metallic sheen; antennae long and thin; wings delicate... family **Chrysopidae** (Figs. **12** and **13**)

Green lacewings are commonly green colored; larvae actively feeding on aphids, mites and other small arthropods.

-. Body a few mm long, with normal wings, sometimes atrophic; several species with body and wings covered by white waxy secretions ....family **Coniopterygidae** (Figs. **14** and **15**)

Dusty wings have body and wings covered with white waxy secretions; larvae sometimes coated with wax and feeding on small arthropods.

16. Forewings with pterostigma; costal cell almost completely reduced; cross-vein 3 r-m missing; hind wings with almost 2 cells; antennae commonly with more than 16 segments, scape reduced; ovopositor often well developed (superfamily **Ichneumonoidea**); forewings with only one or no recurrent veins; hind trochanters with two segments..................................family **Braconidae** (Fig. **29**)

Some species lay eggs in their hosts by bending the abdomen under the body, bringing it forward until it reaches the host; larvae form a cocoon inside (or below) the host, which hardens and turns into a "mummy"

-. Forewings without pterostigma (except in Proctotrupidae and other groups); hind wings without cells (except in some groups); antennae maximum 16-segmented.. 17

17. Forewings with proximal part of costal vein not developed and cell R1 indistinct; ovipositor often obvious; usually the pronotum not reaching laterally the tegulae; gaster rarely compressed; cercoids present .... superfamily **Chalcidoidea**...18

-. Forewings with costal vein well developed, wing venation otherwise reduced (superfamily **Proctotrupoidea**); forewings without venation; antennae 10-segmented (rarely 7-9-segmented); posterior tarsi usually 5-segmented.. family **Platygastridae** (Fig. **30**)

Little, blackish colored, with antennae at rest bent as Z, variously colored like the legs; parasitoids of whiteflies, scale insects and others.

18. Hind femora greatly enlarged and with a row of teeth or serrations along the lower margin; posterior tibiae arched; forewings not bent longitudinally; ovipositor not bent on dorsum of body; mesosoma robust with very strong sculpturing... family **Chalcididae**[1]

-. Hind femora normal, rarely with row of teeth.. 19

19. Basal segments of tarsi II with teeth; antennae without annuli but with the funiculus maximum 6-articulated; forewings with short marginal veins; mesoscutum without parapsidial furrows (or rarely present)... family **Encyrtidae** (Figs. **33-35**)

Body strong, of small and medium size; parasitoids of scale insects, leaf miner and other insects.

-. Tarsi II not proximally enlarged... 20

20. Forewings with postmarginal vein indistinct or barely visible, but with marginal vein and a short stigmatic vein; antennae maximum 9-segmented (usually 6-8-segmented); tarsi 4-5-segmented;

[1] This family has been included to facilitate the construction of the key. It includes important parasitic species, but none play a key role in the biological control of citrus pests.

mesoscutum with parapsidial furrows; body without metallic colors..... .... family **Aphelinidae** (Figs. **36** and **37**)

Small, not metallic colored; gaster broadly sessile, not narrowed between gaster and propodeum; parasitoids of scale insects.

-. Forewings with well developed postmarginal and stigmatic veins... 21

21. Tarsi 4-segmented; tibiae II with normal spur, short and thin; wings normal; body often with metallic colors ....family **Eulophidae**

Body long and often of a metallic hue, antennae with at most 10 segments; parasitoids of various citrus pests.

-. Tarsi 5-segmented; forewings with postmarginal and stigmatic veins well developed; antennae 8- 13-segmented; posterior margin of pronotum distinct from mesoscutum; anterior tibiae with a well developed and bent spur.................. ... . family **Pteromalidae** (Figs. **31** and **32**)

Body metallic green or bluish, some body parts yellow or blackish; oophagous as well as ectoparasitic, ectoparasitoidic and endophagous.

2.2. PATHOGENS

Pathogens affecting citrus pests include viruses, rickettsiae, bacteria and fungi. Under specific environmental conditions viruses control pest populations, as in the case of citrus red mite, *Panonychus citri* (McGregor) [4]. Bacteria and fungi are common pathogens in the Eumycotina division. Among Bactericeae is the well known *Bacillus thuringiensis* Berliner, a sporigenous species that is nowadays produced by the industry and its different strains are sold for the control of various pest insects, such as moths (Lepidoptera), beetles (Coleoptera) and flies (Diptera). The most important pathogenic fungi belong to the subdivisions Deuteromycotina, Mastigomycotina and Ascomycotina [5]. The Deuteromycotina include pathogens of pest mites, such as eriophyiids and spider mites and insect pests (Coccoidea, Aleyrodidae, Aphididae). The main species are *Beauveria bassiana* Vuillemin, *Metharhizium anisopliae* (Metch.) Sorok., *Paecilomyces* sp., *Verticilium lecanii* (Zimmerman) Viegas, *Aschersonia aleirodidis* Webber [6], and *Hirsutella thompsonii* Fisher [7]. The Mastigomycotina include the genus *Entomophthora*, with about 150 species reported as pathogens of mites and insects. Different species of the Exobasidiomycetidae (*Meira argovae* Boekhout *et al.*, *M. geulakonigii* Boekhout *et al.*, *Acaromyces ingoldii* Boekhout *et al.*) attack the citrus rust mite, *Phyllocoptruta oleivora* (Ashmead), and some spider mites [8]. For more information see Evans and Prior [9].

2.3. ARANEAE

Araneae (Fig. **1**) or spiders are an order in the class Arachnida, whose members are very common in natural environments and crops. Their body is 0.5 mm to 0.9 cm long, variously colored and divided into an anterior prosoma and a posterior opisthosoma. The prosoma bears 2 chelicerae, 2 palps, 4 pairs of legs, and commonly 8 eyes. The opisthosoma is peduncolate and posteriorly bears 1-4 spinnerets, organs that emit silken webs [10]. Spiders build webs with the silk filaments in order to capture prey, which is commonly killed by poison secreted through cheliceral glands. The Araneae number about 35,000 species in 80 families. Dozens of families and species are known from citrus in the Mediterranean Region [11-13]. They are generalist predators that affect the stability of pests populations and represent interesting biological indicators of the integrity of the ecosystems.

2.4. ACARI

The Acari, or mites, range from 300 to 500 μm in body length, except for some Eriophyoids that are approximately 100-200 μm long, or for certain ticks whose female measures about 30,000 μm.

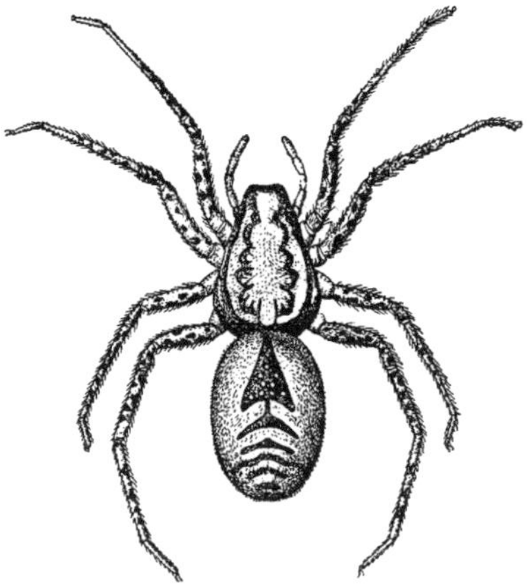

Fig. (1). Araneae. *Lycosa tarentula* Linnaeus [14].

They are a subclass of the class Arachnida of the subphylum Chelicerata, and are distinguished from insects. Although sometimes confused, the Acari are recognizable by their undivided body, by lacking antennae, mandibles and maxillae, and by the adults bearing four pairs of legs (except the Eriophyoids, which have only two pairs of legs). The mites differ from the spiders by the absence of spinnerets and by their undivided body. At present approximately 55,000 mites have been described, but it is believed that the group will grow to number over 1,000,000 species [15]. The subclass numbers about 5,500 genera and 550 families [16].

The families Phytoseiidae, Cheyletidae, Stigmaeidae, Cunaxidae and Hemisarcoptidae are common on citrus and several species contribute to biological control of various pests [3, 17].

Phytoseiidae (Figs. **2-5**) are rounded or oblong mites, white grey and sometimes orange-red or brown in color. The dorsal plate is entire and possesses fewer than 23 pairs of setae. The stigmata are set between legs III and IV and the metasternal plates are small. Leg grooves are absent, leg I has ambulacra, and the digits of the male spermadactyls are free distally [3]. The family consists of the subfamilies Amblyseiinae, Phytoseiinae, Typhlodrominae and more than 2,250 described species [18]. Basically they are adapted to prey on other mites (spider mites, tydeids, tarsonemids, *etc.*), and few species prey on thrips, scale insects and whiteflies, and may supplement their diets with pollen, fungi or other plant substances. Some species have a specialist feeding habit but most are generalist. About 46 species were collected from citrus in different countries of the Mediterranean Region [17].

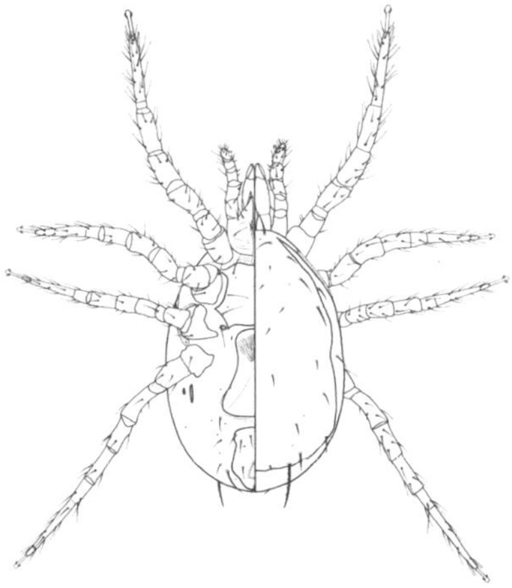

Fig. (2). Dorsal (at right) and ventral view (at left) of a female of *Euseius stipulatus* (Athias Henriot) [19].

Relevant studies on these mites were conducted in Israel [20], Spain [21] and Italy [19]. Different phytoseiid species dominate on citrus in the western and in the eastern Mediterranean parts [22].

Fig. (3). Dorsal (at right) and ventral view (at left) of a female of *Phytoseiulus persimilis* (Athias Henriot) (from Vacante and Nucifora, partially modified) [19].

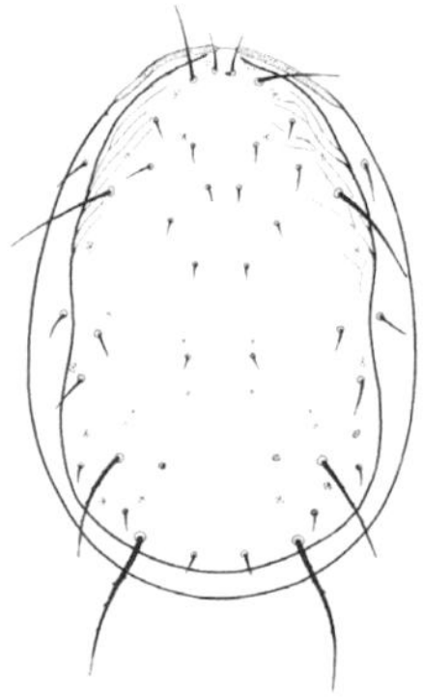

Fig. (4). Female of *Amblyseius swirskii* (Athias Henriot) [19].

Euseius stipulatus (Athias Henriot) (Fig. **2**) is the most common phytoseiid occurring on citrus from Turkey through Greece and Italy to Spain [19, 21, 23]. It is especially active in the control of citrus red mite, *Panonychus citri*, followed in Italy by *Phytoseilus persimilis* Athias Henriot (Fig. **3**), commonly feeding on *Tetranychus* spp., while in Israel and Egypt *Amblyseius swirskii* (Athias-Henriot) (Fig. **4**) and *Amblyseius athiasae* Porath and Swirski (Fig. **5**) appear to be dominant [20, 24]. The development of their populations is related to several environmental factors, especially the use of selective chemicals; their introduction into an environment with classic biological strategy has a concrete meaning, whereas their use in inundative biological control [25] proved to be ecologically and economically unproductive. For a catalogue of the world species of Phytoseiidae, see de Moraes *et al.* [18].

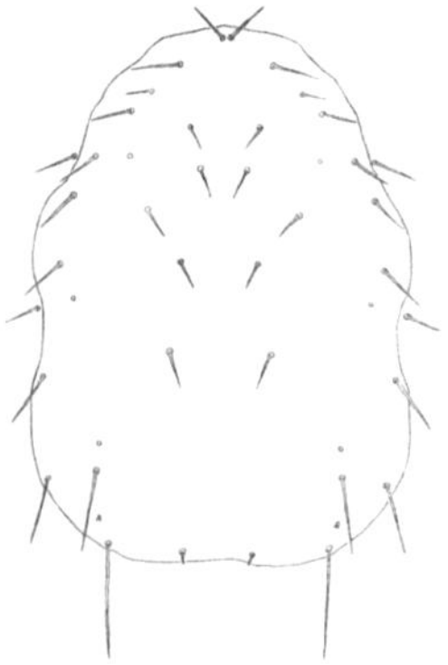

Fig. (5). Female of *Amblyseius athiasae* Porath and Swirski [20].

Cheyletidae. These mites (Fig. **6**) have a rounded or oblong body, usually intense yellow to orange-red in color. They have a prominent gnathosoma and a strong thumb-claw complex; the palptarsus usually bears sickle- and/or comb-like sensillae. The dorsal setae are simple, hair-like, clavate, shell-like or staghorn-like. The males may occur as homeomorphs, similar to the females, or as heteromorphs, dissimilar from females [3]. The family consists of about 400 described species in 80 genera [26]. Eight species were recorded from Mediterranean citriculture; *Cheletogenes ornatus* (Canestrini and Fanzago) (Fig. **6**) is common in Italy, Israel, Egypt, Lybia and Spain [17]. In Italy this predator was found in colonies of *Tetranychus urticae* Koch on citrus leaves and in lemon buds infested by *Aceria sheldoni* (Ewing), and in association with armored scale insects [27].

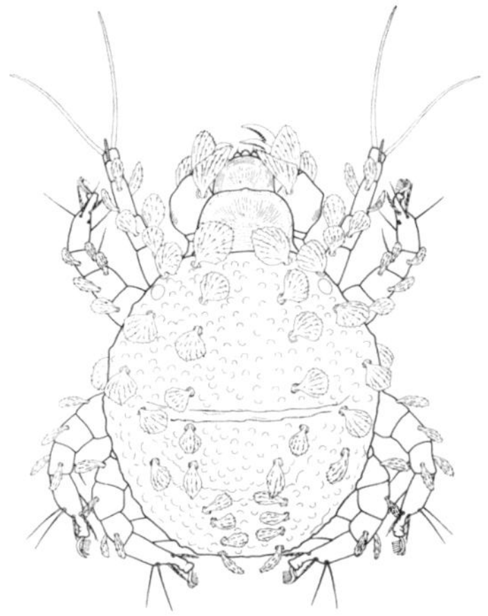

Fig. (6). Female of *Cheletogenes ornatus* (Canestrini and Fanzago) [30].

In Israel, it is a year-round resident in chaff scale (*Parlatoria perganii* Comstock) colonies on citrus bark and represents a "second line" predator [28]. In Egypt the predator feeds on the eggs of scale insects and mites [29]. *Cheletomimus berlesei* (Oudemans) was found in Italy, Israel and Egypt, whereas other species, as *Hemicheyletia bakeri* (Ehara) seem confined to more restricted sites (Middle East) [17]. For detailed information see [3, 26].

Stigmaeidae (Fig. **7**) are ovoid or elongate mites, yellow, orange to red colored, often covered by one or more plates. No obvious stigmata or peritremes are present. The chelicerae are short and stylet-like, the palpi bear a thumb-claw complex and the tarsi often carry a distal three-tined seta [3]. At present the family includes about 300 species and 25 genera [31]. About 14 species are known from the Mediterranean citriculture [17].

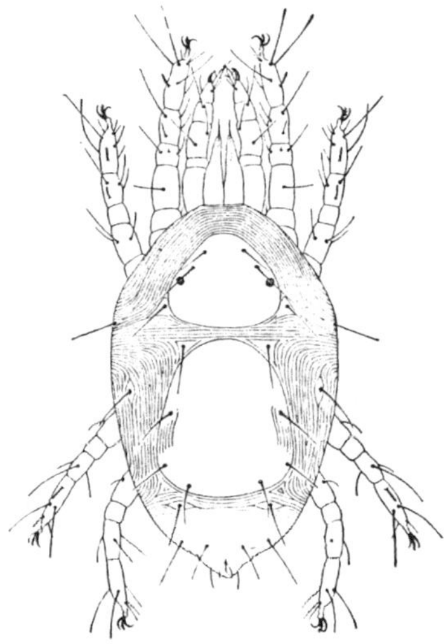

Fig. (7). Adult of *Zetzellia mali* (Ewing) [35].

They include *Zetzellia mali* (Ewing) collected in Italy and Yugoslavia, *Z. graeciana* Gonzalez, found in Italy in the buds and on leaves of lemon and other citrus, where it feeds on *Aceria sheldoni* and other pest mites [27], *Agistemus fanari* Dosse, feeding on spider mites and tenuipalpids in Lebanon [32], *A. exsertus* Gonzalez Rodriguez found in Egypt, where it is recorded as the year-round, key acarine predator of mites and some scale insect pests [33], and in Israel [34].

Cunaxidae (Fig. **8**) are rounded or oblong mites variously colored, brown to brown-red or brown green, sometimes yellow brown red. The palpi are three to five segmented, with strong spines, spurs or apophyses, and commonly ending with a strong claw. The prodorsum bears two pairs of sensillae, and tibiae IV has one pair. The family numbers about 200 species in less than 20 genera [3]. Species found in the Mediterranean citrus include *Cunaxa capreolus* (Berlese) in Italy [17] and Egypt [36], *C. setirostris* (Hermann) in Italy, Egypt and Spain, *Cunaxoides olivieri* (Schruft) in Italy and *Cunaxoides americanus* Baker in Spain [17]. They feed on all stages of mite pests and other little arthropods.

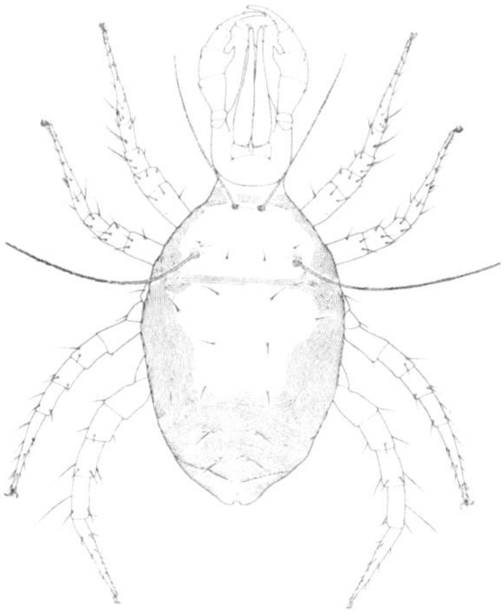

Fig. (8). Female of *Cunaxa capreolus* (Berlese) [30].

Hemisarcoptidae (Fig. **9**) are small, whitish mites. They have a prodorsal sclerite and genital and anal openings set beyond coxae IV, confluent in the female, separated in the male. The latter has a median ventral sucker plate anterior to the genital area. The tarsi end with an elongate, sucker-like ambulacrum. The hypopodes (deutonymphs) are dark brown and heavily sclerotised. Species of *Hemisarcoptes* feed on armored scale insects (Diaspididae) and their hypopodes are phoretic on predatory ladybirds of the genus *Chilocorus* (Coccinellidae), which feed on the same prey [3].

Fig. (9). Female of *Hemisarcoptes malus* (Shimer) [37].

Hemisarcoptes coccophagus Meyer, found in Israel [38] and Spain [21], and *H. malus* (Shimer) collected in Italy [27] and Lybia [39] occur on citrus in the Mediterranean Region. These mites may contribute considerably to the control of various diaspidids, such as the California red scale, *Aonidiella aurantii* (Maskell) [38] and the chaff scale [40].

2.5. INSECTS

2.5.1. Hemiptera

The Hemiptera have piercing and sucking mouthparts, consisting of two mandibular and two maxillary stylets lying in the lower labrum (rostrum). Members of the suborder Heteroptera include predators of small and medium-sized insects and are characterized by the texture of the front wings, which are leathery near the base and membranous at the apex [41]. Species of the families Anthocoridae and Miridae are common on citrus.

Anthocoridae (Fig. **10**) or minute pirate bugs, are 2 to 5-6 mm long and red to dark black in color. The head has two ocelli, antennae with four segments and the rostrum 3-segmented. The membranous part of the forewings has no cells, and bears a short basal vein. The family consists of about 450 zoophagous species in 80 genera, usually feeding on small arthropods [42]. A few are phytophagous, or supplement their diets with pollen or plants. The common species belong to the genera *Orius* and *Anthocoris*.

Miridae (Fig. **11**) or plant bugs, are 2-4 mm long and green to brown in color. The head has no ocelli; the antennae are 4-segmented and the rostrum is 4-segmented. The membranous portion of the forewings bears 1 or 2 closed cells, and the thickened portion has a *cuneus*, a somewhat triangular-shaped area, adjacent to the membranous area. The tarsi are 3-segmented [42]. Usually they prey on small arthropods, but some species are phytophagous or feed on animals as well as plants. Species of the genera *Macrolophus*, *Nesidiocoris*, and *Deraeocoris* are common on citrus.

2.5.2. Neuroptera

Neuroptera are 5-150 mm long, slender and delicate. The antennae are long, thin, with many segments. They carry two pairs of membranous, transparent wings, similar in size and rich in veins, which at rest form a roof. Usually, they have chewing mouthparts, with mandibles and maxillae. The larvae possess three pairs of legs and mouthparts with mandibles and maxillae; in some groups mouthparts are modified, to bite and suck. The Neuroptera belong to the superorder Neuropteroidea, and include about 5,000 species [44]. Species of the families Chrysopidae and Coniopterygidae are common on citrus.

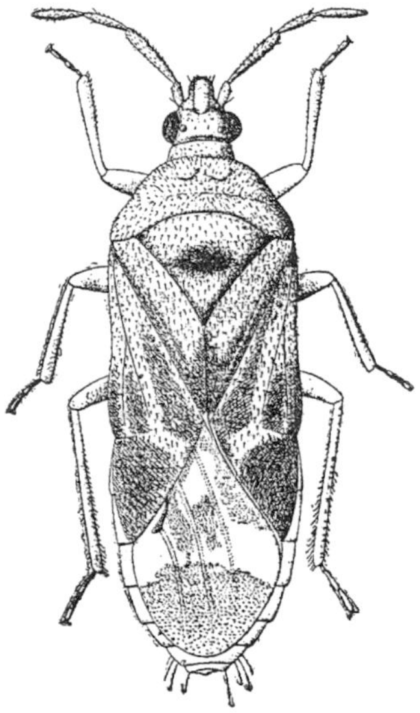

Fig. (10). Female of *Anthocoris nemoralis* (Fabricius) [43].

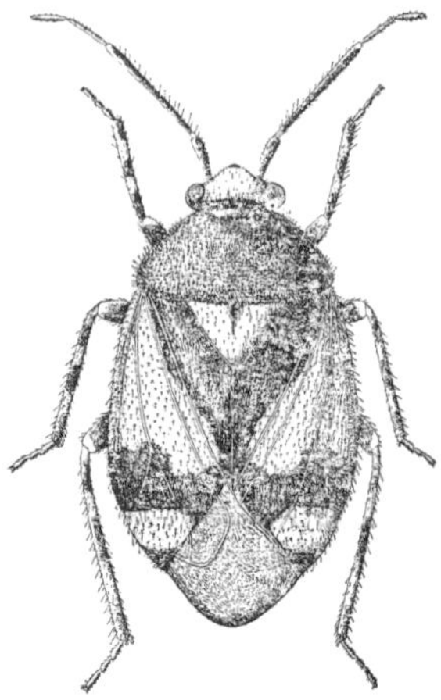

Fig. (11). Female of *Deraeocoris schach* (Fabricius) [43].

Chrysopidae (Figs. **12** and **13**) or green lacewings have a medium to large size, and are green or brown in color. They possess two pairs of similar, membranous wings, with many veins; the forewings with cross veins in the costal area that are not bifurcate. The eyes are pronounced with a metallic sheen and there are no ocelli. The antennae are long and thin. The larva is campodeiform, with long mouthparts and setae, preying actively on aphids, mites, scale insects and other suitable prey. Most species are predatory as adults, also feeding on pollen or liquid sugars of plant origin. The more common species belong to the genera *Chrysopa*, *Chrysoperla* and *Mallada* [45, 46].

12

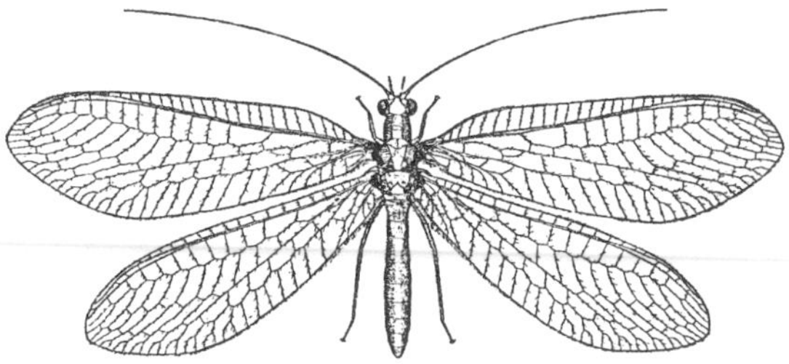

13

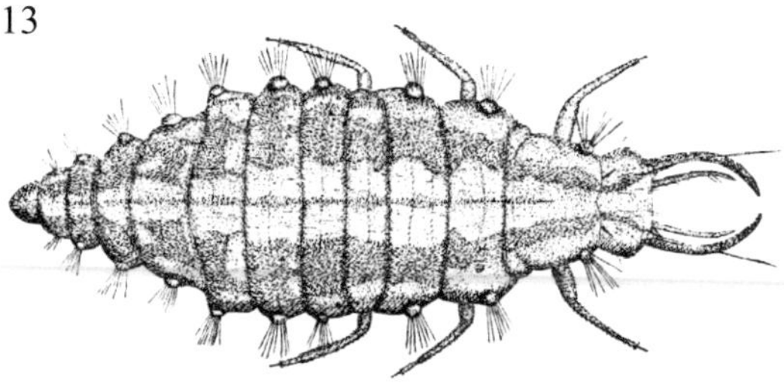

Figs. (12 and 13). Female (12) and larva (13) of *Chrysopa carnea* Stephens [43].

Coniopterygidae (Figs. **14** and **15**) or dustywings are a few mm long; they have normal wings (sometimes atrophic) whose the costal area has no more than two veins. Body and wings are covered with white waxy secretions (like the whiteflies, Aleyrodidae). The larva is campodeiform, fusiform, and sometimes coated with wax and weaves a cocoon on leaves. It preys small arthropods (scale insects, aphids, whiteflies, mites and others). The adults feed on pollen or liquid sugars of plant origin. *Semidalis aleyrodiformis* Stephens and *Conwentzia psociformis* (Curtis) are common [45, 47].

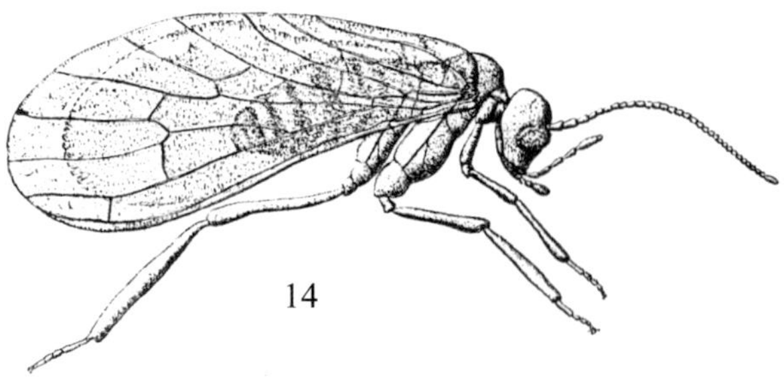

14

15

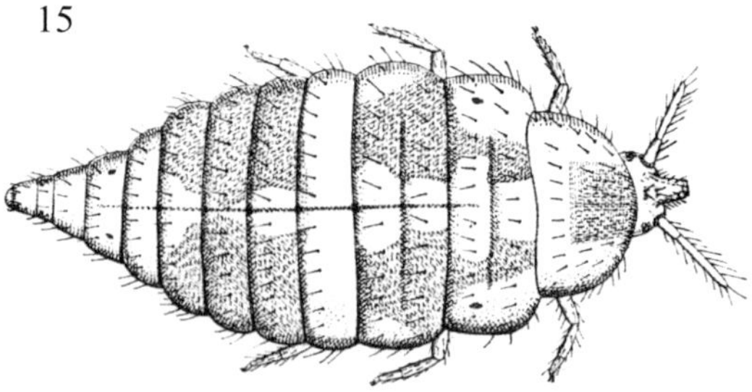

Figs. (14 and 15). Female (14) and larva (15) of *Semidalis aleyrodiformis* Stephens [43].

2.5.3. Lepidoptera

Lepidoptera or butterflies and moths, have wings covered with flattened scales. Mouthparts are adapted to sucking or piercing and sucking, and the external lobes (galeae) of maxillae I form a proboscis. Larvae possess 3 pairs of legs and additional abdominal prolegs. These insects are usually phytophagous during the larval stage, with the exception of the family Erebiidae, which includes zoophagous species, like *Eublemma scitula* Rambur, feeding on scale insects in the Mediterranean Region (Figs. **16** and **17**) [48].

2.5.4. Diptera

Diptera or flies have hind wings modified into halteres, small, club-like structures that function as balancing organs during flight. Mouthparts are adapted for sucking or piercing and sucking. Larvae vermiform, without legs. The flies comprise 12-15% of all animal species and their systematic is complex [49]. A traditional system divides the Diptera into the suborders Nematocera and Brachycera, and this latter into the sections Orthorrhapha and Cyclorrhapha.

16

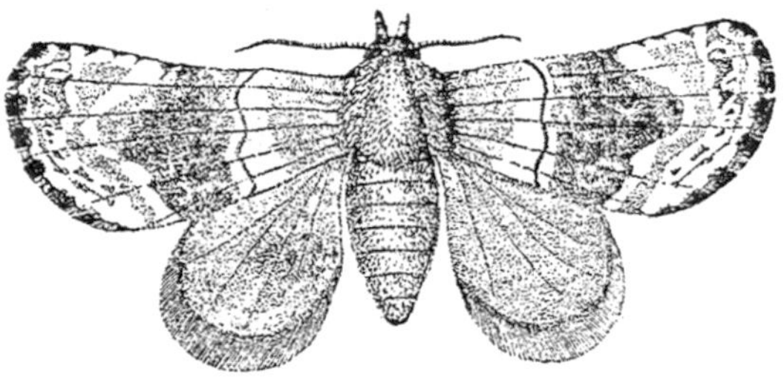

17

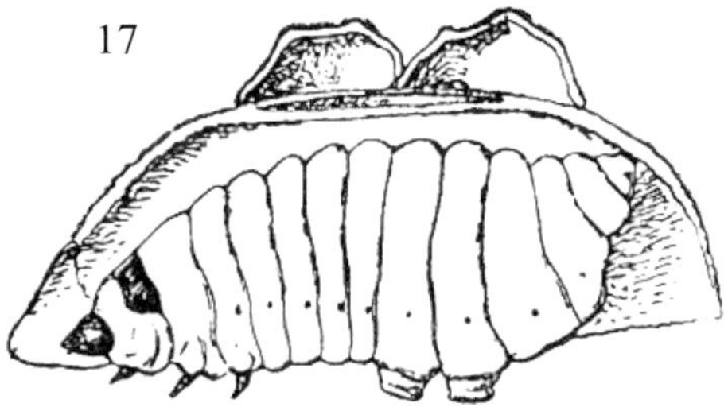

Figs. (16 and 17). Adult (16) [50] and larva (17) of *Eublemma scitula* Rambur protected by its cocoon (from Rouzaud, in Berlese) [14].

The larvae of the Nematocera have a well-developed head, antennae of several segments and mouthparts whose mandibles moving horizontally. The head of the Orthorrhapha larvae is tucked into the thorax, they have very small mandibles that move in a vertical plane, and antennae are sometimes developed. The Cyclorrhapha larvae lack a distinct head or are microcephalic, with inconsistent antennae and mouthparts consisting of two hook-like sclerites. Many Diptera are known predators and parasitoids. The Brachycera comprise various families (Muscidae, Syrphidae, Cryptochaetidae, Chamaemyiidae) that include important species in biological control. Some, like members of the Asilidae, prey in the larval and adult stages. Adults of the parasitoids feed on sugar liquids produced by other insects or on vegetable liquids. The larval bio-ecology indicated separating the families to predators (Syrphidae), mainly predators (Cecidomyiidae, Asilidae, Anthomyiidae), parasites (Tachinidae), and mainly parasitoids (Sarcophagidae) [51].

There are about 200 species in the Cryptochaetidae, including *Cryptochaetum iceryae* Will., an important natural enemy of the cotton cushion scale, *Icerya purchasi* Maskell, which was introduced by Koebele in 1888 from Australia into California [52].

The Chamaemyiidae prey during their larval stages on aphids and armored scale insects, whereas the adults feed on sugar liquids [53]. *Leucopis* spp. are common on citrus. The most important biological control species are in the families Cecidomyiidae, Muscidae, and Syrphidae.

Cecidomyiidae (Figs. **18** and **19**), or gall midges, are in the Nematocera and include about 4,500 described species. They are small, with a wing length of 1-5 mm; wings are broad, with R5 unbranched and venation commonly reduced to three main veins. Antennae long, with 14 or more segments and the flagellomeres with thin-walled sensoria, long and thread-like loops [55].

Some species are used in biological control, such as the predatory *Feltiella acarisuga* (Vallot), feeding on spider mites [56], *Aphidoletes aphidimyza* (Rondani), feeding on aphids [54], *Dicrodiplosis pseudococci* (Felt), feeding on citrus mealybug *Planoccocus citri* (Risso) [57], as well as *Lestodiplosis* spp., predators of various coccids [55].

Muscidae or flies, have antennae with an arista, a simple or variously modified apical or subapical bristle corresponding to the last three articles of the flagellum, usually plumose for its entire length. They also bear a hypopleuron (a sclerite on the thorax just above hind coxae), usually without bristles; and generally more than one sternopleural bristle.

18

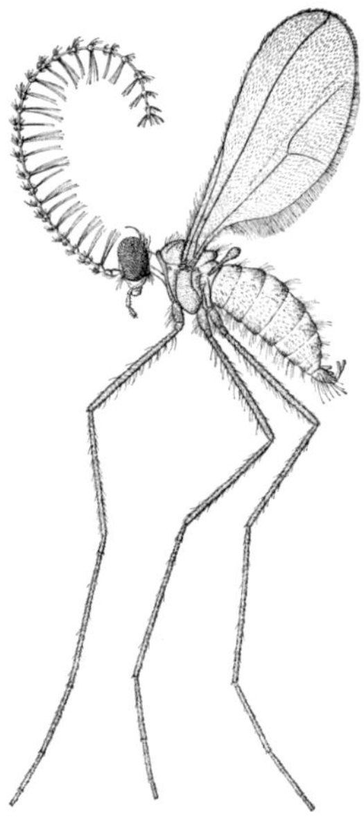

19

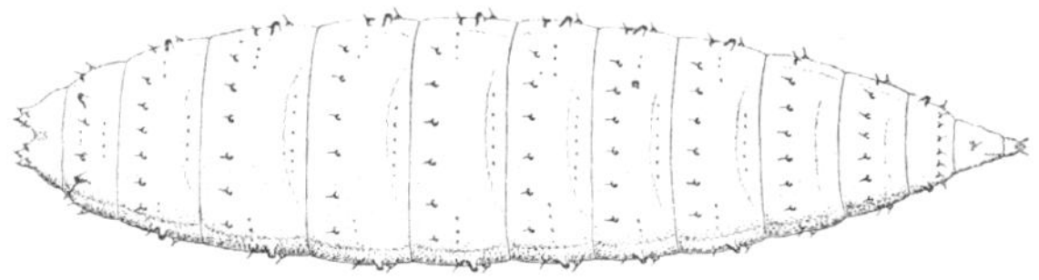

Figs. (18 and 19). Male (18) and larva (19) of *Aphidoletes aphidimyza* (Rondani) [54].

The R5 cell on the wings is either parallel-sided or narrowed distally; vein 2A short and not reaching the wing margin [58]. The hunter fly, *Coenosia attenuata* Stein, which preys on small flying arthropods (*e.g.* Aleyrodids), is active in biological control. For more information on the biology of the Muscidae, see [59].

Syrphidae (Figs. **20** and **21**) or hover flies, have antennae with arista, the wings with a spurious vein, located parallel to the fourth longitudinal wing vein. The bodies are densely covered by hairs; the adults feed mainly on nectar and pollen [61].

They hover around flowers, thus their common name. Over 4,000 species are known, of which about 500 belong to the subfamilies Syrphinae and Chilosiinae, preying on aphids [62]. Species in the genera *Episyrphus*, *Metasyrphus*, *Syrphus*, *Melangyna*, *Pipiza*, *Epistrophe* and others occur on citrus.

20

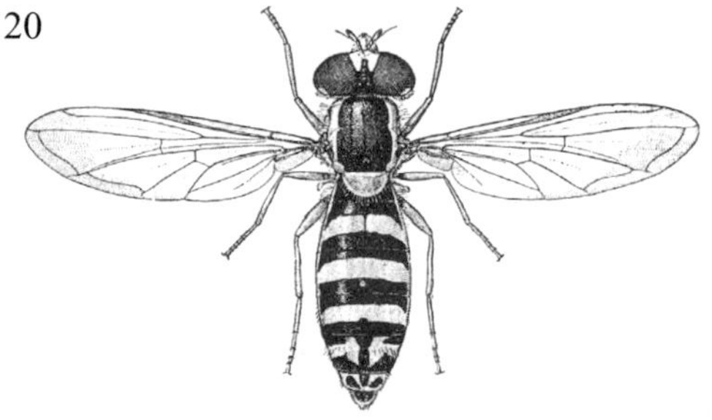

21

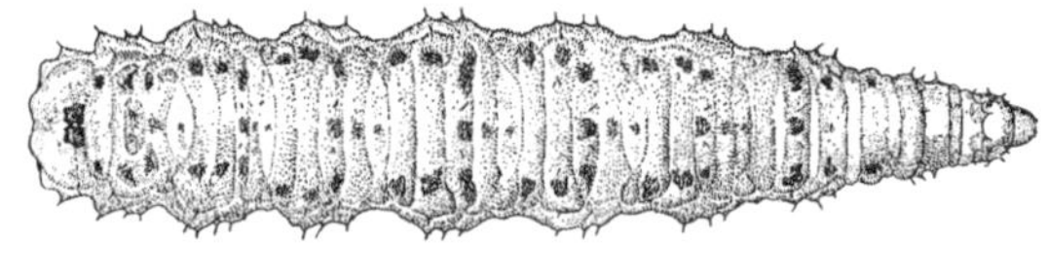

Figs. (20 and 21). Adult of Syrphidae (20) (from Gridelli, in Grandi) [60] and larva of *Metasyrphus* (21) (from Metcalf, in Silvestri) [43].

2.5.5. Coleoptera

Body of beetles is from less than 1 mm to several cm in length. Mouthparts possess mandibles and maxillae; forewings (elytrae) are hardened and hind wings are membranous. The larvae are campodeiform or vermiform, with mouthparts like the adult. The Coleoptera, with about 300,000 species and 150 families, include two suborders, Polyphaga and Adephaga. The Polyphaga include several families with members active in pest control, such as the Alleculidae, Anthribidae, Cantharidae, Coccinellidae, Cybocephalidae, Endomychidae, Nitidulidae, Staphylinidae, and Tenebrionidae [63, 64]. The most important biocontrol species are in the Coccinellidae.

Several Staphylinidae prey on fly larvae in organic substrates. They include *Belonuchus formosus* Grav, introduced into Italy by Silvestri [66] for the biocontrol of the Tephritidae. The Cybocephalidae include generalist and specialized predators such as *Cybocephalus* spp., whose larvae and adults feed on various coccids.

Coccinellidae (Figs. **22-28**) number over 5,000 species in about 500 genera. Their body is oval, from less of 1 mm to about 2 cm long. The apical segment of the maxillary palpus is commonly securiform and often conical or parallel-sided. Antennae are short, club-like and commonly 11-segmented (sometimes 7-segmented). Elytrae never striated. Legs are cursorial [67]. Larvae are elongated, campodeiform and sometimes covered with a white waxy substance. A key to the genera, including those of species feeding on armored scale insects, is by Drea and Gordon [67] and a key to identifying the larvae by Hodek [68].

More than half the species prey on aphids, whereas others feed on small arthropods, such coccids, mites and others. The family is divided into several subfamilies: the Scymninae, Chilocorinae, Coccidulinae, and Coccinellinae, all of which include polyphagous zoophagous species [67, 69, 70, 71].

Coccinellinae (Figs. **22** and **23**) are over one cm in length with a smooth dorsum. Antennae 11-segmented, the clypeus is normal and the dental surface of the mandibles is variously divided. This subfamily has the largest number of species, including the common *Adalia bipunctata* (Linnaeus), feeding on aphids (e.g *Myzus persicae* Sulzer) and *Coccinella septempunctata* Linnaeus, feeding on *M. persicae*, *Aphis fabae* Scopoli and *A. gossypii* Glover and other aphids. Other Coccinellinae include the genera *Hippodamia* and *Harmonia.*

Scymninae (Fig. **24**) are small or medium sized, with a pronounced dorsum, short antennae, clypeus extended laterally, last segment of the maxillary palpus not strongly triangular, and the lateral sides of elytra narrow and short. *Stethorus punctillum* (Weise) feeds on spider mites, *Clitostethus arcuatus* (Rossi), preys on various small species as whiteflies, and both *Cryptolaemus montrouzieri* (Mulsant) and *Nephus includens* (Kirsch) prey on citrus mealybugs.

Chilocorinae (Figs. **25** and **26**) have the clypeus strongly expanded laterally, the last segment of the maxillary palpus is subcylindrical, and the antennal bases are not visible in dorsal view. They include important species as *Chilocorus bipustulatus* (Linnaeus), a predator of soft and armored scale insects, *C. kuwanae* Silvestri also feeding on armored scale insects, and *Brumus quadripustulatus* (Linnaeus) preying on soft scales.

Coccidulinae (Figs. **27** and **28**) possess antennae about half as long as the width of head and the body dorsum is expanded. They include *Rodolia cardinalis* (Mulsant), a predator of *Icerya purchasi*, *Rhyzobius lophantae* (Blaisdell), feeding on diaspidids.

22

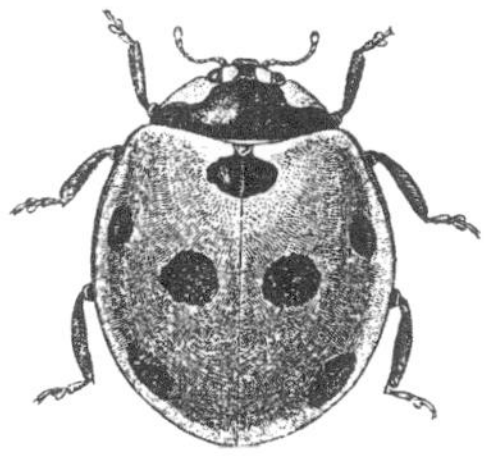

23

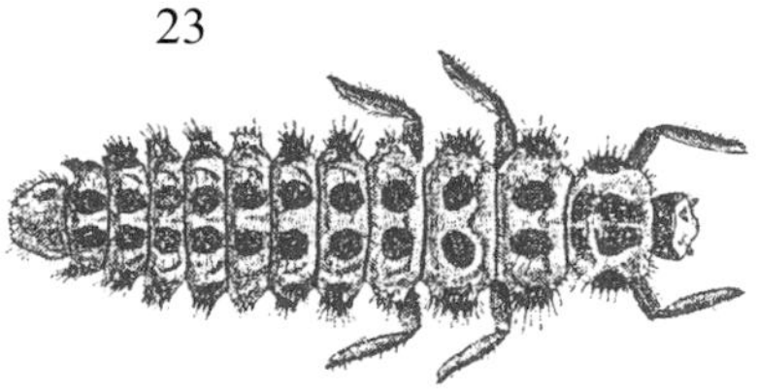

Figs. (22 and 23). Adult (22) and larva (23) of *Coccinella septempunctata* Linnaeus [65].

2.5.6. Hymenoptera

Members of the order Hymenoptera have chewing mouthparts (except in bees where the maxillae and labium form a proboscis for sucking nectar), and well developed compound eyes. The wing are membranous with relatively few veins, forewings larger of the hind wings, and both wings are held together by small hooks (hamuli).

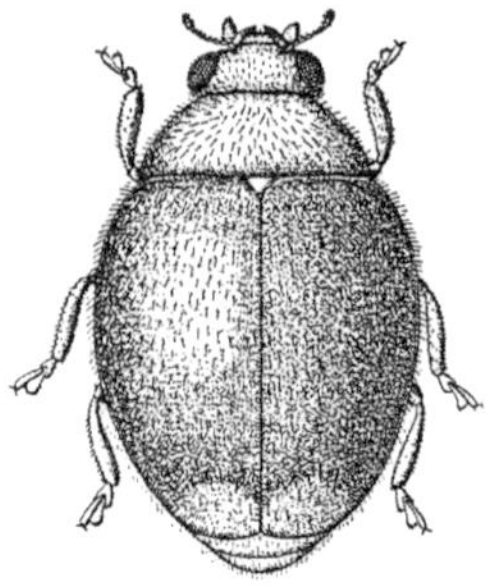

Fig. (24). *Cryptolaemus montrouzieri* (Mulsant) [43].

Fig. (25). *Chilocorus bipustulatus* (Linnaeus) [50].

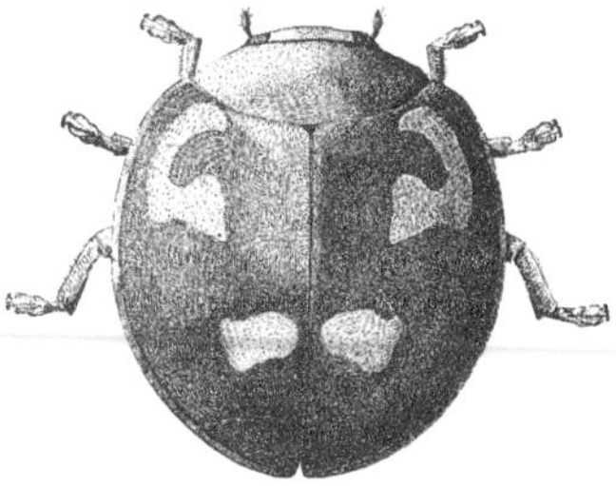

Fig. (26). *Brumus quadripustulatus* (Linnaeus) [50].

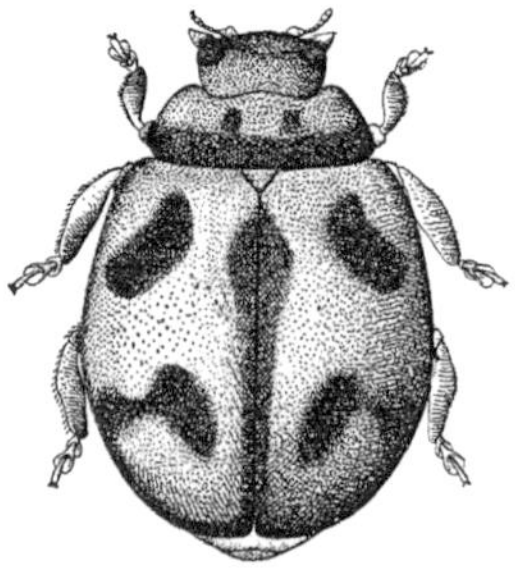

Fig. (27). *Rodolia cardinalis* (Mulsant) [43].

The Hymenoptera include over than 200,000 species [72] organized into the suborders Symphyta and Apocrita.

The Symphyta are less advanced and include phytophagous species. The more advanced Apocrita are characterized by the narrow waist or petiole, a structure deriving from the first two abdominal segments. The first abdominal segment, the propodeum, is fused to the thorax. The Apocrita include many predatory and parasitoid species and members of some families, like the Braconidae, Platygastridae, Pteromalidae, Encyrtidae, Eulophidae, and Aphelinidae, play a very important role in the biological control of citrus pests. A key for the identification of the families is by Goulet and Huber [73], and a catalogue of the species in the superfamily Chalcidoidea, which includes the most important families, is by Noyes [74].

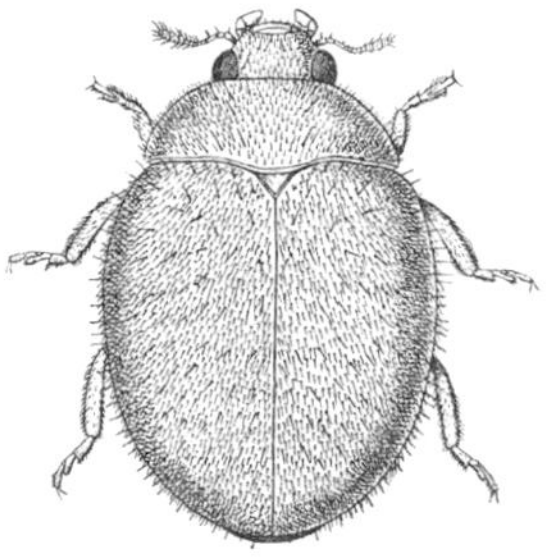

Fig. (28). *Rhyzobius lophantae* (Blaisdell) [43].

Braconidae (Fig. **29**) are commonly black-brown (sometimes with reddish markings) insects, although some species exhibit striking coloration patterns. The wings have one or no recurrent veins, and wing venation patterns are also divergent to apparent randomness. The antennae have 16 or more segments; the hind trochanters are 2-segmented. Females commonly with long ovipositors. About 45 subfamilies are known, 1,000 genera and 40,000-50,000 species [75, 76].

The subfamily Aphidiinae includes species important in the control of citrus pests. They are small, dark, and brown, yellowish or variously colored, whose forewing veins are reduced. They are solitary koinobiont endoparasitoids of aphids, currently considered to include approximately 50 genera and 400 species [77, 78]. The females lay eggs in the host by bending their abdomen under its body, pushing it forward until reaching the aphid. The mature larvae form a cocoon inside the body of the host, (*e.g. Lysiphlebus* spp.), or below, as with *Praon* spp. Once the cocoon is formed inside the host body, the latter loses its segmentation, hardens and turns into a "mummy". *Lysiphlebus fabarum* (Marshall), *L. testaceipes* (Cresson) are parasitoids of aphids in the genera *Aphis*, *Toxoptera* and others. *Aphidius matricariae* Haliday is a parasitoid of *Myzus* and other citrus aphids.

Platygastridae (Fig. **30**) are small and blackish, with 10-segmented (rarely 7-9-segmented) antennae, bent at rest as a Z, and like the legs, variously colored. The wings usually lack veins, except for the submarginal vein. About 4200 species are known, which are usually endoparasitoids (sometimes oophagous) of the juveniles of Aleyrodidae and Coccidae [79, 80]. A catalogue of the Platygastridae of the world is by Vlug [81].

The subfamily Sceliotrachelinae includes *Amitus hesperidum* Silvestri, introduced from the Far East into the American continent to control the whitefly *Aleurocanthus woglumi* Ashby, and *A. spiniferus* (Brèthes), introduced into the Mediterranean Region to control *Aleurothrixus floccosus* (Maskell).

Pteromalidae (Figs. **31** and **32**) are 1-7 mm in length, metallic green or bluish in color, sometimes with body parts yellow or blackish. Head and thorax sculptured, reticulate. Antennae 8 to13-segmented. The family includes over 3,000 species, 600 genera, and 32 subfamilies; information on the palearctic species is in De Vere Graham [83]. The Pteromalidae are predatory oophagous as well as ectoparasites, ectoparasitoids and endophagous. The subfamily Eunotinae includes parasitoids or hyperparasitoids of Coccoidea, or species that develop as predators on coccoid eggs within the host-female's body [84, 85].

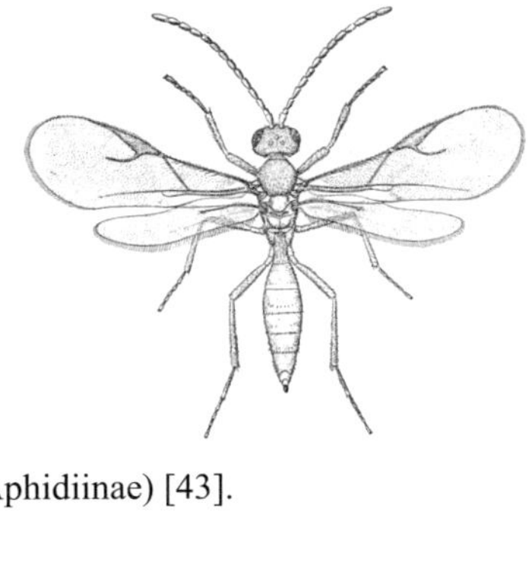

Fig. (29). Adult of *Lysiphlebus* (Braconidae Aphidiinae) [43].

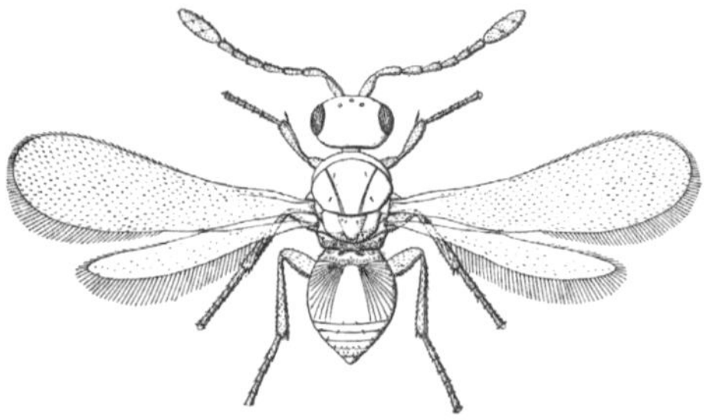

Fig. (30). Sceliotrachelinae. *Amitus hesperidum* Silvestri [82].

Several species have been transported from one country to others in biological control programs [86]. Species common in citriculture include the cosmopolitan *Scutellista caerulea* Fonscolombe, which is a predator and parasite of soft scales (e.g *Ceroplastes*, *Coccus*, *Lichtensia*, *Saissetia*) [87], and *Moranila californica* (Howard), common as an enemy of *Saissetia oleae* (Olivier) in some areas of the Mediterranean Region.

31

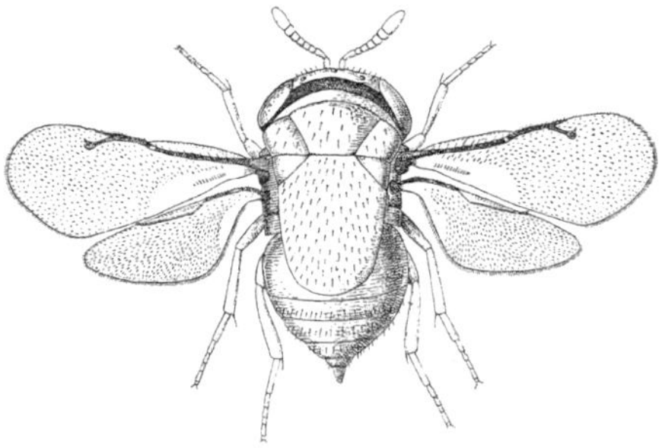

32

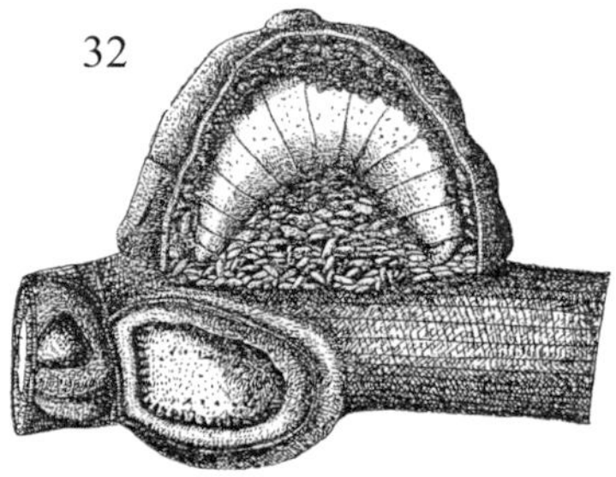

Figs. (31 and 32). Adult (31) and a larva (32) of *Scutellista caerulea* Fonscolombe within the body of *Ceroplastes rusci* Linnaeus [87].

Encyrtidae (Figs. **33-35**) have strong bodies and are at most 3.5 mm long, with antennae of 13 or less segments. Mandibles usually with 4 teeth. Fore wings with short marginal and submarginal veins. The family includes over 3,000 species and 450 genera, and is divided into the subfamilies Tetracneminae and Encyrtinae [88], members of both are important in the control of citrus pests. The Tetracneminae (Figs. **33** and **34**) includes parasitoids of mealybugs, like *Anagyrus pseudococci* (Girault), *A. fusciventris* (Girault), *A. greeni* Howard, *Leptomastidea abnormis* (Girault), *Leptomastix dactylopii* Howard, *Tetracneumoidea brevicornis* (Girault), and *T. peregrina* (Compere). The Encyrtinae (Fig. **35**) include several major natural enemies of *Saissetia oleae* and other soft scales. The parasitoids are *Metaphycus lounsburyi* (Howard), *M. bartletti* Anneke and Mynhardt, *M. flavus* (Howard), *M. helvolus* (Compere), and *Diversinervus elegans* Silvestri. *Comperiella bifasciata* Howard is a parasitoid of Diaspididae, including *Aonidiella aurantii*, whereas *Ageniaspis citricola* Logvinovskaya is an ectoparasitoid of the citrus leaf miner, *Phyllocnistis citrella* Stainton.

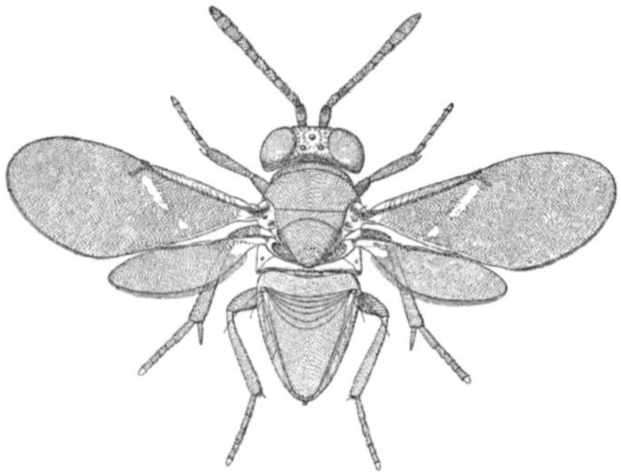

Fig. (33). *Anagyrus greeni* Howard [43].

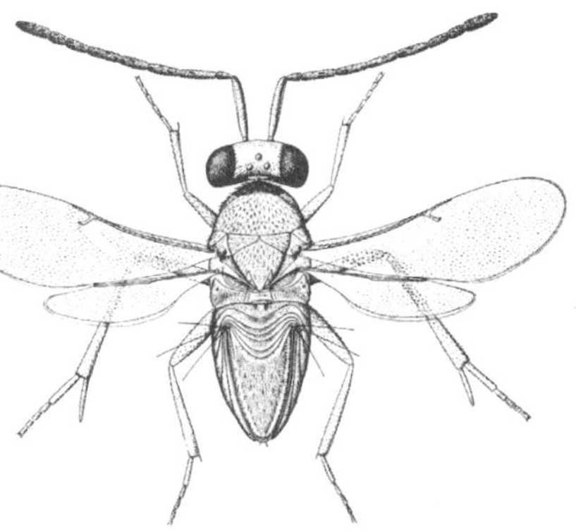

Fig. (34). *Leptomastix dactylopii* Howard [89].

Eulophidae number over 3,400 species in about 330 genera. Their bodies are longer than 6 mm and are often of a metallic hue. Antennae with at most 10 segments, including a funicle of 2-4 segments. The tarsi are 4-segmented. Other characters are the structure of the thorax and abdomen. They are parasitoids of various insects [90] and are ascribed to five subfamilies [91]. *Cirrospilus pictus* (Nees) and *Pnigalio agraules* (Walker) are common parasitoids of the citrus leaf miner.

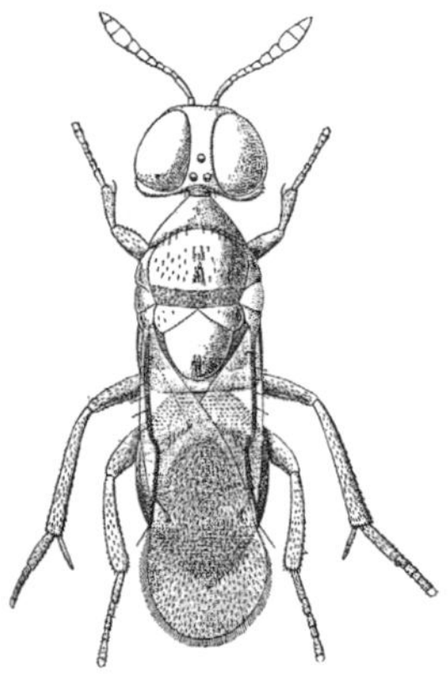

Fig. (35). *Diversinervus elegans* Silvestri [43].

Aphelinidae (Figs. **36** and **37**) are about 0.6-1.4 mm long and their bodies lack a metalic hue. The gaster is broadly sessile and not narrowed between gaster and propodeum. Notaular lines are present as deep, straight grooves. The antennae are 5-8 segmented, the tarsi 4- or 5-segmented, and there are cerci at the apex of gaster.

The family includes over 1,000 species, 50 genera [92] and 7 subfamilies. Members of the genus *Aphytis* are very important in the control of Diaspididae, including *A. chilensis* Howard, a parasitoid of *Aspidiotus nerii* (Bouchè), *A. chrysomphali* (Mercet), attacking *Chrysomphalus dictyospermi* (Morg.), *A. lingnanensis* Compere and *A. melinus* DeBach, parasitoids of *Aonidiella aurantii*, and *A. hispanicus* developing on *Parlatoria pergandii*, and many others. Other interesting species belong to the genus *Coccophagus*, as *Coccophagus lycimnia* (Walker), *C. flavoscutellatum* Ashmead, *etc.*, developing on various Coccidae.

Species parasitizing the Aleyrodidae include *Encarsia lahorensis* (Howard), a parasitoid of *Dialeurodes citri* (Ashmead), and *Encarsia meritoria* Gahan, attacking *Parabemisia myricae* (Kuwana) and other aleyrodids. In the same genus *E. citrina* (Craw), *E. lounsburyi* (Berlese and Paoli), *E. fasciata* (Malenotti), *E. herndoni* (Girault), *E. inquirenda* (Silvestri), which parasitize different Diaspididae. *Eretmocerus debachi* Rose and Rosen develops on *P. myricae*, and *Cales noacki* Howard attacks *Aleurothrixus floccosus* and other whiteflies. Both species provide good control.

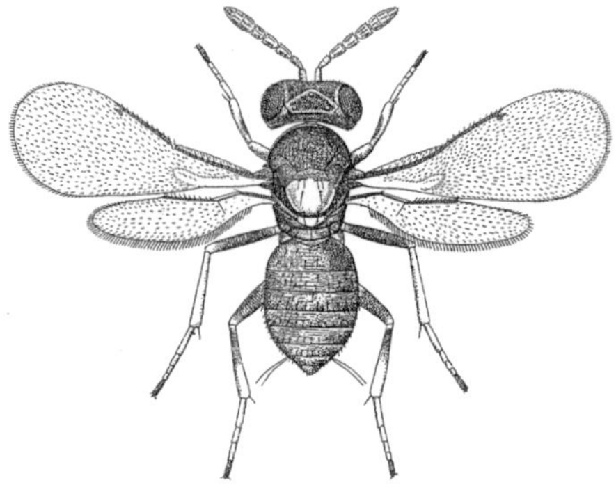

Fig. (36). *Coccophagus flavoscutellatum* Ashmead (from Masi, in Silvestri) [43].

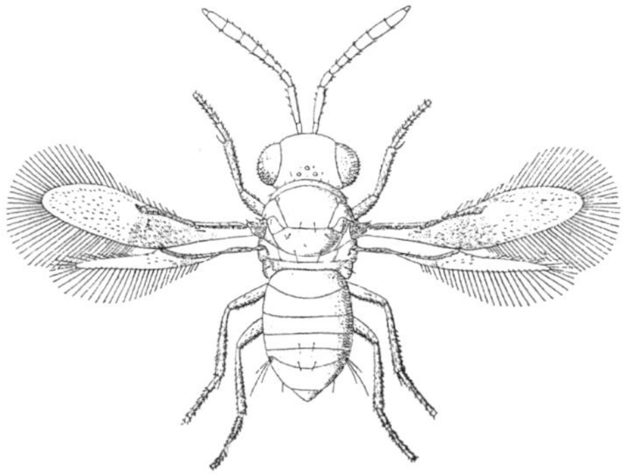

Fig. (37). *Encarsia citrina* (Craw) [43].

REFERENCES

[1] Rosen D. Methodologies and strategies of pest control in citriculture. In: Cavalloro R, Di Martino E, Eds. Proceedings of the Experts' Meeting " Integrated pest control in citrus-orchards"; 1985: Rotterdam: Balkema 1986; pp. 521-30.

[2] Rosen, D. (ed.) Armored scale insects their biology, natural enemies and control. Vol. B. Amsterdam: Elsevier 1990.

[3] Gerson U, Smiley RL, Ochoa R. Mites (Acari) for Pest Control. Oxford: Blackwell Publishing 2003.

[4] Reed DK, Tashiro H, Beevers JB. Determination of mode of transmission of the citrus red mite virus. J Invert Path 1975; 26: 239-46.

[5] Samson RA, Evans HC, Latge JP. Atlas of entomopathogenic fungi. Berlin, Heidelberg, New York: Ed. Springer Verlag 1988.

[6] Samson RA, McCoy CW. *Aschersonia aleyrodis*, a fungal pathogen of whitefly I. Scanning electron microscopy of the development on the citrus whitefly. Z Angew Entomol 1983; 96(1-5): 380-6.

[7] Bell JV. Mycoses. In: Cantwell GE, Ed. Insect Diseases. New York: Marcel Dekker 1974; pp. 185-236.

[8] Paz Z, Gerson U, Sztejnberg A. Assaying three new fungi against citrus mites in the laboratory, and field trial. BioControl 2007; 52: 855-62.

[9] Evans HC, Prior C. Entomopathogenic fungi. In: Rosen D, Ed. Armored scale insects their biology, natural enemies and control. Vol. B. Amsterdam: Elsevier 1990; pp. 3-17.

[10] Barnes RB. Zoologia: gli invertebrati. Padova: Piccin editore 1972.

[11] Benfatto D, Di Franco F, Vacante V. Spiders of Italian Citrus Groves. In: Tribulato E, Gentile A, Reforgiato G, Eds. Proc. Int. Soc. Citriculture. Vol. 3. Acireale, March 8-13, 1992; pp. 938-42.

[12] Ribeiro C, Cardoso P, Franco JC. Abundance and diversity of spiders in lemon orchards with different weed management systems. Bull IOBC/WPRS 2006; 29(3): 167-78.

[13] Monzó C, Mollà O, Montón H, *et al.* Ground-dwelling spiders (Araneae) in citrus orchards in Spain. Bull IOBC/WPRS 2008; 38: 267.

[14] Berlese A. Gli Insetti. Vol. II. Milano: Società Editrice Libraria 1925.

[15] Walter DE, Proctor HC. Mites – Ecology, Evolution and Behaviour. Sydney and Wallingford: University of New South Wales Press and CAB International 1999.

[16] Krantz GW. Introduction. In: Krantz GW, Walter DE, Eds. Acarology. Lubbock: Texas Tech University Press 2009; pp. 1-2.

[17] Vacante V, Nucifora A, Tropea Garzia G. Citrus mites in the mediterranean Area. In: Goren R, Mendel K, Eds. Proc VI Int Citrus Congress, Tel Aviv, 6-11 marzo 1988; pp. 1325-34.

[18] de Moraes GJ, McMurtry JA, Denmark HA, Campos CB. Zootaxa, A revised catalog of the mite family Phytoseiidae. Auckland: Magnolia Press 2004.

[19] Vacante V, Nucifora A. Gli Acari degli Agrumi in Italia. II. I Gamasidi (Phytoseiidae Berlsese e Ascidae Voigts & Oudemans): chiave per il riconoscimento dei generi e delle specie e note di biologia. Boll Zool agr Bachic ser. II 1987; 19: 145-84.

[20] Porath A, Swirski E. A survey of phytoseiids mites (Acarina: Phytoseiidae) on citrus with a description of one new species. Israel J Agric Res 1965; 15: 87-100.

[21] Garcia-Marì F, Ferragut F, Marzal C, Costa Comelles J, Laborda R. Acaros que viven en las hojas de los citricos espanoles. Investigacion Agraria Produccion y Proteccion Vegetales 1986; 1(2): 219-50.

[22] Gerson U, Vacante V. The use of indigenous acarine predators to control citrus mite pests. Bull OILB SROP 1993; 16(7): 115-19.

[23] Vacante V. Influence of white mineral oil treatments on the population dynamics of some mites in a lemon orchard in Eastern Sicily. In: Cavalloro R, Di Martino E, Eds. Proceedings of the Experts' Meeting "Integrated pest control in citrus-groves"; 1985: Acireale, 26-29 March: Rotterdam: Balkema 1986; pp. 423-31.

[24] Rasmy AH, Zaher MA, Albagoury ME. Mites Associated with Citrus in the Nile Delta (U.A.R.). Z Angew Entomol 1972; 70: 183-6.

[25] Abad-Moyano R, Urbaneja A, Schausberger P. Intraguild interactions between *Euseius stipulatus* and the candidate biocontrol agents of *Tetranychus urticae* in Spanish clementine orchards: *Phytoseiulus persimilis and Neoseiulus californicus.* Exp Appl Acarol 2010; 50: 23-34.

[26] Gerson U, Fain A, Smiley RL. Further observations on the Cheyletidae (Acari), with a key to the genera of the Cheyletinae and a list of all known species in the family. Bull Inst R Sci Nat Belgique 1999; 69: 35-86.

[27] Vacante V, Nucifora A, 1986. A first list of the mites in citrus orchards in Italy. In: Cavalloro R, Di Martino E, Eds. Proceedings of the Experts' Meeting "Integrated pest control in citrus-groves"; 1985: Acireale, 26-29 March: Rotterdam: Balkema 1986; pp. 177-88.

[28] Avidov Z, Blumberg D, Gerson U. *Cheletogenes ornatus* (Acarina: Cheyletidae), a predator of the chaff scale on citrus in Israel. Israel J Entomol 1968; 3: 77-93.

[29] Zaher MA, Yousef AETA, Kandil MM. Effect of food on the biology of *Cheletogenes ornatus* (C. & F.) (Acari: Prostigmata: Cheyletidae). Acarologia 1981; 22: 361-66.

[30] Vacante V, Nucifora A. Gli Acari degli agrumi in Italia. I. Specie rinvenute e chiave per il riconoscimento degli ordini, dei sottordini e delle famiglie. Boll Zool agr Bachic ser. II: 115-66.

[31] Tseng YH. Mites of the family Stigmaeidae of Taiwan, with key to genera of the world (Acarina: Prostigmata). Phytol Entomol Nat Taiwan Univ 1982; 9: 1-52.

[32] Dosse G. Schädmilben des Libanons und ihre Prädatoren. Z Angew Entomol 1967; 59: 16-48.

[33] Elbadry EA, Abo Elghar MR, Hassan SM, Kilany SM. Life history studies on the predatory mite *Agistemus exsertus*. Ann Entomol Soc Am 1969; 62: 649-51.

[34] Gerson U. The mites associated with Citrus in Israel. Israel J Ent 1971; 6: 5-22.

[35] Baker EW, Wharton GW. An Introduction to Acarology. New York: The MacMillan Company 1952.

[36] Zaher MA, Soliman ZR, El-Bishlawy SM. Feeding habits of the predaceous mite, *Cunaxa capreolus* (Acarina: Cunaxidae). Entomophaga 1975; 20: 209-12.

[37] Vacante V. Acari simbionti di Diaspididae (Rhynchota, Homoptera) in Sicilia: *Hemisarcoptes malus* (Shimer) (Acarina, Hemisarcoptidae). II Contributo. Atti XIV Congresso nazionale italiano di Entomologia, Palermo-Erice-Bagheria, 28 maggio-1 giugno 1985; pp.749-58.

[38] Gerson U. Observations on *Hemisarcoptes coccophagus* Meyer (Astigmata: Hemisarcoptinae) with a new synonimy. Acarologia 1967; 9: 632-8.

[39] Vacante V, Di Martino E. Reperti sull'acarofauna degli agrumi in Libia. Atti del Convegno "Il recente contributo della ricerca allo sviluppo dell'agrumicoltura italiana", Cagliari, 29 aprile-30 maggio 1986: Roma: A. Delfino Ed.1987; pp. 559-63.

[40] Izraylevich S, Gerson U. Population dynamics of *Hemisarcoptes coccophagus* Meyer (Astigmata: Hemisarcoptidae) attacking three species of armored scale insects (Homoptera: Diaspididae). Exp Appl Acarol 1993; 17: 877-88.

[41] Miller NCE. The biology of Heteroptera. London: Leonard Hill 1956.
[42] Pericart J. Hémiptères Anthocoridae, Cimicidae, Microphysidae de l'Ouest-Paléarctique. Paris: Masson 1972.
[43] Silvestri F. Compendio di entomologia applicata. Vol. 1-2. Portici, Napoli: Tipografia Bellavista 1934-51.
[44] Aspöck H, Aspöck U, Hölzel H (1980) Die Neuropteren Europas. Vol 1-2. Krefeld: Goecke & Evars, 1980.
[45] McEwen P, New TR, Whittington A E. (Eds) Lacewings in the Crop Environment. Cambridge: Cambridge University Press 2001.
[46] Brooks SJ, Barnard, PC. The green lacewings of the world: a generic review (Neuroptera Chrysopidae). Bull Br Mus (Nat Hist) Entomol 1990; 59: 117-286.
[47] Meinander M. A revision of the family Coniopterygidae (Planipennia). Acta Zoologica Fennica 1972; 136: 1-357.
[48] Kravchenko VD, Fibiger M, Mooser J, Junnila A, Müller GC. The Eublemminae of Israel (Lepidoptera Erebiidae). SHILAP, Revista de Lepidopterología 2007; 35(140): 513-9.
[49] Yeates KD, Wiegmann BM, Courtney GW, Meier R, Lambkin C, Pape T. Phylogeny and systematics of Diptera: Two decades of progress and prospects. Zootaxa 1668, 2007; 565–90 www.mapress.com/zootaxa/
[50] Martelli G. Osservazioni fatte sulle Cocciniglie dell'olivo e loro parassiti in Puglia ed in Calabria. Boll Lab Zool Portici 1908; vol. II: 217-96.
[51] Clausen CP. Entomophagous insects. New York, London: McGraw-Hill Book Company 1940.
[52] DeBach P. Biological Control of insects pests and weeds. London: Chapman and hall Ltd 1964.
[53] Tanasiychuk VN. Fauna USSR: Diptera. Vol. 14, part 7: Mukhi-Serebryanki (Chamaeiidae) Nauka Leningrad 1986.
[54] Roberti D. La *Phaenobremia aphidimyza* (Rond.) (Diptera Cecidomyidae) predatrice di *Aphis* (*Doralis*) *frangulae* Koch. Boll Ist Ent Univ Bologna 1946; 15: 233-56.
[55] Harris KM. Cecidomyiidae and Other Diptera. In: Rosen D, Ed. Armored scale insects their biology, natural enemies and control. Vol. B. Amsterdam: Elsevier 1990; pp. 61-6.
[56] Roberti D. I simbionti degli acari fitofagi. I. *Therodiplosis persicae* Kieffer. Boll lab Ent agr Filippo Silvestri 1954; 13: 3-20.
[57] Bodenheimer FS. Citrus Entomology in the Middle East. The Hague: Junk W Pub. 1951.
[58] Séguy E. Diptera, family Muscidae. In: Wystmann P, Ed. Genera Insectorum. Brussels 1937; 205.
[59] Skidmore P. The biology of the Muscidae of the world. Series entomologica, 29. Dordrecht: Junk 1985.
[60] Grandi G. Introduzione allo studio della Entomologia. Vol. II. Bologna: Edagricole 1965.
[61] Speight MCD. External morphology of adult Syrphidae (Diptera) - Tijdschr Entomol 1987; 130: 141-75.
[62] Schneider F. Bionomics and physiology of aphidiphagous Syrphidae (Dipt.). Annu Rev Entomol 1969; 14: 104-24.
[63] Balduf WV. The bionomics of entomophagous Coleoptera. St. Louis 1935.
[64] Drea JJ. Other Coleoptera. In: Rosen D, Ed. Armored scale insects their biology, natural enemies and control. Vol. B. Amsterdam: Elsevier 1990; pp. 41-9.
[65] Menozzi C. Animali e vegetali dannosi alla barbabietola da zuchero e mezzi per combatterli. Bologna: Soc. Tip. Bolognese 1947.
[66] Silvestri F. Descrizione e biologia del coleottero stafilinide *Belonuchus formosus* Grav. introdotto in Italia per la lotta contro ditteri tripaneidi. Boll Lab R Entomol Agr Portici 1945; 5: 312-26.
[67] Drea JJ, Gordon RD. Coccinellidae. In: Rosen D, Ed. Armored scale insects their biology, natural enemies and control. Vol. B. Amsterdam: Elsevier 1990; pp. 19-40.
[68] Hodek I. Biology of Coccinellidae. The Hague: Dr. W Junk NV Publ. 1973.
[69] Sasaji H. Coccinellidae (Insecta: Coleoptera). Fauna Japonica. Japan: Acad Press 1971.
[70] Iablokoff-Khnzorian SM. Les Coccinelles Coléoptères-Coccinellidae Tribu Coccinellini des regions Palearctique et Orientale. Paris: Boubée 1982.
[71] Hodek I, Honek A. Ecology of Coccinellidae. Dordrecht: Kluwer Academic Publishers 1996.
[72] Gauld I, Bolton B. The Hymenoptera. Oxford: Oxford University Press 1988.
[73] Goulet H, Huber JT (Eds). Hymenoptera of the World: an identification guide to families. Ottawa: Research Branch, Agriculture Canada 1993.
[74] Noyes JS. Catalogue of Chalcidoidea of the World. CD-ROM Series, ETI, Amsterdam 1998.
[75] Van Achterberg C. Essay on the phylogeny of Braconidae (Hymenoptera: Ichneumonoidea). Entomol Tidskr 1984; 105: 41-58.
[76] Van Achterberg C. Braconidae: an illustrated key to all subfamilies. CD-ROM produced by ETI and available from Springer-Verlag, NY 1997.
[77] Mackauer M, Stary P. World Aphidiidae (Hymenoptera: Ichneumonoidea). Paris: Le Francois 1967.

[78] Stary P. Aphididae. In Minks AK, Harrewijn P, Eds. Aphids, Their Biology, Natural Enemies and Control. Vol. 2B. Amsterdam: Elsevier 1988; pp. 171-84
[79] Masner L, Huggert L. 1989. World review and keys to genera of the subfamily Inostemmatinae with reassignment of the taxa to the Platygastrinae and Sceliotrachelinae (Hymenoptera: Platygastridae). Mem Entomol Soc Can 1989; 147: 1-214.
[80] Masner L. Superfamily Platygastroidea. In: Goulet H, Huber J, Eds. Hymenoptera of the World: an identification guide to families. Ottawa: Research Branch, Agriculture Canada 1993; pp. 558-65.
[81] Vlug H J. Catalogue of the Platygastridae (Platygastroidea) of the world (Insecta: Hymenoptera). Hymenopterorum Catalogus 1995; 19: 1-168.
[82] Silvestri F. Contribuzione alla conoscenza degli Aleurodidae (Insecta: Hemiptera) viventi su citrus in Estremo Oriente e dei loro parassiti. Boll Lab Zool Portici 1927; 21: 1-60.
[83] Graham MVR de Veze. The Pteromalidae of North-Western Europe (Hymenoptera: Chalcidoidea). Bull Br Mus (Nat Hist) Entomol 1969; suppl. 16.
[84] Smith HS, Compere H. Preliminary report on the insects parasites of the black scale, Saissetia oleae (Bern.). Univ Calif Publ Entomol 1928; 4: 231-334.
[85] Flanders SE. Predatism by the adult hymenopterous parasite and its role in biological control. J Econ Entomol 1953; 46: 541–44.
[86] Burks BD. Pteromalidae. In: Krombein *et al.* Eds. Catalog of Hymenoptera in America North of Mexico. Vol. 1. Symphyta and Apocrita (Parasitica). Washington: Smithsonian Institution Press1979; pp. 768-835.
[87] Silvestri F, Martelli G. La Cocciniglia del Fico (*Ceroplastes rusci* L.). Boll Lab Zool Portici 1908 ; pp. 297-358.
[88] Trjapitzin VA. Parasitic Hymenoptera Encyrtidae of Palaearctics. Leningrad: Ac Sc USSR, Zool Institute 1989 (in Russian).
[89] Zinna G. Ricerche sugli insetti entomofagi. I. Specializzazione entomoparassitica negli Encyritidae: Studio morfologico, etologico e fisiologico del *Leptomastix dactylopii* Howard. Boll Lab Ent agr Filippo Silvestri 1959; 18: 1-147.
[90] Bouček Z, Askew RR. Hym. Chalcidoidea - Palearctic Eulophidae (excl. Tetrastichinae). Index of entomophagous insects. Paris: Ed. Delucchi and Remaudière, Le François 1968.
[91] Peck O, Bouček Z, Hoffer A. Keys to the Chalcidoidea of Czechoslovakia (Insecta: Hymenopetra). Mem Entomol Soc Can 1964; 34: 1-120.
[92] Hayat M. The genera of Aphelinidae (Hymenoptera) of the world. Syst Entom 1983; 8: 63-102.

CHAPTER 6

Acari

Uri Gerson[1,*] and Vincenzo Vacante[2]

[1]*Department of Entomology, Faculty of Agricultural, Food and Environmental Quality Sciences, The Hebrew University of Jerusalem, P.O. Box 12, Rehovot, 76-100, Israel and* [2]*Dipartimento Patrimonio Architettonico ed Urbanistico (PAU), Mediterranean University of Reggio Calabria, Via Melissari, Località Feo di Vito, 89100, Reggio Calabria, Italy*

Abstract: Eleven mite (Acari) species in four families are serious pests of citrus in the Mediterranean Region. These include *Aceria sheldoni* (Ewing), *Phyllocoptruta oleivora* (Ashmead) and *Aculops pelekassi* (Keifer) in the family Eriophyidae; *Brevipalpus californicus* (Banks), *B. lewisi* McGregor, *B. obovatus* Donnadieu and *B. phoenicis* (Geijskes) in the Tenuipalpidae; *Polyphagotarsonemus latus* (Banks) in the Tarsonemidae and *Eutetranychus orientalis* (Klein), *Panonychus citri* (McGregor) and *Tetranychus urticae* (Koch) in the family Tetranychidae. Diagnostic characters and illustrations are provided for each species, along with data on their life history, economic importance and management.

Keywords: Citriculture, Mediterranean Region, citrus pest mites, bio-ecology, damage, control.

1. INTRODUCTION

Pest mites play an important ecological role in citrus groves and their population increases, under suitable conditions, can result in heavy economic losses. Eleven species are important pests of citrus in the Mediterranean Region. They are ascribed to four families: Eriophyidae [*Aceria sheldoni* (Ewing), *Phyllocoptruta oleivora* (Ashmead) and *Aculops pelekassi* (Keifer)]; Tenuipalpidae [*Brevipalpus californicus* (Banks), *B. lewisi* McGregor, *B. obovatus* Donnadieu, and *B. phoenicis* (Geijskes)]; Tarsonemidae [*Polyphagotarsonemus latus* (Banks)], and Tetranychidae [(*Eutetranychus orientalis* (Klein), *Panonychus citri* (McGregor), *Tetranychus urticae* Koch)].

The implementation of an Integrated Pest Management (IPM) program for these Acari requires, as with insect pests, much basic biological data. Thus we provide keys for the identification of the relevant species, their diagnostic characters, life histories, economic importance and suggestions for management. Much information was gleaned from two major books on mite pests [1, 2], and they will not be individually cited below. In each legend of the figures the author of the design is listed.

2. ERIOPHYIDAE

The Eriophyidae, the rust and gall mites, is a large family of over 3,000 described species placed in about 230 genera; all feed on plants. They are very small (usually 0.15-0.25 mm in length), with a worm-like body that bears up to five pairs of prodorsal setae and at most seven pairs of opisthosomal setae. The eriophyiids have only two anteriorly-located pairs of legs, which bear empodial featherclaws but no real claws.

They reproduce by arrhenotoky; males deposit spermatophores that are taken up by the females. Dispersal is often by winds or on infested plant material. Several species (none infesting citrus in the Mediterranean Region) transmit plant viruses.

Three Eriophyidae are wide-spread, serious pests of citrus in the Mediterranean Region (*Aceria sheldoni*, *Phyllocoptruta oleivora*, *Aculops pelekassi*). They can be separated with the following key.

1. Body with fewer, but larger, opisthosomal rings (annuli) on dorsum than on venter, usually found on leaves or fruit... 2

***Address correspondence to Uri Gerson:** Department of Entomology, Faculty of Agricultural, Food and Environmental Quality Sciences, The Hebrew University of Jerusalem, P.O. Box 12, Rehovot, 76-100, Israel; E-mail: Gerson@agri.huji.ac.il

Vincenzo Vacante and Uri Gerson (Eds)

-. Body with a similar number of dorsal and ventral rings (annuli) on opisthosoma, usually found in buds or flowers..***Aceria sheldoni***

2. Body color yellow-brown; dorsal shield setae placed ahead of rear shield margin, projecting upwards; a broad longitudinal trough located on the abdominal dorsum…......................***Phyllocoptruta oleivora***

-. Body color pink; dorsal shield setae placed near rear shield margin, projecting backwards; without a broad longitudinal trough located on abdominal dorsum.... ....***Aculops pelekassi***

2.1. *Aceria sheldoni* (Ewing)

Diagnostic characters: The body of this pest, known as the citrus bud mite, is yellow to pink; about 0.17 mm in length, cylindrical, with a similar number (70-80) of dorsal and ventral opisthosomal rings (annuli) (Fig. **1**). The prodorsum (prodorsal shield) (Fig. **2**) bears a pair of backwards-pointing setae and two parallel ridges along its entire length. The featherclaw (tarsal empodium) is five-rayed (Fig. **3**).

Life history: The citrus bud mite reproduces by arrhenotoky, each female depositing about 50 eggs, and its entire cycle (egg to egg) requires ca 15 days in summer, twice that long in winter. Thus the pest completes about 20 annual generations. The sex ratio is male-biased in spring, female-biased during fall [4].

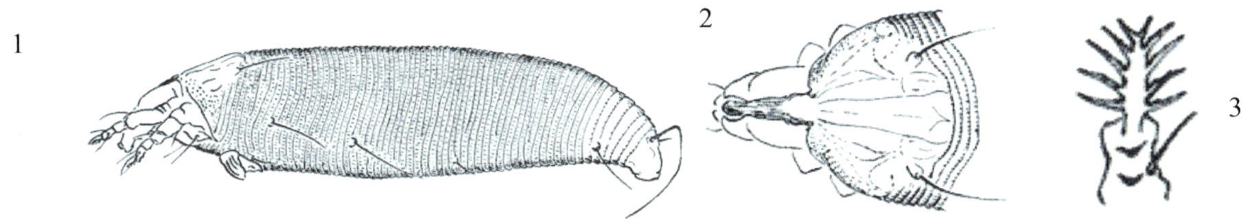

Figs. (1-3). *Aceria sheldoni* (Ewing). Lateral view of adult (1); prodorsal shield (2) and featherclaw (3) [3].

Although the mite lacks eyes, it is attracted to yellow and red, not to blue or green surfaces [5]. The pest develops throughout the year, mostly in milder and more humid regions, with peaks in spring and autumn [6]. Dispersal takes place during the spring growth of most citrus species, throughout the year on the several growth flushes of lemons.

Fig. (4). Blossoms of lemon deformed by *Aceria sheldoni* (Ewing).

Young, elongating twigs are rapidly colonized by the mite, its infestations increasing as the twigs age. Temperatures below 8°C or above 35°C and relative humidities below 40% are lethal to the mite.

Fig. (5). Twigs of lemon foreshortened by *Aceria sheldoni* (Ewing).

Economic importance: This mite occurs almost wherever citrus is grown, attacking mostly lemons. It lives in and feed on the buds, blossoms, flowers and fruits of the host, and may occur in very large numbers. Mite feeding kills the cells and causes bud blackening and irregular growth of leaves, blossoms (Fig. **4**), flowers, branches (Fig. **5**) and fruit (Fig. **6**).

Fig. (6). Fruit of lemon deformed by *Aceria sheldoni* (Ewing).

The damaged buds proliferate and produce irregular and twisted growth of leaf clusters. Twigs are foreshortened (Fig. **5**) and broomed, and the fruit deforms into various distorted shapes (Fig. **6**), including finger-like forms. Further damage is caused by fruit drop. Although the pest attacks all citrus varieties, it causes the most serious damage to lemons, whose yield may be seriously affected [7].

There is however some doubt whether this mite is actually an economic pest [8, 9].

Management: When necessary, the mite can be controlled by most acaricides as well as by white oils, but the cost of such control may exceed its economic benefit. Several predatory mites were found to be associated with the citrus bud mite, but due to the pest being protected within buds, none appear to affect the populations or the damage that it causes.

Website: http://www.viarural.com.ar/viarural.com.ar/agricultura/frutales/plagas/aceria-sheldoni.htm

2.2. *Aculops pelekassi* (Keifer)

Diagnostic characters: The body of this pest, known as the pink citrus rust mite, is pink to reddish in color, about 0.14-0.15 mm long, with a distinctive rounded body. The prodorsal setae are backwards-pointing and the opisthosoma bears about 35 dorsal and 50 ventral rings (annuli) (Figs. **7** and **8**). The featherclaw is four-rayed (Fig. **9**).

Life history: This mite is a subtropical species, whose optimal conditions for development are at 22-27°C and 70-80% RH [11]. In Croatia, *A. pelekassi* completes 18-22 annual generations between May and October [12]. A generation was completed in 6.3 days at 30°C and in 15 at 20°C (68°F) [13]. Maximal oviposition (at 25°C) was 21.8 eggs/female, ceasing when temperatures dropped to 15°C [13]. The females overwinter in citrus terminal and lateral buds, begin to lay eggs on sprouting buds and newly emerging leaves in spring, and then move onto developing leaves and fruit [12, 13]. *Aculops pelekassi* is a vagrant species usually located along the leaf margins of young leaves. The adults show strong responses to light, causing them to move rapidly. They tend to aggregate at leaf edges in large numbers, thus being more easily dispersed by winds. The mite prefers citrus species and varieties of loose skin, such as clementines and mandarins.

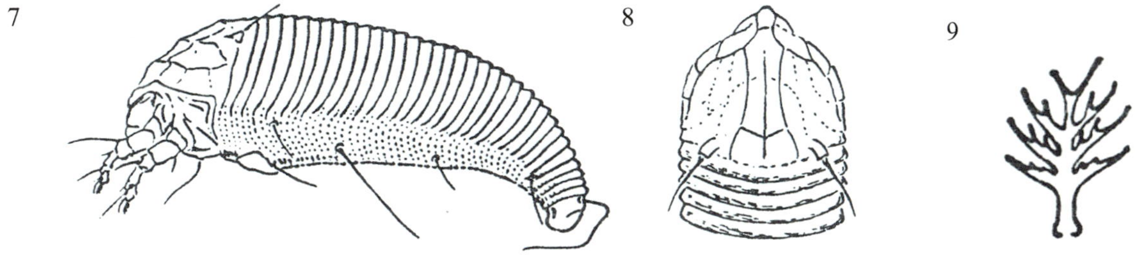

Figs. (7-9). *Aculops pelakassi* (Keifer). Lateral view of adult (7); prodorsal shield and first opisthosomal annuli of male (8); featherclaw (9) [10].

Economic Importance: Infestations of the pink citrus rust mite bring about quantitative and qualitative yield losses. This includes leaf (Fig. **10**) and fruit russeting (Fig. **11**) that results in increased cull rate at harvest, distortion of new growth (Fig. **12**) when large populations infest the trees [14], and even in leaf drop. Some of the damage is similar to that caused by plant viruses.

Management: Most common acaricides (*e.g.* abamectins, wettable sulphur, spirodiclofen, *etc.*) may control the pest. The first treatment should be applied when the mites migrate from the leaves to the fruits and a second, if necessary, about two months later [12]. A binomial sampling program to estimate mite densities on orange fruit was developed in the USA [15]. Several predatory mites of the family Phytoseiidae feed on the pink citrus rust mite [16], but their controlling effect is not known.

Fig. (10). Lower leaves surface of lemon damaged by *Aculops pelekassi* (Keifer).

Fig. (11). Fruit and leaves of orange damaged by *Aculops pelekassi* (Keifer).

Website: http://www.cababstractsplus.org/abstracts/Abstract.aspx?AcNo=19750523248

2.3. *Phyllocoptruta oleivora* (Ashmead)

Diagnostic characters: The body of this pest, known as the citrus rust mite or silver mite, is wedge-shaped, yellowish-brown, about 0.15 mm in length. The prodorsum has a six-sided cell pattern and a pair of very short setae that are not backwards oriented (Fig. **14**). The opisthosoma is marked by a distinct dorsal furrow and bears about 30 dorsal and 60 ventral rings (annuli) (Fig. **13**). The featherclaw is five-rayed (Fig. **15**).

Fig. (12). Distorsion of young leaf and brownig of young stem of lemon produced by *Aculops pelekassi* (Keifer).

Life history: The citrus rust mite lives only on its host plants, which besides *Citrus* spp. include *Swinglea glutinosa* (Blanco) Merr. and *Fortunella* spp. Its optimal conditions for development are around 25°C and very high relative humidities. The mite is resistant to sub-freezing temperatures but susceptible to warm and dry winds [17, 18]. Thus its populations (unless reduced by hot winds) increase during spring, remain high in summer and decline in autumn. Before fruits are about 1.5 cm in diameter, the vagrant mites mostly occur on the underside of young leaves. Later they colonize the fruits, initially feeding at protected sites, and then on their entire surfaces.

The pest develops throughout the year, its entire cycle (egg to egg) requiring about a week in summer, several weeks in winter, completing up to 30 annual generations [17]. Reproduction is indirect; spermatophores are deposited by males on the leaves and are taken up by the females. The species is arrhenotokous, each female depositing about a dozen eggs at 20-31°C. Mite dispersal (mostly females) between patches is on infested plant material and by being wind-borne [19].

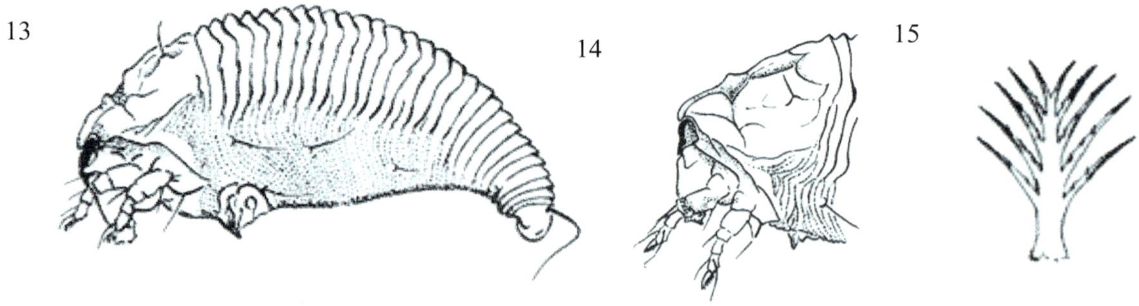

Figs. (13-15). *Phyllocoptruta oleivora* (Ashmead). Lateral view of adult (13); lateral view of prodorsum (14); featherclaw (15) [3].

Economic importance: The citrus rust mite is a major pest of citrus; unless controlled it can reduce yield by 70-100%. The mite feeds on the fruit, leaves and green twigs, piercing the epidermal cells and killing them. Rust-like blemishes appear on the affected areas on orange, grapefruit and mandarin fruits; on lemons and citron the blemishes are silvery. As infestation proceeds the surface of the fruit becomes russeted and bronzed, rough and scaly; peels of young fruit lose their elasticity and may shrivel. Heavily infested fruit is smaller, its quality is reduced and it may drop; such fruit is rejected as fresh product as well as by the industry. Dark spots appear on affected leaves, which can lose their gloss, and their lower surfaces become rust-like. Infested green twigs may show blackish areas. Attack on leaves and twigs can affect next year's growth.

Management: Due to the minute size of the pest, it may not be noticed until damage has already been inflicted; its sampling is therefore necessary. Orchards, or parts thereof where the pest had occurred in former years, should be sampled early in the season. An examination of 1 cm^2 per fruit (with hand lens of 10X or a larger magnification) of four fruits, taken from around the tree will provide a reasonable estimate of infestation, provided 10 or more mites are present in each sample. A rapid estimate of the population density can also be obtained by using standardized visual keys based on estimating infestation rates per 1 cm^2 of plant surface and comparing them to the keys [20]. However, as the mite cannot feed on the damaged fruit skin, it will not be found there; the areas to be examined are the margins of the affected area and the sepals.

Efforts to explore the relative tolerance of citrus species and varieties to the pest showed that lemons, oranges, tangerines and rough lemon are very susceptible, whereas pummelo and sour orange seem to be resistant [21].

Due to the rapid rate of increase in mite populations during early summer, the pest should be chemically treated as soon as it is recognized, especially at sites that had experienced damage in the past. A repeat spray may be needed within 4-6 weeks. Fruit intended for processing can tolerate some cosmetic damage and may require fewer chemical treatments than fruit sold as fresh product. Acaricides in current usage include abamectin, fenbutatin oxide (which is being phased out) and liquid sulphur. Field resistance to zineb, formerly in common use, was reported from Florida, Cyprus and Israel [22].

Predatory mites of the families Phytoseiidae and Stigmaeidae prey on the citrus rust mite, and may be able to control it together, especially if active during early spring. None, however, appear to be either specific or capable of predictably controlling the pest [23]. Another option is the acaropathogenic fungus *Hirsutella thompsonii* Fisher, which adversely affects the mite in Florida [24]. This fungus, which requires saturation humidities in order to develop, failed to affect the pest in Israel due to environmental aridity [25]. The recently discovered endophytic fungus *Meira geulakonigii* Boekhout *et al.* may protect grapefruits from the pest in Israel [26].

Website: http://www.ipm.ucdavis.edu/PMG/r107400511.html

3. TENUIPALPIDAE

The family Tenuipalpidae, known as the false spider mites or flat mites, consists of about 900 species in 34 genera [27, 28]. All are phytophagous and several are pests of crops, including citrus and grapes. Their feeding causes leaf mottling and necrosis, fruit blistering and deformation, and may reduce yields. In rare cases false spider mites induce the formation of galls [29]. In Florida and in Central and South America several species of the genus *Brevipalpus* transmit plant viruses (e.g. leprosies) to citrus trees [30]. Tenuipalpids are usually red to yellow, with flat bodies, about 0.3-0.4 mm in length. They bear strongly-recurved whip-like chelicerae, and palpi with 1-5 segments, whose tarsi lack a thumb-claw complex. These mites have short legs whose empodia may be modified. The dorsum usually bears a reduced number of setae and two pairs of eyes. The Tenuipalpidae are sensitive to arid conditions, usually preferring to feed on the lower sides of leaves near the veins, at protected sites on fruits, and also on tree bark and even in galls. Most species have a narrow host plant range, but a few are very polyphagous. Most investigated species require 3-4 weeks at ca 25°C to complete a generation, and usually deposit about 30 to 50 eggs/female [31]. Reproduction is usually by arrhenotoky, although a few species are parthenogenetic. One cosmopolitan species, *Brevipalpus phoenicis*, appears to consist only of haploid females [32].

Websites: www.sel.barc.usda.gov/acari/content/brevipalpus/brevi1.html; http://en.wikipedia.org/wiki/Tenuipalpidae

Four species of false spider mites are pests of citrus in the Mediterranean Region: *Brevipalpus californicus, B. lewisi, B. obovatus* and *B. phoenicis.* A key using microscopic characters, intended to separate these species follows; in the field they are very similar and thus no field key is added.

1. Hysterosoma with six lateral setae (*c3*, *d3*, *e3*, *f3*, *h2*, *h1*).. 2

-. Hysterosoma with seven lateral setae (*c3*, *d3*, *e3*, *f2*, *f3*, *h2*, *h1*).. 3

2. Tarsus II with a single solenidion... ***Brevipalpus obovatus***

-. Tarsus II with two solenidia... ***Brevipalpus phoenicis***

3. Tarsus II with a single solenidion .... ***Brevipalpus lewisi***

-. Tarsus II with two solenidia.. ***Brevipalpus californicus***

3.1. *Brevipalpus californicus* (Banks)

Diagnostic characters: The body of this pest, known as the citrus flat mite, is oblong, red, about 0.25-0.30 mm in length. The entire dorsum is covered by reticulations. The opisthosoma bears 7 pairs of very short lateral setae and tarsus II has two distal sensory setae (solenidia).

Life history: This cosmopolitan, polyphagous species (having been found on 316 host plants [33], is a slow-moving mite that usually reproduces by parthenogenesis. Development from egg to first egg requires about 30 days [34]. Females on orange or lemon fruit produced 15-33 progeny each, dependent on the Egyptian climate, and completed 11 annual generations there [35]. The mites feed on all above-ground parts of all citrus varieties, seeming to prefer limes and bitter oranges [36]. On the fruits they prefer rind depressions [37]. The populations peak in mid-summer and they favor the eastern parts of the canopy, the lower branches and frequently the fruits. In Egypt the mite was present throughout the year, in larger populations on oranges than on mandarins [38].

Economic importance: The damage caused by this mite differs on different citrus species and varieties [36]. Feeding results in leaves with yellow-brown spots around punctures and in the skins of fruits, tending to crack at the damaged sites, causing their downgrading. Heavy infestations also result in leaf drop [39].

In South Africa the feeding of *B. californicus* produces brownish irregular scab-like spots on the rind of sweet oranges and corky blemishes on Valencia oranges [37]. In Greece *B. californicus* is a serious pest, causing silvering of lemon and mandarin fruits, and a brown speckling of oranges [40]. In the western Mediterranean Region this mite is regarded as a secondary pest of citrus. Although another species of *Brevipalpus* is known to transmit viral citrus diseases, in the Mediterranean Region *B. californicus* does not transmit such diseases.

Management: Due to the fact that this mite is regarded as a secondary pest in the Mediterranean Region, no special control measures are usually applied; rational horticultural practices and pesticides used against other pests suffice. In Egypt the need to control *B. californicus* is stronger, and it is recommended [38] that acaricides (*e.g.* abamectin and petroleum oils) should be applied to the lower tree levels and to their eastern aspects, before August [41]. Mite outbreaks on lemons were observed in Egypt after applications of organophosphorus pesticides [42]. Releases of the phytoseiid predator *Euseius scutalis* (Athias-Henriot) in groups of 50 per tree on orange trees in Egypt resulted in about 90% control [39].

Website: http://www.google.co.il/search?hl=en&q=Brevipalpus+californicus&btnG=Search&meta=&aq=f&oq=

3.2. *Brevipalpus lewisi* McGregor

Diagnostic characters: The body of this pest, known as the citrus flat mite or the citrus scab mite, is oblong, red, about 0.25 mm in length. The dorsal reticulations do not meet midway on the propodosoma. The opisthosoma bears 7 pairs of very short lateral setae and tarsus II has only a single sensory seta (solenidion).

Life history: This is a cosmopolitan polyphagous species (collected from over 30 plant species). Because little is known about this pest infesting citrus in the Mediterranean Region, data from other Regions will be used. The mite is usually found on the stem-end of fruits, near or under the button.

Green fruits are preferred over mature fruits (especially on the trees' canopy), and both sites over twigs and leaves. The pest prefers warm and dry conditions, its populations peaking during the warmest and driest months [43]. When reared on grapevine leaves, the mite completed a generation in about five weeks at 28°C and deposited up to 19 eggs [44]. Very few males are known, suggesting that it mostly reproduces by parthenogenesis.

Economic importance: Damage to citrus usually consists of irregular, corky, silvery or brown, scab-like scars at the sites of feeding, which are slightly depressed and appear on the exposed fruit surfaces. At maturity, as the fruits attain their orange color, these spots may disappear; if they persist, such fruit could be downgraded. Damage may occur on lemons, oranges and (especially) on tangerines (of which the cull might reach 25%), but not on grapefruits, despite the latter being sometimes heavily infested. The damage caused by *B. lewisi* has been attributed to the wounding of the oil glands on the fruit peel [45]. Mite feeding on the glands releases chemicals, which cause the damage. Grapefruit peel has fewer peel glands and *B. lewisi* may be able to feed around them, which could explain the relative tolerance of grapefruit to the mite. In Greece *B. lewisi* is a serious pest of citrus, causing scab-like scars on the fruits' skin [40].

Although another species of *Brevipalpus* transmits viral diseases of citrus, in the Mediterranean Region *B. lewisi* is not known to transmit such diseases.

Management: No data are available about the chemical or biological management of *B. lewisi* in the Mediterranean Region, where it is, at most, a minor citrus pest. Organophosphorous acaricides seem to be ineffective [43] but dicofol and Envidor (spirodiclofen) controlled the pest on citrus in Arizona for one month [46]. On grapevines in Spain several acaricides are recommended during June [47]. In south-eastern Australia *B. lewisi* is controlled on grapevines by two phytoseiid predators [48]. Other natural enemies have also been listed [2], but their effects on pest populations are little known.

Website: http://www.google.co.il/search?hl=en&q=Brevipalpus+lewisi&btnG=Search&meta=&aq=f&oq=

3.3. *Brevipalpus obovatus* Donnadieu

Diagnostic characters: The body of this pest, known as the privet mite, is oblong, red, about 0.25 mm in length. The striae are irregular mid-dorsally on the propodosoma and on the opisthosoma, which bears 6 pairs of very short lateral setae; tarsus II has only a single sensory seta (solenidion) (Fig. **16**).

Life history: This cosmopolitan, polyphagous species (obtained from 450 host plants [33]), can raise a generation in about three weeks at 27°C and produce about 50 eggs during five weeks on violet leaves [49]. Female longevity came to about five weeks on citrus at 25°C [50]. In Egypt *B. obovatus* occurs on citrus throughout the year, its populations increasing in late winter, peaking in spring, declining during summer and rising again in the fall [50]. The mites prefer mature leaves to young ones and hidden fruit regions.

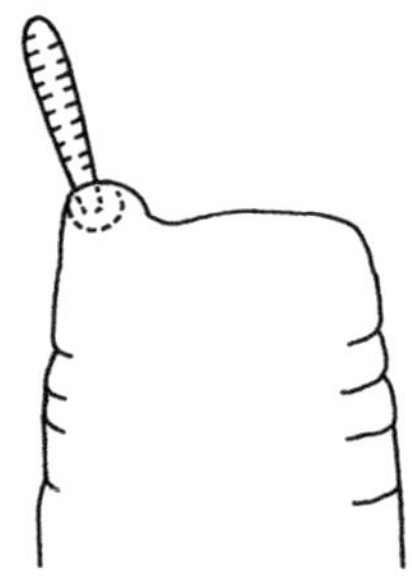

Fig. (16). Distal solenidion of tarsus II (without other setae) of *Brevipalpus obovatus* Donnadieu.

Winter is usually passed in the adult stage, the mites hiding in sheltered sites on the plants. The color of the female may change from light orange to dark red, dependent of the host plant. This thelytokous species seems to consist of several parthenogenetic strains, males being very rare [51].

Economic importance: Mite feeding sites on citrus leaves and fruit become brown and rusty, causing leaf drop [40, 50]. On the whole, *B. obovatus* is usually only of minor economic importance. Although another species of *Brevipalpus* transmits viral diseases of citrus, in the Mediterranean Region *B. obovatus* is not known to transmit such diseases.

Management: Different acaricides reduce mite populations on citrus [50]. A clear prey-predator relationship between the cheyletid mite *Cheletogenes ornatus* (Canestrini and Fanzago) and *B. obovatus* was observed on citrus trees in Egypt [52].

Website:http://www.google.co.il/search?hl=en&q=Brevipalpus+obovatus&btnG=Search&meta=&aq=f&oq=

3.4. *Brevipalpus phoenicis* (Geijskes)

Diagnostic characters: The body of this pest is oblong, red, about 0.25-0.30 mm in length. It is known as the reddish black flat mite, due to the irregular dark-red color patterns on its dorsum [53]. The mid-dorsal area is covered by irregular, rounded striae, without reticulation, an area that may show some variability, even in mites obtained off the same host plant [54, 55]. The opisthosoma with six pairs of short lateral setae (Fig. **17**) and tarsus II bears two sensory setae (solenidia) (Fig. **18**). Data from mitochondrial COI sequences of *B. phoenicis* were not consistent with its taxonomy based on morphology, suggesting that a revision of the entire *Brevipalpus* group is needed [56].

Life history: This cosmopolitan, polyphagous species, which was found on 486 host plants [33], is thelytokous, males being very rare. It completed ten annual generations on citrus in Egypt, a generation requiring about three weeks at 27°C. Besides temperature, duration of development is also strongly dependent on the host plants and the season. At 25°C a generation was completed in 19 days on citrus fruit, but required 29 days on papaya fruit [58].

17 18

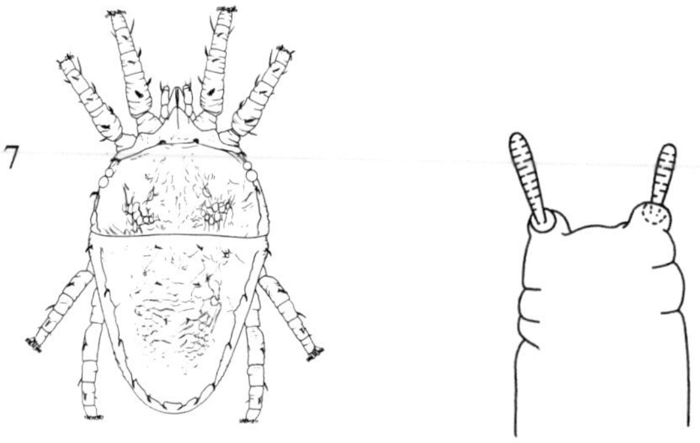

Figs. (17 and 18). Dorsal view of adult of *Brevipalpus phoenicis* (Geijskes) (17) [57]; distal solenidia of tarsus II (without other setae) (18).

Fecundity depended on the season, as 16.3 eggs were deposited on citrus in the Egyptian summer, only 12.5 in winter [59]. On other host plants more eggs may be produced; it produced an average of 57 progeny per female when reared on papaya fruit [53]. The mite prefers the lower sides of mature leaves and aggregates in protected invaginations, in the pedicle and at the button of fruits [59].

Economic importance: During feeding the mite injects a toxin into the host, causing a variety of symptoms on different plants, including citrus. In Venezuela *B. phoenicis,* infesting sour lemon seedlings cause a disorder called "halo scab", occurring when the mite settles in scabs caused by a fungus; affected seedling may drop their leaves and even die. Galls may also be formed on the stems of such seedlings [1]. Oranges damaged by *B. phoenicis* usually weigh less, loss that was inversely proportional to the degree of infestation. Furthermore, affected trees may lose 50% of their yield [60]. In Italy pest feeding resulted in greyish scabby patches and medial cracks on the apical epidermis of mandarin fruit. Many lesions were located on the oleiferous glands, which are emptied and dried; affected orange fruits show rounded reddish patches [61]. In Greece the mite infests all above-ground citrus parts, causing the leaves to become brownish, dry and to drop during heavy infestations [40].

The damage caused by leprosis, transmitted by this mite, including yield reductions and the cost of acaricide applications, was recently calculated to cost the Brazilian citrus industry $62 million annually, constituting 35% of total agrochemical costs and 14% of total production costs in mature orchards [62]. These data, probably representing only part of the total cost, are from South America. No such damage has yet been reported in the Mediterranean Region.

Management: Outbreaks of *B. phoenicis* on citrus in Brazil were controlled by the acaricides hexythiazox and fenbutatin oxide, but the large number of applications is promoting the development of mite resistance to these chemicals [63, 64]. Reports from the same region indicated that neem oil, at 1.5% v/v, successfully controlled the mite on orange fruits [65]. Another option, intended to reduce the damage caused by mite-vectored leprosis, is to explore the tolerance or resistance of citrus species and varieties to the disease [66]. The large numbers of the pest found on hedges, windbreak plants and weeds near or around citrus groves suggested that such plants represent threats to the crop [67]. Efforts to control the pest by biological means consisted of assaying fungi and predators. Of 52 isolates of various fungi tried against *B. phoenicis* infesting citrus fruit in the laboratory, *Hirsutella thompsoni* provided over 90% mite mortality, indicating the potential of this fungus to control the pest [68]. Several predators of the families Phytoseiidae and Stigmaeidae reproduced best and survived longest when feeding on *B. phoenicis* in the laboratory [69]. Nothing is known about similar efforts from the Mediterranean Region.

Website: http://www.google.co.il/search?hl=en&q=Brevipalpus+phoenicis&btnG=Search&meta=&aq=f&oq=

4. TARSONEMIDAE

This family, which includes mites of very different feeding habits, consists of about 550 species in 40 genera. Tarsonemids are small mites (ca 0.2-0.3 mm in length) with broad to elongate oval bodies covered by a hard and shiny integument. Their setation is reduced and females carry a pair of prodorsal sense organs (pseudostigmatic organs) that are lacking in males. Legs IV differ from other legs; in females they end with apical or subapical whip-like setae, (Fig. **19**) whereas in males they usually terminate as large claws. Tarsonemids reproduce by arrhenotoky and the sex ratio is variable (but usually heavily female-biased). Tarsonemids undergo a foreshortened life cycle, with only two active stages, larvae and adults; generation time may be one week or less. The larvae have a brief feeding period and molt directly to adults. They are dispersed by winds, on insects and on infested plant material. Many species are saprophagous, fungivorus, parasitic [e.g. the bee tracheal mite, *Acarapis woodi* (Rennie)] or predatory, but only a few feed on green plants and may become pests. Only one, *Polyphagotarsonemus latus*, is a pest of citrus in the Mediterranean Region.

4.1. *Polyphagotarsonemus latus* (Banks)

Diagnostic characters: The body of this pest, known on citrus as the citrus silver mite, but usually as the broad mite, is white to yellow, oval, about 0.20 mm in length. The dorsal setae are short, venter of metapodosoma with 3 pairs of coxal setae and leg I with a strong claw (Fig. **19**). The male has a button-like claw on leg IV, and the eggs are covered with small white tubercles.

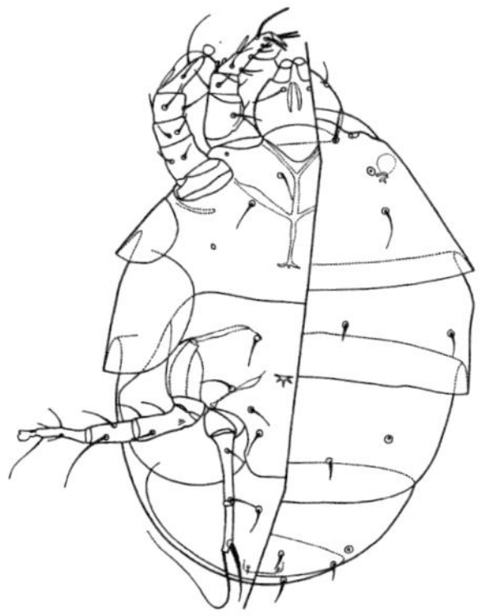

Fig. (19). Dorsal (at right) and ventral view (at left) of female of *Polyphagotarsonemus latus* (Banks) [71].

Life history: This cosmopolitan, polyphagous species (having been found on plants assignable to almost 60 families) [70], can raise a generation in one week at 27°C and produce about 40 eggs during five weeks. Low (12-14°C) or high (33-35°C) temperatures and dry conditions adversely affect reproduction, but there is no diapause. Males appear first and locate the pharate females (known as 'diapausing larvae', 'pupae' or 'quiescent nymphs'), placing these female on the terminal pads of their genital capsule and then moving upwards, to the plants' apical parts. This movement causes the pest's populations to ascend to the upper parts of host plants. Mating occurs shortly after the female molts, and the sex ratio is 4:1, female-biased. The males live for one week, females a few days more. In warm and wet climates *P. latus* is most active, and very large populations may develop rapidly and unexpectedly under such conditions [72].

The mite may be dispersed from area to area and from country to country through the transfer of infested plants. Dispersal within a habitat can be effected by various insects, especially whiteflies (Hemiptera: Aleyrodidae), such as the tobacco whitefly, *Bemisia tabaci* Gennadius. In citrus groves the citrus whitefly, *Dialeurodes citri* (Ashmead), is a suitable carrier [73].

Economic importance: The mite occurs on all species and varieties of citrus but serious damage has usually been reported from the yellow varieties, especially lemons and limes. In Nigeria, however, sweet oranges were also severely attacked [74]. Young, developing leaves and fruit are usually preferred. Leaf blistering, crinkling and narrowing, as well as various distortions, such as downwards curling, along with flower drop may also follow [75]. Fruits react by developing silvery (on lemons and limes) (Fig. **20**) or rust-colored (on oranges) blemishes with a "sharkskin" or corky texture that may extend over the entire surface [76]; most of such yield may be lost. The damage (especially to the leaves) is probably due to feeding and the injection of toxic substances. Citrus silver mite damage is somewhat similar to that caused by the citrus rust mite.

Management: This minute pest may often remain unnoted until its injury becomes evident, necessitating adequate sampling. This is facilitated by its clumped distribution on the young leaves and smaller fruits in the shaded parts of its host plants. On lime trees in Florida the pest was often concentrated in the middle canopy, and its numbers per cm^2 fruit surface had a close relationship with the total number of mites present on such fruits. An area of one cm^2 fruit surface, taken from the middle tree stratum, thus sufficed for reliable sampling. In addition, due to the good fit between the percentages of fruit surfaces damaged and percentages of fruit damage per tree, either could be used to estimate broad mite injury to limes. The aggregated occurrence of *P. latus* on lime fruit further facilitated sampling, as fewer fruits per orchard were needed for an adequate estimation of its populations [77]. No similar data are available from the Mediterranean Region.

Fig. (20). Young fruits of lemon damaged by *Polyphagotarsonemus latus* (Banks).

Most available acaricides provide satisfactory silver mite control, but pyrethroids are ineffective. The application of these chemicals against cotton pests in Africa were followed by outbreaks of *P. latus* [78]. Damage of this pest on citrus was first noted in Morocco after spraying with synthetic pesticides [79], suggesting that these chemicals may have killed the local natural enemies of *P. latus*.

Many predatory mites, especially of the family Phytoseiidae, control the pest in greenhouses [80]. Satisfactory control was obtained in Florida on limes with *Neoseiulus californicus* (McGregor), acting along with a complex of indigenous predacious mites [81]. In Nigeria, mites of the family Bdellidae were the main predators on sweet oranges [74]. Efforts to control the pest with several acaropathogenic fungi were reported [82], but it was concluded that fluctuating relative humidities in the field would detract from the efficacy of these fungi. On the other hand, a formulation of the Beta-exotoxin of *Bacillus thuringiensis* Berliner provided consistent control for nine days post-application. No data on the biological control of *P. latus* infesting citrus in the Mediterranean Region are available.

Websites: http://images.google.com/images?hl=en&lr=&ie=UTF-8&oe=UTF-8&q=Polyphagotarsonemus+latus&sa=N&tab=wi

http://www.entnemdept.ufl.edu/creatures/orn/broad_mite.htm

5. TETRANYCHIDAE

The family Tetranychidae, or spider mites, consists of ca 1,260 described species in about 70 genera [83], and is divided into two subfamilies, the "primitive" Bryobiinae and the "advanced" Tetranychinae. They can be recognized by bearing a thumb-claw complex on their tarsi, two pairs of eyes and usually two pairs of duplex setae on tarsi I. All spider mites are phytophagous and several polyphagous species are major pests of many crops, including apples, cassava, citrus, cotton, roses and citrus. They can be destructive plant feeders that cause serious economic losses and even plant death. They feed by wounding plant tissues with their chelicerae, which are then inserted into the cells to suck out their contents. As the affected cells die they turn yellow, imparting a mottled appearance to the affected leaves, which may shrivel. Injury is also due to the dense webbing produced by many members of the Tetranychinae, which may totally cover entire or parts of the host plants. These webs are used for dispersal, protection, pheromone carriers and colony definition. Dispersal is often by "ballooning": the mites produce a single long silken thread that is taken up by winds, carrying them along [84]. Many spider mites have a very high rate of increase, being capable of completing a life cycle within 2 weeks at 25°C and reproducing (by arrhenotoky) the year around.

Spider mites were formerly considered to be pests of secondary importance, but after the Second World War they attained the status of major pests. This change was due to several factors that probably compounded each other. First and foremost was the destruction of their natural enemies by synthetic pesticides (most of which are less toxic to the pests).

The elimination of competitors was a second factor, along with the emergence of resistance to many pesticides and acaricides [85]. Other reasons were the continuous, year-round cultivation of many crops under glass, the world-wide introductions of many new, susceptible commercial plants and outbreaks engendered by pesticides, such as the pyrethroids [86].

Three species of spider mites are pests of citrus in the Mediterranean Region: *Eutetranychus orientalis, Panonychus citri* and *Tetranychus urticae.* Another secondary species, *Eutetranychus banksi* (McGregor), lives in a restricted area of the Iberian peninsula, where it has not so far caused serious damage.

A key using microscopic characters to separate these species follows, along with a key for their field recognition.

1. Empodial claw present.. 2

-. Empodial claw absent.. ***Eutetranychus orientalis***

2. With 2 pairs of para anal setac .... ***Panonychus citri***

-. With a single pair of para anal setae.. ***Tetranychus urticae***

A key for recognizing these species in the field

1. Dorsal setae not set on tubercles; color dark-green or red, with dark spots along body .... ***Tetranychus urticae***

-. Dorsal setae set on tubercles.. 2

2. Body red; dorsal setae longer than setae next in line.. .***Panonychus citri***

-. Body dark brown yellow, dorsal setae shorter than setae next in line ....***Eutetranychus orientalis***

5.1. *Eutetranychus orientalis* (Klein)

Diagnostic characters: The oval body of this pest, known as the oriental spider mite or citrus brown mite, is about 0.4-0.5 mm in length, yellow-brownish, and legs that are about as long as the body. The dorsal setae are short, spatulate and placed on small tubercles (Fig. **21**). The smaller male is reddish, its hysteronotum is pointed posteriorly and the legs are longer than the body.

Life history: This cosmopolitan, polyphagous species occurs on about 220 host plants [83]. It colonizes the upper side of leaves whereon its eggs are deposited along the midrib. The threshold of development is at 10°C and development requires about two weeks at around 25°C [88]. Each female produces 30-40 eggs during two-three weeks, being most fecund and living longest on lemons, probably due to their thin cuticle [89]. In the East Mediterranean the pest raises about 20 annual overlapping generations and its sex ratio is usually 0.7, female-biased.

Economic importance: This spider mite is the most persistent citrus pest in Upper Egypt and an occasional serious pest in Israel, especially in the drier areas. Mite feeding results in yellow-grey stippled spots on the leaves that cause their wilting and drop. Heavy infestations may incur fruit drop and the death of upper branches, and following-years' blossoming can be much affected. Seedlings may be severely damaged. Lemon trees of all ages are more damaged than other citrus species and varieties, whereas oranges are not much affected. Reduced irrigation increases the damage.

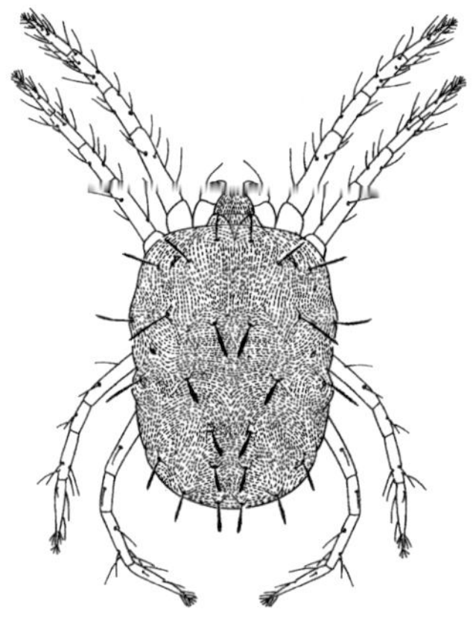

Fig. (21). Dorsal view of female of *Eutetranychus orientalis* (Klein) [87].

Management: Autumn pruning of heavily damaged branches can reduce some damage [88]. As far as known, the pest has not yet developed any resistance to pesticides and may still be controlled by acaricides as fenbutatin oxide, hexythiazox and others, and even by petroleum oils [90-92]. Nevertheless, continuous monitoring for resistance is necessary. The pest is attacked by many indigenous predatory mites of the families Phytoseiidae and Stigmaeidae [93]. The phytoseiid *Euseius scutalis* regulates the numbers of the citrus brown mite in Jordan [94], but little is known about the ability of other predators to reduce pest populations.

Websites: http://www.springerlink.com/content/k5r73tv37m586067/

5.2. *Panonychus citri* (McGregor)

Diagnostic characters: The body of this pest, known as the citrus red mite, is about 0.4 mm in length. The dorsum bears 13 pairs of mostly long, tapering setae that are set on reddish tubercles (Fig. **22**). The tarsi carry claw-like empodia with three pairs of ventrally-directed hairs (Fig. **23**). All stages of the citrus red mite, including the males and eggs, are dark red.

Life history: This cosmopolitan, polyphagous species occurs on about 110 host plants [83], but is mainly known as a pest of citrus. Development on leaves of sweet orange [*Citrus sinensis* (Linnaeus) Osbeck] requires about 10 days and an average of 26 eggs per female was produced at 25°C and 65% RH. When feeding on fruits the fecundity of the mite may be much higher, coming to 77 eggs per female [95]. The female development threshold was at about 9.5°C [96], and at 26°C a generation can be completed in a fortnight. Eggs are usually placed along the main vein on the upper sides of young leaves. Citrus red mite populations are more numerous on citrus during early winter, least abundant in summer, but they continue to reproduce throughout the year.

Mite susceptibility to high temperatures (32 to 40°C) and low humidities explains its population decline in mid-summer. The pest often disperses by "ballooning", floating on air streams by holding on to a silken thread.

Economic importance: Mite feeding on the leaves causes stippling spots (Fig. **24**) that combine to cause yellow or silvery areas (Fig. **25**). Heavy infestations result in leaf drop, twig dieback, low-quality fruit and even tree death. The expression of damage is compounded by dry, hot winds [1]. Furthermore, the citrus red mite is a serious allergen [98].

22

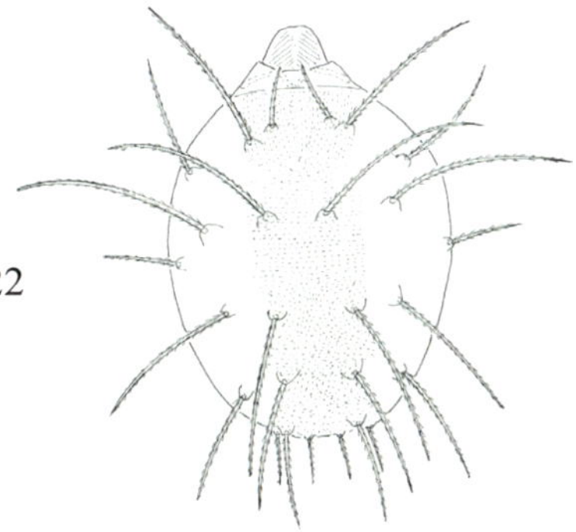

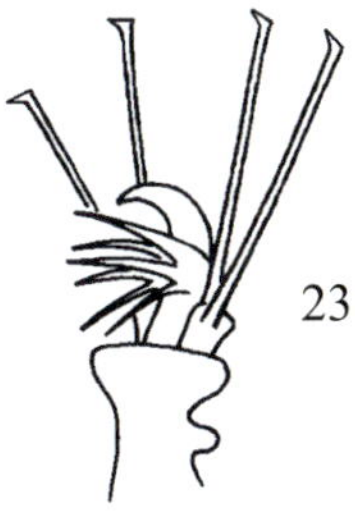

23

Figs. (22 and 23). Dorsal view (22) [97] and pretarsus I of the female of *Panonychus citri* (McGregor) [2].

Management: Sampling may be conducted by counting mites on entire leaves, usually 20 per tree (five from each aspect), taken off ten trees. Sample size however depends on weather and region, because more damage can be expected from fewer individuals during warm and dry periods. In Italy a visual sampling system, based on observations on 4 shoots of the most recent vegetative growth from 10% of trees chosen at random, has been positively verified. The mite has developed wide-spread resistance to most pesticides [99]; resistant strains may however have shortened lives and reduced fecundity [100].

Fig. (24). Stippling spot on leaves of lemon due to feeding activity of *Panonychus citri* (McGregor).

In most orange orchards the pest is usually maintained below its economic injury level by natural enemies, predatory mites of the families Phytoseiidae [101] and Stigmaeidae, as well as by coccinellid beetles [102] and coniopterygids.

Fig. (25). Severe attack of *Panonychus citri* (McGregor) on a lemon leaf.

In the western Mediterranean countries the phytoseiid *Euseius stipulatus* (Athias Henriot) provides excellent control of the pest, bud it is necessary to preserve their populations by using selective acaricides and/or petroleum oils [103]. Also the coniopterygid *Conwentzia psociformis* (Curtis) limits mite outbreaks. Pest outbreaks sometimes follow the application of pesticides [86, 104]. The application of selective chemicals that are harmless to the natural enemies, could thus be the key to citrus red mite management. In California the pest is partially controlled by a virus disease [105].

Website: http://www.inra.fr/Internet/Produits/HYPPZ/RAVAGEUR/6pancit.htm

5.3. *Tetranychus urticae* Koch

Diagnostic characters: The body of this pest, known as the two-spotted spider mite, is about 0.4 mm long. The dorsum bears 13 pairs of mostly long, slender setae and there is only one pair of para-anal setae (Fig. **26**). The tarsal empodium bears three pairs of ventrally-directed hairs (Fig. **27**). The body color of the female may be green or red, often depending on the host plant, and with two darker internal spots on each side; diapausing females are yellow. The males and young stages are green-yellow. This species occurs in two forms, green and red [the latter often listed as *Tetranychus cinnabarinus* (Boisduval)] [106].

Life history: At 24°C and 40% RH, a generation develops in about two weeks, each female depositing over 100 eggs [107]. In the Middle East the mite may raise 20-25 annual generations. Prior to the last molt (between second nymph and adult) the female covers itself with a few silken threads, which attract the males. Mating takes place as soon as the emerging female frees itself from the webs [84].

The two-spotted spider mite lives in webbed colonies, and its webbing is seen on the lower sides of leaves. Most webbing is produced under the same environmental conditions (24°C and 40% RH) that promote its population increase [108]. The webs are used to construct mite colonies, to protect themselves and their eggs from predators and unsuitable humidities, to discourage competitors, and as a carrier for their sex pheromone. In addition, the webs serve for dispersal, which usually takes place when the host plant begins to wither as a result of intensive feeding [84]. Mated females disperse by ballooning and by forming webbed "balls" that contain all their stages; such balls become detached from the host plants by winds and are then blown about.

Tetranychus urticae (green form) undergoes a reproductive diapause, in contrast to the red form, which may reproduce throughout the year.

Economic Importance: This cosmopolitan, polyphagous species occurs on over 1000 host plants [83], being a pest of many crops besides citrus. The mites usually colonize and feed on the lower sides of leaves where they suck out cell constituents. The resulting yellow or silvery spots fuse to form wide-spread stippling (Figs. **28** and **29**), resulting in chlorophyll content and photosynthetic losses, leaf wilt and even death (especially of seedlings).

In eastern Mediterranean citrus orchards the two-spotted spider mite is usually only of minor economic importance, in contrast to the situation in the eastern Mediterranean Region, where it can seriously damage leaves and fruits of lemon (Fig. **30**), orange (Fig. **31**), mandarin and clementine.

Management: Sampling the mite on clementine mandarins in Spain showed that symptomatic leaves (showing typical injury spots) carried 57.1% of the total mite populations and represented the mites' dynamics on all leaves. On fruits mite feeding activity produces russeted areas around the navel end that may spread over the whole fruit; on lemons damage commonly appears at the blossom end of the fruit. The two-spotted spider mite develops rapid and extensive resistance to most pesticides in use.

26 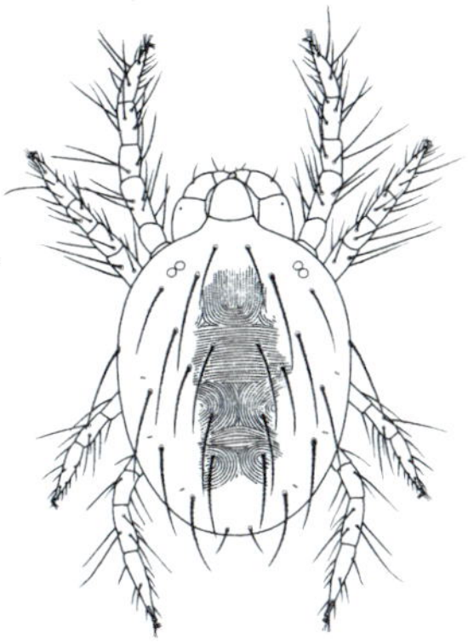

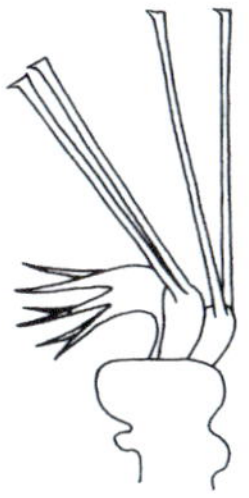 27

Figs. (26 and 27). Dorsal view (26) and pretarsus (27) I of female of *Tetranychus urticae* Koch [2].

The chemicals still available in Mediterranean citrus groves include mineral oil, tebufenpyrad, clofentezine, fenazaquin [110], spirodiclofen [111] and bifenazate [112]. In the citrus orchards the pest may be controlled by many indigenous and especially introduced predatory mites of the family Phytoseiidae [113]. Among insects coccinellids and cecidomyiids are also common.

Fig. (28). Leaf of lemon damaged by *Tetranychus urticae* Koch.

Fig. (29). Leaves of clementine attacked by *Tetranychus urticae* Koch.

Fig. (30). Blossom end of lemon damaged by *Tetranychus urticae* Koch.

The large variation in mite numbers per leaf necessitated enumerative sampling, whereas on fruits binomial sampling was the only valid method for estimating pest numbers [109]. From a practical point of view it is sufficient to sample 10% of plants at random, observing four shoots of the recent vegetative growth, and 20 fruits per plant. Sampling and monitoring should be conducted every two weeks from June to October.

Fig. (31). Oranges damaged by *Tetranychus urticae* Koch.

Websites: http://creatures.ifas.ufl.edu/orn/twospotted_mite.htm

http://www.inra.fr/Internet/Produits/HYPPZ/RAVAGEUR/6teturt.htm

REFERENCES

[1] Jeppson LR, Keifer HH, Baker EW. Mites Injurious to Economic Plants. Berkeley: University of California Press 1975.

[2] Vacante V. Citrus Mites. Identification, Bionomy and Control. Wallingford: Cabi Publishing 2010.

[3] Keifer HH. Eriophyid studies. Bull Dept Agric, California, 1938; XXVII, 2: 181-206.

[4] Sternlicht M, Goldenberg S. Fertilization, sex ratio and postembryonic stages of the citrus bud mite *Aceria sheldoni* (Ewing) (Acarina, Eriophyidae). Bull Entomol Res 1971; 60: 391-7.

[5] Sternlicht M. Effect of different wave lengths of light on the behaviour of an eriophyid bud mite, *Aceria sheldoni*. Entomol Exp Appl 1969; 12: 377-82.

[6] Sternlicht M. Contribution to the biology of the citrus bud mite *Aceria sheldoni* (Ewing) (Acarina: Eriophyidae). Ann Appl Biol 1970; 65: 221-30.

[7] Walker GP, Voulgaropoulos AL, Phillips PA. Effect of citrus bud mite (Acari: Eriophyidae) on lemon yields. J Econ Entomol 1992; 85: 1318-29.

[8] Hare JD, Rakha M, Phillips PA. Citrus bud mite (Acari: Eriophyidae): an economic pest of California lemons? J Econ Entomol 1999; 92: 663-75.

[9] Vacante V, Bonsignore C, Federico S. Indagini sul ruolo ecologico di *Aceria sheldoni* su limone. Italus Hortus 2007; 14(6), 103-5.

[10] Keifer HH. Eriophyid studies XXVII. Occasional Papers. California Dept Agric.

[11] Ebrahim HM. Influence of temperature and relative humidity on the biology and life table parameters of *Phyllocoptruta oleivora* and *Aculops pelekassi* (Acari: Eriophyidae) on "Hamlin" orange in Central Florida. Egypt J Agricul Res 2000; 78: 143-61.

[12] Mijuskovic M, Kosac D. Control of *Aculops pelekassi* Keifer (Acarina, Eriophyidae), an important mite injurious to Citrus on the coast of Montenegro. Jugoslovenskog Vocarstva 1972; 21/22: 835-42.

[13] Seki M. Life cycle of the pink citrus rust mite, *Aculops pelekassi* (Keifer). In: Proc Int Soc Citriculture 1981. Vol 2; pp. 656-8.

[14] Burditt AK, Reed DK, Crittenden CR. Observations on the mites *Phyllocoptruta oleivora* (Ashmead) and *Aculus pelekassi* Keifer under laboratory conditions. Fl a Entomol 1963; 46: 1-5.

[15] Hall DG, Childers CC, Eger JE. Binomial sampling to estimate rust mite (Acari: Eriophyidae) densities on orange fruit. J Econ Entomol 2007; 100: 233-40.

[16] Kim SS, Paik CH. Predation of five species of phytoseiid mites on *Panonychus citri* and *Aculops pelekassi*. Korean J. Entomol. 1999; 29: 261-4.

[17] Swirski E. Contribution to the knowledge of the fluctuations in populations of the citrus rust mite (*Phyllocoptruta oleivora* Ashm.) in the coastal plain of Israel. Israel J Agricul Res 1962; 12: 175-87.

[18] Allen JC, Yang Y, Knapp JL. Temperature effects on development and fecundity of citrus rust mite (Acari: Eriophyidae). Environ Entomol 1995; 24: 996-04.

[19] Bergh JC. Ecology and aerobiology of dispersing citrus rust mites (Acari: Eriophyidae) in central Florida. Environ Entomol 2001; 30: 318-26.

[20] Rogers JS, McCoy CW, Manners MM. Standardized visual comparison keys for rapid estimations of citrus rust mite (Acari: Eriophyidae) populations. J Econ Entomol 1994; 87: 1507-12.

[21] Yue B, Sun M, Qiu A. A comparison of citrus resistance to citrus rust mite (Acari: Eriophyidae) in the laboratory. Syst Appl Acarol 1998; 3: 59-62.

[22] Swirski E, Kehat M, Gruenberg S, Dorzia N, Amitai S. Trials for control of the citrus rust mite (*Phyllocoptruta oleivora* Ashm.). Israel J Agricul Res 1967; 16: 115-24.

[23] Argov Y, Domeratzky S, Gerson U, Steinberg S, Palevsky E. Augmentation and conservation of indigenous generalist acarine predators for the control of citrus rust mite, *Phyllocoptruta oleivora*, in Israel. IOBC/WPRS Bull 2007; 30(5): 9-15.

[24] Childers CC. In: Rosen D, Bennett FD, Capinera JL, Eds. Pest Management in the Subtropics: Biological Control - A Florida Perspective. Andover: Intercept 1994; pp. 255-88.

[25] Gerson U, Kenneth R, Muttath TI. *Hirsutella thompsonii*, a fungal pathogen of mites. II. Host-pathogen interactions. Ann Appl Biol 1979; 91: 29-40.

[26] Paz Z, Burdman S, Gerson U, Sztejnberg A. Antagonistic effects of the endophytic fungus *Meira geulakonigii* on the citrus rust mite *Phyllocoptruta oleivora.* J Appl Microbiol 2007; 103: 2570-9.

[27] Meyer MKP (Smith). The Tenuipalpidae (Acari) of Africa with keys to the world genera. Entomol Mem, Dept Agricul Techn Serv, Republic of South Africa 1979; 50.

[28] Mesa N, Ochoa R, Welbourn WC, Evans GA, De Moraes G. A Catalog of the Tenuipalpidae (Acari) of the world, with a key to genera. Zootaxa 2098: New Zealand: Magnolia Press 2009.

[29] Childers CC, French JV, Rodrigues JCV. *Brevipalpus californicus, B. obovatus, B phoenicis,* and *B. lewisi* (Acari: Tenuipalpidae): a review of their biology, feeding injury and economic importance. Exp Appl Acarol 2003a; 30: 5-28.

[30] Childers CC, Derrick KS. *Brevipalpus* mites as vectors of unassigned rhabdoviruses in various crops. Exp Appl Acarol 2003; 30: 1-3.

[31] Gerson U. The Tenuipalpidae: an under-explored family of plant-feeding mites. Syst Appl Acarol 2008; 13: 83-01.

[32] Weeks AR, Marec F, Breeuwer JAJ. A mite species that consists entirely of haploid females. Science 2003; 292: 2479-82.

[33] Childers CC, Rodrigues JCV, Welbourn WC. Host plants of *Brevipalpus californicus, B. obovatus* and *B. phoenicis* (Acari: Tenuipalpidae) and their potential involvement in the spread of viral diseases vectored by these mites. Exp Appl Acarol 2003b; 30: 29-105.

[34] Manglitz GR, Cory EN. Biology and control of *Brevipalpus australis*. J Econ Entomol 1953; 46: 116-9.

[35] Attiah HH, Soliman AA, Wahba ML. On the biology of *Brevipalpus californicus* Banks. Proc 3rd Int Congr Acarol. The Hague: W. Junk; pp. 187-91.

[36] Sadana GL, Joshi R. Comparative susceptibility of different varieties of citrus to the attack of the phytosphagous mite *Brevipalpus californicus*. Sc Cult 1979; 45: 413-4.

[37] Schwartz A. Flat mites of citrus. Farming in South Africa, Citrus series H.13 1977a; 1-3.

[38] El-Halawany ME. Some factors affecting the distribution of *Brevipalpus californicus* (Banks) on citrus trees. Egyptian J Agricul Res 1991; 69: 185-92.

[39] El-Halawany ME, Abd El Samed MA, Ibrahim GA, Radwan Z, Ebrahim HM. Evaluating the efficiency of different levels of *Euseius scutalis* (Athias-Henriot) released for the control of *Brevipalpus californicus* (Banks) on baladi orange trees. Menofiya J Agricul Res 1993; 18: 2697-2707.

[40] Hatzinikolis EN. The genus *Brevipalpus* in Greece (Acari: Tenuipalpidae). Entomol Hellen 1986; 4: 37-48.

[41] El-Kady MH, Nassar MES, Heykal I H. The effect of certain insecticides on the mite *Brevipalpus californicus* on citrus trees (Acarina, Tenuipalpidae). Agricul Res Rev 1977; 55: 95-7.

[42] Attiah HH, Wahba ML. Phosphorus compounds as a cause of an increase in numbers of the flat mite. Proc 3rd Int Congr Acarol. The Hague: W. Junk 1971; pp. 193-6.

[43] Elmer HS, Jeppson LR. Biology and control of the citrus flat mite. J Econ Entomol 1957; 50: 566-70.

[44] Buchanan GA, Bengston M, Exley EM. Population growth of *Brevipalpus lewisi* McGregor (Acarina: Tenuipalpidae) on grapevines. Austral J Agric Res 1980; 31: 957-65.

[45] Elmer HS. Citrus flat mite injury patterns on some citrus varieties. J Econ Entomol 1968; 61: 1689-92.

[46] Kerns D. Efficacy of envidor on mites of citrus, 2005. Arthrop Manag Tests 2006, 2006; 31: D5.

[47] Rodriguez JA, Arias A, Santiago R, Nieto J. Observaciones sobre la biologia de *Brevipalpus lewisi* (McGregor) en vinedos de la Comarca de Guarena (Badajoz), 1984-1986. Bol Sanid Veg Plagas 1987; 13: 249-59.

[48] James DG, Whitney J. Mite populations on grapevines in south-eastern Australia: Implications for biological control of grapevine mites (Acarina: Tenuipalpidae: Eriophyidae). Exp Appl Acarol 1993; 17: 259-70.

[49] Morishita FS. Biology and control of *Brevipalpus inornatus* (Banks). J Econ Entomol 1954; 47: 449-56.

[50] Rezk HA. The false spider mite, *Brevipalpus obovatus* Donnadieu (Acari: Tenuipalpidae): host-related biology, seasonal abundance, and control. In: Halliday *et al.*, Eds. Proceedings of the 10th International Congress of Acarology; 2001: Camberra, Australia: CSIRO Publishing; pp. 291-4.

[51] Helle W, Bolland HR, Heitmans WRB. Chromosomes and types of parthenogenesis in the false spider mites (Acari: Tenuipalpidae). Genetica 1980; 54: 45-50.

[52] Rezk HA, Gadelhak GG. Relationship between phytophagous and predatory mites in citrus orchards and the effect of acaricides on their populations. Alexandria J Agricul Res 1996; 41: 217-24.

[53] Haramoto FH. Biology and control of *Brevipalpus phoenicis* (Geijskes) (Acarina: Tenuipalpidae). Hawaii Agricul Exp St, Tech Bull 1969; 68: 1-63.

[54] Manson DCM. Mites of the families Tenuipalpidae and Tetranychidae intercepted entering New Zealand from overseas. NZ J Sc 1967; 10: 664-74.

[55] Gonzalez RH. Revision of the *Brevipalpus phoenicis* "complex" with descriptions of new species from Chile and Thailand (Acarina, Tenuipalpidae). Acarologia 1975; 17: 81-91.

[56] Groot TVM, Breeuwer JAJ. *Cardinium* symbionts induce haploid thelytoky in most clones of three closely related *Brevipalpus* species. Exp Appl Acarol 2006; 39: 257-71.

[57] Vacante V, Nucifora A. Gli Acari degli Agrumi in Italia. I. Specie rinvenute e chiave per il riconoscimento degli ordini, dei sottordini e delle famiglie. Boll Zool agr Bachic, ser. II, 1985; 18: 115-66.

[58] Chiavegato LG. Biologia do acaro *Brevipalpus phoenicis* em citros. Pesquisa Agropecuaria Brasileira 1986; 21: 813-16.

[59] Zaher MA, Wafa AK, Yousef AA. Biology of *Brevipalpus phoenicis* (Geijskes) in Egypt. Bull Soc entomol Egypte 1970; 54: 177-83.

[60] Rodrigues JCV, Kitajima EW, Childers CC, Chagas CM. Citrus leprosis virus vectored by *Brevipalpus phoenicis* (Acari: Tenuipalpidae) on citrus in Brazil. Exp Appl Acarol 2003; 30: 161-79.

[61] Di Martino E. In: Spina P, Di Martino E, Eds. Trattato di agrumicoltura. Bologna: Edagricole 1985: pp. 113-95.

[62] Bassanezi RB, Laranjeira FF. Spatial patterns of leprosis and its mite vector in commercial citrus groves in Brazil. Pl Path 2007; 56: 97-06.

[63] Oliveira CAL, Campos Neto RR, Fernandes CB. Efeito de diferentes volumes de calda no controle do ácaro-da-leprose *Brevipalpus phoenicis* (Geijskes) em citros. Anais Soc Entomol Brasil 1998; 27: 117-54.

[64] Campos FJ, Omoto C. Estabilidade da resistência de *Brevipalpus phoenicis* (Geijskes) (Acari: Tenuipalpidae) a hexythiazox em pomares de citros. Neotrop Entomol 2006; 35: 840-8.

[65] Wagner J, Pereira MFA, de Souza Amorim LC, Goes Maciel CD. Efficiency of neem oil on citrus leprosies mite *Brevipalpus phoenicis* (Geijskes, 1939) control. Pesquisa Agropecuária Tropical 2009; 39(1): 38-42.

[66] Bastianel M, de Oliveira AC, Cristofani M *et al.* Inheritance and heritability of resistance to citrus leprosis. Phytopathology 2006; 96: 1092-6.

[67] Maia OMA, Oliveira CAL. Capacidade de colonização de *Brevipalpus phoenicis* (Geijskes) (Acari: Tenuipalpidae) em cercas-vivas, quebra-ventos e plantas invasoras. Neotrop Entomol 2004; 33: 625-9.

[68] Rossi-Zalaf LS, Alves SB. Susceptibility of *Brevipalpus phoenicis* to entomopathogenic fungi. Exp Appl Acarol 2006; 40: 37-47.

[69] De Vis RMJ, Moraes GJ de, Bellini MR. Initial screening of little known predatory mites in Brazil as potential pest control agents. Exp Appl Acarol 2006; 39: 115-25.

[70] Gerson U. Biology and control of the broad mite *Polyphagotarsonemus latus* (Banks) (Acari: Tarsonemidae). Exp Appl Acarol 1992; 13: 163-78.

[71] Nucifora A, Vacante V. Citrus mites in Italy. VII. The family Tarsonemidae. Species collected and notes on ecology. Acarologia 2004; 44 (1/2): 49-67.

[72] Jones VP, Brown RD. Reproductive responses of the broad mite, *Polyphagotarsonemus latus* (Acari: Tarsonemidae), to constant temperature-humidity regimes. Ann Entomol Soc America 1983; 76: 466-9.

[73] Palevsky E, Soroker V, Weintraub P, Mansour F, Abo-Moch F, Gerson U. How species-specific is the phoretic relationship between the broad mite, *Polyphagotarsonemus latus* (Acari: Tarsonemidae), and its insect hosts? Exp Appl Acarol 2001; 25: 217-24.

[74] Umeh V, Amih C, Kolade J, Onukwu D. Damage on sweet orange fruit by the mite *Polyphagotarsonemus latus* (Banks) and the population of its predatory mites in Southwestern Nigeria. Fruits 2007; 62: 149-56.

[75] Uygun N, Ulusoy MR, Karaca I. A citrus pest in the East Mediterranean Region of Turkey, *Polyphagotarsonemus latus* (Banks) (Acarina, Tarsonemidae). Turk J Entomol 1995; 19: 1-4.

[76] Schwartz A. Citrus silver mite, *Hemitarsonemus latus* (Banks). Farm South Africa, Citrus Series H.14 1977b; 1-5.

[77] Peña JE, Baranowski RM. Dispersion indices and sampling plans for the broad mite (Acari: Tarsonemidae) and the citrus rust mite (Acari: Eriophyidae) on limes. Environ Entomol 1990; 19: 378-82.

[78] Vaissayre M. Chemical control of the mite *Polyphagotarsonemus latus* (Banks) in cotton fields. Coton Fibres Tropicales 1986; 41: 38-43.

[79] Delucchi VL. Un acarien nuisible aux agrumes au Maroc *Hemitarsonemus latus* Banks (Acarina, Tarsonemidae). Al Awamia 1963; 6: 17-29.

[80] Weintraub PG, Kleitman S, Mori R, Shapira N, Palevsky E. Control of the broad mite (*Polyphagotarsonemus latus* (Banks)) on organic greenhouse sweet peppers (*Capsicum annuum* L.) with the predatory mite, *Neoseiulus cucumeris* (Oudemans). Biol Cont 2003; 27: 300-9.

[81] Peña JE, Osborne L. Biological control of *Polyphagotarsonemus latus* (Acarina: Tarsonemidae) in greenhouses and field trials using introductions of predaceous mites (Acarina: Phytoseiidae). Entomophaga 1996; 41: 279-85.

[82] Peña JE, Osborne LS, Duncan RE. Potential of fungi as biocontrol agents of *Polyphagotarsonemus latus* (Acari: Tarsonemidae). Entomophaga 1996; 41: 27-36.

[83] Migeon A, Dorkeld F. Spider Mites Web: a comprehensive database for the Tetranychidae [Internet] - Available from:http://www1.montpellier.inra.fr/CBGP/spmweb/ 2009.

[84] Gerson U. In: Rodriguez JG, Ed. Recent Advances in Acarology. New York: Academic Press Vol. 1 1979; pp. 177-88.

[85] McMurtry JA, Huffaker CB, Van de Vrie M. Ecology of tetranychid mites and their natural enemies: a review. I. Tetranychid enemies: their biological characters and the impact of sprays practices. Hilgardia 1970; 40 (11): 331-90.

[86] Gerson U, Cohen E. Resurgences of spider mites (Acari: Tetranychidae) induced by synthetic pyrethroids. Exp Appl Acarol 1989; 6: 29-46.

[87] Gutierrez J, Helle W. Deux nouvelles espèces du genre *Eutetranychus* Banks (Acariens: Tetranychidae) vivant sur plantes cultivèes à Madagascar. Ent Ber Amst 1971; 31: 45-60.

[88] Klein Z. Contributions to the knowledge of the red spiders in Palestine. 1. The Oriental red spider, *Anychus orientalis* Zacher. Hadar 1936; 9: 107-11, 126-32.

[89] Al-Gboory JJ. In: Dusbabek F, Bukva V., Eds. Modern Acarology: The Hague: W. Junk 1991; pp. 607-11.

[90] Osman MS. Petroleum oils as a component of integrated pest management of phytophagous mites. Arab J Sc Res 1997; 15: 125-35.

[91] Drishpoun Y. A Guide to the Control of Citrus Pests. Israel Ministry Agricul Rural Dev 2000.

[92] Marquez AL, Wong E, Garcia E, Olivero J. Efficacy assay of different phytosanitary chemicals for the control of *Eutetranychus orientalis* (Klein) (Oriental Spider Mite) on Fine lemon and Valencia-Late orange crops. Bull IOBC/WPRS 2006; 29 (3): 305-10.

[93] Momen F, El-Borolossy M. Suitability of the citrus brown mite, *Eutetranychus orientalis* as prey for nine species of phytoseiid mites (Acari: Tetranychidae, Phytoseiidae). Acarologia 1999; 40: 19-23.

[94] Tanigoshi LK, Bahdousheh M, Babcock JM. *Euseius scutalis* (Athias-Henriot) a predator of *Eutetranychus orientalis* (Klein) in Jordan: toxicity of some acaricides to *E. orientalis*. Arab J Pl Prot 1990; 8: 114-20.

[95] Chiavegato LG. Biologia do acaro *Panonychus citri* (McGregor, 1916) (Acari: Tetranychidae) praga dos citros. Cientifica Jaboticabal 1988; 16: 79-84.

[96] Kasap I. The biology and fecundity of the citrus red mite *Panonychus citri* (McGregor) (Acari: Tetranychidae) at different temperatures under laboratory conditions. Turkish J Agricul Forest 2009; 33: 593-600.

[97] Meyer M K P (Smith). African Tetranychidae (Acari: Prostigmata) - with reference to the world genera. Republic of South Africa Department of Agriculture and Water Supply Entomology Memoir 69, 1987.

[98] Lee MH, Cho SH, Park HS *et al.* Citrus red mite (*Panonychus citri*) is a common sensitizing allergen among children living around citrus orchards. Ann Aller Asth Immunol 2000; 85(3): 200-04.

[99] García-Marí F. Resistencia dc *Tetranychus urticae* y *Panonychus citri* a acaricidas en el cultivo de los cítricos. Phytoma Espana 2005; 173:71-8.

[100] Yamamoto A, Yoneda H, Asada M. Influence of hexythiazox resistance on life-history parameters in the citrus red mite, *Panonychus citri* (McGregor). J Pest Sc 1995; 20: 521-7.

[101] Swirski E, Gokkes M, Amitai S. Phenology and natural enemies of the citrus red mite, *Panonychus citri* (McGregor) in Israel. Israel J Entomol 1986; 20: 37-44.

[102] Jamieson LE, Chhagan A, Charles JG. Predation of citrus red mite *Panonychus citri* by *Stethorus* sp. and *Agistemus longisetus*. NZ Pl Prot 2008; 61: 317-21.

[103] Vacante V. Influence of white mineral oil treatments on the population dynamics of some mites in a lemon orchard in Eastern Sicily. In: Cavalloro R., Di Martino E., Eds. Proceedings Experts' Meeting "Integrated pest control in citrus groves". Acireale: Balkema, Rotterdam 1986; pp. 423-31.

[104] Childers CC, Abou-Setta MM. Yield production in "Tahiti" lime from *Panonychus citri* feeding injury following different pesticide treatment regimes and impact on the associated predacious mites. Exp Appl Acarol 1999; 23: 771-83.

[105] McMurtry JA, Shaw JG, Johnson HG. Citrus red mite populations in relation to virus disease and predaceous mites in Southern California. Environ Entomol 1979; 8: 160-4.

[106] Ben-David T, Melamed S, Gerson U, Morin S. ITS-2 sequences as barcodes for identifying and analyzing spider mites (Acari: Tetranychidae). Exp Appl Acarol 2007; 41: 169-81.

[107] Hazan A, Gerson U, Tahori AS. Life history and life table of the carmine spider mite. Acarologia 1973; 15: 414-40.

[108] Hazan A, Gerson U, Tahori AS. Spider mite webbing. I. The production of webbing under various environmental conditions. Acarologia 1974; 16: 68-84.

[109] Martínez-Ferrer MT, Jacas JA, Ripollés-Molesa JL, Aucejo-Romero S. Approaches for sampling the two spotted spider mite (Acari: Tetranychidae) on clementines in Spain. J Econ Entomol 2006; 99: 1490-99.

[110] Urbaneja A, Pascual Ruiz S, Pina T, Abad-Moyano R *et al.* Efficacy of five selected acaricides against *Tetranychus urticae* (Acari: Tetranychidae) and their side effects on relevant natural enemies occurring in citrus orchards. Pest Manag Sc 2008; 64(8): 834-42.

[111] Tropea Garzia G, Mazzeo G, Suma P. Experimental evaluation of spirodiclofen efficacy in the control of spider mites and armored scales in Sicilian citrus orchards. Bull. WPRS/SROP 2008; 38: 256-60.

[112] Khajehali J, van Leeuwen T, Tirry, L. Susceptibility of an organophosphate resistant strain of the two-spotted spider mite (*Tetranychus urticae*) to mixtures of bifenazate with organophosphate and carbamate insecticides. Exp Appl Acarol 2009; 49: 185-92.

[113] Abad-Moyano R, Urbaneja A, Schausberger P. Intraguild interactions between *Euseius stipulatus* and the candidate biocontrol agents of *Tetranychus urticae* in Spanish clementine orchards: *Phytoseiulus persimilis and Neoseiulus californicus*. Exp Appl Acarol 2010; 50: 23-34.

CHAPTER 7

Thripidae

Rita Marullo* and Alessandra De Grazia

[1]*Department of Agricultural and Forest Management Systems (GESAF) (Section of Entomology), Mediterranean University of Reggio Calabria, Località Feo di Vito, 89060 Reggio Calabria, Italy*

Abstract: Eleven Thysanoptera (thrips) species, all included in the family Thripidae, are listed as pests on citrus in the Mediterranean region, and a diagnostic dichotomic key is provided for them. For each species the main morphological characters, biological features and data on the economic importance are reported. Amongst them, five species *i.e. Chaetanaphothrips signipennis* Bagnall, *Frankliniella bispinosa* Morgan, *Scirtothrips aurantii* Faure, *S. citri* (Moulton) and *S. dorsalis* Hood are quarantine species that are not yet present in the Mediterranean countries; however, *Heliothrips haemorrhoidalis* (Bouchè), *Pezothrips kellyanus* Bagnall, *Thrips flavus* Schrank and *T. tabaci* Lindeman are common pests on citrus. Data on pest activity due to *F. occidentalis* (Pergande) are not yet known, whereas literature reports on its spread onto new hosts, have only very recently become available from a few Mediterranean countries.

Keywords: Citriculture, Mediterranean Region, citrus thrips, bio-ecology, damage, control.

1. INTRODUCTION

The thrips (Thysanoptera) species known to be citrus pests in the Mediterranean region belong to the Thripidae, a large family in the sub-order Terebrantia. Two subfamilies are recognized in this family, Thripinae and Panchaetothripinae; about 1700 species in 260 genera are currently known, with many remaining undescribed [1]. The thripids are small to medium in body size (0.5 to 2.5 mm long), variable in colour, from the palest yellow or white, through shades of brown, to almost black. The body surface (particularly the legs) of the Panchaetothripinae is strongly reticulated. The antennae usually have seven or eight segments (nine or six segments in a few taxa); the terminal segments are usually small, and segments III and IV bear one or two slender, usually forked, emergent trichomes. The forewings (when present) are slender and more or less pointed, with two longitudinal veins, each usually bearing a more or less complete row of setae. The abdomen of the adult females has a serrated ovipositor which is generally turned down at the apex; adult males are usually smaller and paler than the females, with a more slender body. Their median sternites often carry one or several glandular areas, and the apical tergites may bear stout setae, tubercles or a pair of lateral horns called drepanae. The pest thripids infesting citrus in the Mediterranean region are oligophagous, and except *Scirtothrips aurantii* Faure (of South Africa origin), are not restricted to *Citrus* spp. A few, like *Frankliniella occidentalis* (Pergande), *Heliothrips haemorrhoidalis* (Bouchè), *Thrips flavus* Schrank and *T. tabaci* Lindeman are highly polyphagous. Moreover, the monocultural characteristic of citrus orchards, the biological habits of thripids and the intensive international trade in plants, may favour the spread and invasion of exotic thrips (quarantine species), such as *F. bispinosa* Morgan and *S. citri* (Moulton) of North American origin [2].

2. IDENTIFICATION KEY TO THE PEST THRIPS SPECIES IN THE MEDITERRANEAN REGION

1. Body strongly reticulate, dark brown when mature with yellow legs; forewings without long setae.. ***Heliothrips haemorrhoidalis***

-. Body never strongly reticulate, yellow or dark brown; if body brown then legs also dark; forewings with prominent setae on veins.. 2

2. Both longitudinal veins of forewing with complete rows of setae; anterior margin of pronotum with two pairs of prominent setae and two pairs of long pronotal posteroangular setae.......... 3

***Address correspondence to Rita Marullo:** Department GESAF (Section of Entomology), Mediterranean University of Reggio Calabria, Località Feo di Vito, I-89060 Reggio Calabria, Italy; E-mail: rmarullo@unirc.it

Vincenzo Vacante and Uri Gerson (Eds)

-. At least the first vein on the forewing with its setal row interrupted, not continuous; anterior margin of pronotum without long setae, two pairs of posteroangular setae present or absent 4

3. Antennal segment III with pedicel margins weakly rounded; posterior margin of abdominal tergite VIII with a complete comb of microtrichia.. ***Frankliniella occidentalis***

-. Antennal segment III pedicel with a sharp-edged ring; posterior margin of abdominal tergite VIII with lateral but no medial microtrichia.. .... ***Frankliniella bispinosa***

4. Body, legs and antennae mainly dark brown, antennal segments III and IV with apex constricted and bright yellow; first vein setal row widely interrupted, with two setae near apex.. ***Pezothrips kellyanus***

-. Body and legs usually yellow, antennal segments III and IV not sharply constricted at apex.. 5

5. Forewings pale with brown cross-bands medially and at base; tergite VIII with a characteristic band of sculptures extending anteriorly from spiracles....... 6

-. Forewings without transverse dark bands; tergite VIII without band of specialized sculptures associated with the spiracles 7

6. Head without a pair of setae anterior to first ocellus, only two pairs of ocellar setae present; pronotum with two pairs of posteroangular setae; abdominal sternite III without a median pore plate........... .... ***Chaetanaphothrips orchidii***

-. Head with a pair of setae anterior to first ocellus, three pairs of ocellar setae present; pronotum with only one pair of posteroangular setae; abdominal sternite III with a median pore plate .... ***Chaetanaphothrips signipennis***

7. Lateral thirds of abdominal tergites without closely spaced rows of microtrichia, tergites V to VIII with pairs of regular ctenidia laterally; head without a pair of setae anterior to first ocellus 8

-. Lateral thirds of abdominal tergites with many closely spaced rows of microtrichia, tergites V to VIII without paired ctenidia; head with a pair of setae anterior to first ocellus.............................. 9

8. Body Region variable, from yellow to brown, ocellar pigment never red; forewing first vein usually with 4 setae on distal half, but varying from 3 to 6; abdominal tergite II with 3 lateral marginal setae; pleurotergites with closely spaced rows of ciliate microtrichia....................... ....***Thrips tabaci***

-. Region of body yellow, ocellar pigment red; forewing first vein with 3 setae on distal half; abdominal tergite II with 4 lateral marginal setae; pleurotergites without closely spaced rows of ciliate microtrichia..***Thrips flavus***

9. Abdominal tergites and sternites yellow; sternites medially with no discal microtrichia.................. .... ***Scirtothrips citri***

-. Abdominal tergites and sternites anteriorly with dark transverse lines; sternites medially almost covered with microtrichia 10

10. Forewing posteromarginal cilia straight; male tergite IX with no drepanae, male hind femora without a comb of setae...................... .... ***Scirtothrips dorsalis***

-. Forewing posteromarginal cilia undulating; male tergite IX with paired drepanae, male hind femora with a comb of setae.................. ....*Scirtothrips aurantii*

3. CITRUS THRIPS SPECIES

3.1. *Chaetanaphothrips orchidii* (Moulton)

Body color: Macropterous female yellow, forewings pale with brown cross-bands at base and medially; antennal segments V-VI with brown apices.

Diagnostic morphological characters: Antennae 8-segmented, III-IV with slender and forked sensory cones, VII-VIII slender. Head wider than long, with only 2 pairs of ocellar setae, pair III within ocellar triangle. Pronotum with 2 pairs of major posteroangular setae, the external pair no longer than the width of antennal segment III. Metanotum weakly reticulated, median setae small and placed behind anterior margin. Forewings slender, first vein with 3 setae on distal half, second vein with 3-4 setae. Tergites weakly sculptured medially, with complete craspedum on posterior margins; tergite VIII with a specialized sculptured area extending up to the anterior margin. Sternites with large lobed craspedum, except medially on VII; median setae on tergite VII arising in front of posterior margin.

Male unknown.

Biological features: Literature is available about the pest's biology in South-East Asia [3] and on the development of its immature stages and adults in temperate regions [4].

Economic importance, host-plant range and distribution: This species is polyphagous and harmful to several crops, such as bananas, citrus, avocado and ornamentals (orchids, anthurium, begonia and bougainvillea). It colonized *Piper* plants in greenhouses in Tuscany (Italy) [4]. On bananas and grapefruits the feeding bites of larvae and adults on the leaves and fruits induced pronounced rusty necroses. Leaves and flowers of damaged orchids show dark or silver pigmented areas. The pest is known in the Mediterranean Region as very damaging to organic grapefruit in Israel [5]. Originally from South Eastern Asia, it is now widely spread in tropical countries. The pest has been introduced into Florida (USA), where it damages citrus, and also into Europe, where it seems to live only in greenhouses.

3.2. *Chaetanaphothrips signipennis* Bagnall

Body color: Body of the macropterous female yellow, forewings pale with brown cross-bands at base and medially, antennal segments V-VI with brown apices.

Diagnostic morphological characters: Antennae 8-segmented, segments III-IV with slender, forked sensory cones, VII-VIII slender. Head wider than long, 3 pairs of ocellar setae present, pair III between anterior margins of hind ocelli. Pronotum with one pair of posteroangular setae. Tergites without median sculpture, their posterior margins with complete craspedum. Sternite III with a small transverse glandular area.

Male similar to female; tergite IX with a pair of stout thorn-like setae with small tubercles behind them; sternites III-VII with transverse glandular areas.

Biological features: The eggs are laid just below the surface of the fruit or pseudostem. They hatch in about 8 days in summer, and the larvae develop in 8-10 days. The pupal stage, which lasts 7-10 days, is spent in the soil at the base of the host plant. Adults move back up to the plant and may live for many weeks, forming colonies under leaf shelters on the pseudostem and in the bunches. A complete life cycle may require up to three months, more in winter.

Economic importance, host-plant range and distribution: This pest harms mainly banana, anthurium and *Dracaena*. Citrus, tomato and various weeds were also recorded as showing feeding necroses and damage [6]. This thrips is reported as an Alert species for *Citrus aurantium* Linnaeus in the Mediterranean Region (EPPO/PQR46) [7]. It is widespread in tropical regions, including Asia (India, Indonesia, Philippines, Sri Lanka), the Americas (Brazil, Costa Rica, Honduras, Mexico, Panama, Puerto Rico, USA), and Oceania (Australia, Fiji, New Guinea).

3.3. *Frankliniella bispinosa* Morgan

Body color: Color of adult female yellow with brown setae and antennal segments VI-VIII, distal half of IV and apices of II and III brown; forewing pale.

Diagnostic morphological characters: Antennae 8-segmented, segment III pedicel with a sharp-edged ring surmounted by a chalice-shaped structure. Head wider than long; 3 pairs of ocellar setae, pair III arising on anterior margins of triangle; postocular setae pair I present, pair IV as long as hind ocelli. Pronotum with 5 pairs of major setae; anteromarginal setae shorter than anteroangulars; one pair of minor setae present medially between posteromarginal median setae. Metanotum with 2 pairs of setae at anterior margin; campaniform sensilla present. Forewings with 2 complete rows of setae. Posteromarginal comb on tergite VIII with a few broadly based microtrichia laterally. Sternites without discal setae.

Male smaller; tergite IX with one pair of short stout setae medially; sternites III-VII with a broad oval glandular area.

Biological features: This is a common flower-living species, damaging citrus and avocado blossoms in Florida. Its life history and development were studied by Childers [8].

Economic importance, host-plant range, distribution: *Frankliniella bispinosa* causes significant damage to navel orange flowers. The first and second instars and the adults feed on the ovary, style, floral disk, petals and anthers of the swollen buds and open flowers. This feeding results in cellular evacuation, necrosis, plasmolysis and collapse of floral and bud tissues, along with the production of ethylene and then ovary abortion [9]. The species also colonizes vegetables (capsicum and tomato) and ornamentals, and transmits TSWV (Tomato Spotted Wilt Virus) to pepper [10]. The thrips is common in Florida and in the south eastern USA (Georgia, Alabama, South Carolina), The Bahamas and Bermuda. At this time *F. bispinosa* is not yet present in the Mediterranean Region, but is known as one of the most invasive pest thrips and has been placed on the quarantine species lists for several countries (*i.e.* Brazil, Australia).

3.4. *Frankliniella occidentalis* (Pergande)

Body color: Color of adult female variable, winter populations are dark brown, spring/summer populations commonly yellow with brown areas medially on tergites; antennal segments II and VI-VIII brown, III-V yellow with apices variably brown; legs yellow-brown; forewings pale with dark setae.

Diagnostic morphological characters: Antennae 8-segmented. Head wider than long with 3 pairs of ocellar setae, pair III long, arising on anterior margins of ocellar triangle, postocular setae pair I present, pair IV longer than distance between hind ocelli. Pronotum with 5 pairs of major setae; one pair of small setae medially between posteromarginal submedian setae. Metanotum with 2 pairs of setae at anterior margin; campaniform sensilla present. Forewings with complete rows of setae. Ctenidia on tergite VIII anterolateral of spiracle; posteromarginal comb on tergite VIII complete, with microtrichia short and slender on triangular bases. Sternites III-VII without discal setae.

Male smaller and paler; tergite VIII without comb; tergite IX with median dorsal setae shorter than lateral pair, posterolateral setae stout in larger males; sternites III-VII with transverse glandular area.

Biological features: The life cycle of this pest, commonly called WFT, lasts about 20 days at mean temperatures of +25°C (but not under +12°C). Several continuous generations can be raised in greenhouses on ornamentals and horticultural crops. Oviposition starts 36 hours after female emergence, each female producing over 40 eggs [11]. Both adults and larvae live inside flowers, avoiding light.

Economic importance, host-plant range and distribution: The pest is highly polyphagous, being particularly harmful to ornamentals, peach, grape and cotton and in nurseries. Symptoms (due to feeding and oviposition wounds) are a necrosis that is often surrounded by whitish halos (on table-grapes, nectarines), rusts on fruits, bronzing fruits (strawberries), distortion of flowers and young leaves or shoots and abortion of flower ovaries. In all cases the attacks cause loss of production. WFT is also a vector of TSWV. The distribution of this pest, originally from the western

USA, is now worldwide. It is very abundant in citrus orchards, in the Valencia region, Spain [12] and in the Antalya region, Turkey [13]. WFT has recently also been discovered in specialized citrus orchards in southern Italy (unpublished data). This pest is on the Alert List of EPPO [7] for *Citrus paradisi* Macfadyen (grape-fruit).

3.5. *Heliothrips haemorrhoidalis* (Bouchè)

Body color: Color of mature female dark brown, golden-yellow in younger individuals, legs yellow, forewings hyaline. Antennal segments I-II light brown, III-V yellow, VI brown in distal half, VII-VIII yellow.

Diagnostic morphological characters: Adult female body surface, including legs, strongly reticulate. Head with slight cheeks, constricted at base. Antennal segment VIII long and slender (stylus). Metanotum triangular, reticulate sculptured, with a posterior projection. Forewings with apex rounded bearing two long cilia, swollen at base and parallel sided. Abdominal tergites II-VIII with a single pair of dominant setae medially, tergite VIII with a long posteromarginal comb of microtrichia; tergite X short with a complete median division.

Males very rare.

Biological features: This species is parthenogenetic (thelytokous) [14]. Immature larvae and pupae live on lower leaf surfaces, and the adults overwinter under dried leaf litter or on tree bark. In the Mediterranean region the greenhouse thrips may complete 5-7 annual generations, from June to late October in open fields, and more than 15 in greenhouses.

Economic importance, host-plant range and distribution: This pest is very polyphagous, infesting many wild and crop plants, such as ornamentals and edibles. Under field conditions the pest usually occurs on plants with rather hard leaves, including tea and Pinus in parts of Africa, southern Europe [15] and *Viburnum* in eastern Australia [16]. Damage to citrus is manifested as silver necroses on leaves, caused by sucking out the contents of leaf tissues, thus allowing air inside. The surfaces of damaged leaves are covered by drops of dark liquid, excretions of the thrips larvae, which dry to become black spots. These symptoms can also be seen on young fruits and branches of citrus trees; plants may subsequently dry. The necroses on citrus fruits can develop into reticulate rust called "white rust" [17]. Originally from Southwestern Brazil [18], the greenhouse thrips is now found throughout tropical and subtropical countries.

3.6. *Pezothrips kellyanus* Bagnall

Body color: Color of adult female generally brown (Fig. **1**), tibiae yellow, forewings pale brown with the bases slightly paler, antennal segments III-IV with apical whitish necks. Ocellar pigment red; major setae dark brown.

Diagnostic morphological characters: Antennae 8-segmented, segments III and IV with constricted apical necks. Head as wide as long, 3 pairs of ocellar setae, pair III inside the anterior margins of the ocellar triangle; postocular setae IV as long as distance between hind ocelli. Pronotum with 2 pairs of posteroangular setae; 5 pairs of setae on posterior margin. Metanotum with median reticulate sculpture, campaniform sensilla present, median setae at anterior margin. Forewing first vein with 2 setae on distal half, second vein with a complete row of setae. Tergites without ctenidia or sculptures between median setae; posteromarginal comb on tergite VIII with a few lateral teeth. Sternites without discal setae.

Male smaller and slender; antennal segment VI long; posterior margin of tergite IX with a pair of small spine-like processes; sternites III-VII each with more than 20 small circular glandular areas.

Fig. (1). Adults of *Pezothrips kellyanus* Bagnall on a lemon flower.

Fig. (2). Orange fruit damaged by *Pezothrips kellyanus* Bagnall.

Biological features: Data on the development of immature stages and adults, life history and field/laboratory number of generations are available in Baker *et al.* [19]; biological observations and a detailed list of common and occasional host plants in South Italy were reported by Marullo [17, 20]. Studies on sex attractants, including new data about the host range in Australia, and a discussion on the pest's country of origin, were provided by Webster *et al.* [21].

Economic importance, host-plant range and distribution: This species is the most abundant and harmful thrips on citrus orchards in some Mediterranean areas, such as Calabria and Sicily (Italy), Antalya (Turkey), Cyprus and the Oeste region (Portugal). It is listed in the quarantine pest species list for the Mediterranean Region [7]. Damage to young citrus fruits is due to the feeding of young and adult thrips: ring necrosis, loss of epicarp color followed by white-yellowish areas around the feeding bites, necroses that develop into brown corky scars (Figs **2** and **3**). The citrus species and varieties navel orange, lemon and bergamot are the preferred host plants. Kelly's citrus thrips also breeds on scented flowers of *Lonicera*, *Pittosporum* and *Jasminum*. The species occurs in southern European countries (Italy, Spain, Greece, Turkey, Portugal, South France), Australia and New Zealand, even reaching Northern Asia. Its geographical origin remains in doubt, being either native to Australia and introduced to the Mediterranean Region, or vice-versa.

Fig. (3). Lemon fruit damaged by *Pezothrips kellyanus* Bagnall.

3.7. *Scirtothrips aurantii* Faure

Body color: Color of adult macropterous female mainly yellow, antecostal ridges of tergites and sternites dark brown with small brown areas; antennal segment I white, II-III grey, V-VIII brown; forewings pale with clavus shaded.

Diagnostic morphological characters: Antennae 8-segmented, antennal segments III-IV with constricted apical necks, sensory cones forked and stout. Head with closely spaced lines of sculpture in postocular area and ocellar triangle, ocellar setae pair III close together between anterior margins, with 4 pair of setae. Metanotum with irregular longitudinal reticulations on posterior half, median setae close to anterior margin. Forewing first vein with 3 setae on distal half, second vein with 3 widely spaced setae; posterior fringe cilia wavy, clavus with 4 setae. Tergites III-VI with small and placed close together median setae, tergite VIII with a complete comb, microtrichia sparse medially; tergite IX without microtrichia. Sternites covered with rows of microtrichia.

Male similar to female but smaller; tergite IX with a pair of stout drapenae on posterior angles and extending across tergite X; hind femora with a comb-like row of stout setae.

Biological features: Larvae and adults feed on young leaves and the apices of young fruits, often concealed under the calyx. Eggs are inserted into the young tissue. There are two nymphal (feeding) stages, followed by two pupal (non-feeding) stages. Pupation occurs on the ground amongst leaf litter; in spring/summer a life cycle may require less than 30 days.

Economic importance, host- plant range and distribution: This species is the most important pest on citrus crops in South Africa and Zimbabwe, reducing production by heavily damaging young leaves and decreasing the proportion of export quality fruits. The thrips is also the primary cause of banana fruit spotting in Yemen. *Scirtothrips aurantii* infests over 50 plant species in different genera and families. It is a pest of citrus species and varieties as well as of several tropical crops such as mango, *Arachis*, *Asparagus*, *Gossypium* and *Vitis*. The thrips is included in the EPPO Alert 1 list; its occurrence in citrus-growing areas with a subtropical or Mediterranean climate suggests that it could probably establish on citrus in southern Europe and the Mediterranean Region. At present it occurs in southern and eastern African Regions, Egypt and Yemen.

3.8. *Scirtothrips citri* (Moulton)

Body color: Color of adult macropterous female yellow, antennal segments III-VIII grey, major setae pale, forewings pale.

Diagnostic morphological characters: Antennae 8-segmented, III-IV with a constricted apical neck, sensory cones forked and stout. Head wider than long, postocular area and ocellar triangle with closely spaced transverse sculpture lines; 3 pairs of ocellar setae present, pair III close together between anterior margins of hind ocelli; 2 pairs of postocular setae present. Pronotum with transverse sculpture lines; posterior margin with 4 pairs of setae. Metanotum with irregular longitudinal reticulation on posterior half, campaniform sensilla absent. Forewing first vein with 3 setae on distal half, second vein with 3 widely spaced setae; posterior fringe cilia wavy; clavus with 3-4 setae. Tergites III-VI with small and closely packed median setae, II-VIII with lateral thirds covered by spaced rows of fine microtrichia, and 4 small discal setae, posterior margins with fine combs; VIII with comb complete across posterior margin; IX with several rows of discal microtrichia. Sternites without discal setae, laterally with rows of microtrichia.

Male similar to female but smaller; tergite IX without drepanae; hind femora without comb-like row of stout setae; sternites without glandular areas.

Biological features: The life history and biology of this thrips are similar to those of *S. aurantii* [22]. There is no development below 14°C, and more than 8 generations may be produced annually.

Economic importance, host- plant range and distribution: Along with *S. aurantii* and *S. dorsalis*, *S. citri* is a pest of citrus and one of the most important species of the genus *Scirtothrips* for international agriculture. Primarily a pest of citrus (mainly navel oranges) in California, it has been collected from 53 different plant species, including cotton (*Gossypium hirsutum* Linnaeus), dates *(Phoenix dactylifera* Linnaeus), grapevine (*Vitis vinifera* Linnaeus), lucerne (*Medicago sativa* Linnaeus), pecans [*Carya illinoensis* (Wangenh.) K. Koch] and ornamentals such as *Magnolia* and *Rosa*. The native host plants are *Quercus* spp., or more likely *Rhus laurina* (Nutt.) Abrams. It occurs in northern Mexico and in some southern states of the USA, like Arizona, California and Florida. The California citrus thrips is not present in the EPPO region but is included in the Alert1 List.

3.9. *Scirtothrips dorsalis* Hood

Body color: Color of adult female yellow, tergal and antecostal ridges brown with brown areas; antennal segments V-VIII brown; forewings weakly shaded.

Diagnostic morphological characters: Antennae 8-segmented. Head wider than long, postocular area and ocellar triangle with closely spaced transverse lines; three pairs of ocellar setae, members of pair III close together, between hind ocelli. Pronotum with transverse sculpture lines and four pairs of posteromarginal setae. Metanotum with parallel longitudinal sculpture lines on posterior half; campaniform sensilla absent; median setae arise behind anterior margin. Forewing first vein with 3 setae on distal half, second vein with 2 spaced setae; posterior fringe cilia straight. Tergites III-VI with median setae, located close together; II-VIII with thirds covered by spaced rows of fine microtrichia; VIII with a complete comb across its posterior margin; IX with several rows of microtrichia. Sternites covered by microtrichia, except anteromedially.

Male smaller; tergite IX without drepanae on posterior angles.

Biological features: The biology of *S. dorsalis* is similar to that of other *Scirtothrips* spp.

Economic importance, host - plant range and distribution: In common with other members of the genus, *S. dorsalis* may severely damage the plants on which it feeds, the main damage being due to severe distortions of young leaves. It is a highly polyphagous species, breeding on young terminal leaves of many different plants, and is especially associated with plants that grow in warm dry conditions. It is a serious pest of herbaceaous crops, mainly edibles and ornamentals. Citrus crops are seriously affected in Japan and Taiwan, the damage being similar to that caused by *S. aurantii*, with distortion of young leaves and fruit scaring, leading to reductions in crop yield and quality. Its native host plants are probably various Fabaceae, such as *Acacia, Brownea, Mimosa* and *Saraca. Scirtothrips dorsalis* is known as a pest of many crops, including *Actinidia*, *Arachis*, *Capsicum*, *Citrus*, cotton, *Fragaria*, grapevine, onions, *Ricinus*, *Rosa*, tamarinde and tea. Widespread across Asia to the subtropical parts of Australia to Hawaii (USA), the pest is now established in Israel and in South Africa. It is absent from the EPPO region but has been intercepted in The Netherlands. On the basis of its geographical and host plant range this pest has been added to the Alert 1 List of EPPO. Although *S. dorsalis* is less likely than *S. aurantii* or *S. citri* to become established on citrus under Mediterranean conditions, there is considerable danger that, due to its wide host plant range that includes many vegetable crops, the pest will invade European glasshouse crops.

3.10. *Thrips flavus* Schrank

Body color: Body of adult female yellow, major setae light brown; antennal segments IV-V brown distally, VI-VII brown, forewings pale.

Diagnostic morphological characters: Antennae 7-segmented. Head wider than long, with two pairs of ocellar setae; pair III small, arising within the ocellar triangle behind fore ocellus. Metanotum with transverse sculpture lines on anterior half, with irregular longitudinal lines or narrow reticulations on posterior half; median setae arising behind anterior margin, campaniform sensilla present. Forewing first vein with 3 setae on distal half. Tergite II with 4 lateral marginal setae; tergite VIII with a complete posterior marginal comb of long and slender microtrichia, pleurotergites without discal setae. Sternites III-VII with 3 pairs of marginal setae; sternites without discal setae.

Male smaller; tergite VIII with a posteromarginal irregular comb; sternites III-VII with large transverse glandular area.

Biological features: Biological and developmental data are available [14].

Economic importance, host-plant range and distribution: This thrips has been recorded from the flowers of many plant species in Europe and across Asia (from England to Taiwan). It is very polyphagous, especially damaging to cotton leaves and to flowers of lucerne, apple and orange, causing their sterility and fruit distortion [14].

3.11. *Thrips tabaci* Lindeman

Body color: Color of adult female varying from yellow to dark brown, ocellar pigment grey, major setae dark.

Diagnostic morphological characters: Antennae 7-segmented. Head wider than long, with 2 pairs of ocellar setae; pair III small, arising on anterior margin or within ocellar triangle. Pronotum with 2 pairs of posteroangular setae; posterior margin with 3 pairs of setae. Metanotum irregularly reticulate medially with lines converging to midpoint of posterior margin; median setae short, arising behind anterior margin. Forewing first vein usually with 4 setae on distal half. Tergite II with 3 lateral marginal setae; V-VIII with lateral ctenidia, on VIII posteromedially to spiracles; posterior margin of tergite VIII with a complete comb of long and slender microtrichia; pleurotergites with sculpture and a row of fine microtrichia. Sternites without discal setae.

Male small and yellow; tergite VIII posterior comb with a few irregular microtrichia; sternites III-V with narrow transverse glandular area.

Biological features: Data on the biology and development of the onion thrips in Europe has been provided by Bournier [14]. The pest may complete 2-3 annual generations in northern Europe, and as many as 15-16 in more southern areas. Two ecotypes are known: one reproducing by thelytoky (parthenogenesis), the other is bisexual, with a sex ratio of 1:1 (males:females) in field populations. The latter is the only ecotype that transmits viruses.

Economic importance, host- plant range and distribution: The host plant range of *T. tabaci* includes more than 150 plant species of different genera and families. Although the pest prefers to breed on onion and garlic, it can develop large populations on many crops, including cotton, wheat, potatoes, strawberry and greenhouse ornamentals. Damage of economic importance is due to feeding, which causes necroses and small de-pigmented areas on the host-plant tissues, and to the transmission of TSWV. The distribution of this species is worldwide, except in the wet tropics. Its origin is the eastern Mediterranean Region.

4. MANAGEMENT

Chemical control and IPM. Management of thrips with pesticides is problematic, because repeated applications have caused high levels of resistance and cross-resistance. The constant threat of migration, the ability to transmit tospoviruses, the propensity of thrips to prefer tight, protected areas on the plant, and the failure of many systematic pesticides to translocate to the flowers where many thrips feed, all combine to make chemical control exceedingly difficult [23]. The aim of IPM strategies is to promote high and “clean” quality production. Field population monitoring and evaluation of economic and action thresholds are essential techniques for every management program intended to control pest thrips on citrus. Data available from citrus crops in the Mediterranean Region pertain mainly to two species, *Heliothrips haemorrhoidalis* [24] and *Pezothrips kellyanus* [25-27]. For exotic/quarantine species, such as *Frankliniella bispinosa* and *Scirtothrips* spp., interested personnel may consult data obtained from North American and tropical citrus growing areas [28].

Biological control. Although much has been written about the natural enemies of thrips, and there are successful control programmes in greenhouses, little has actually been done concerning the management of thrips pests that infest field crops. Considering the economic importance of citriculture in the Mediterranean Region, natural enemies, particularly anthocorid predators like *Orius* spp., and predatory mites of the genus *Amblyseius*, may play an interesting role, but rarely do they do enough to prevent crop damage. Much more useful for controlling of thrips infestations might be to combine rational cultural practices and the application of appropriate (low impact) insecticides. The actual role and effectiveness of any native natural enemies might be discovered in some situations by the careful use of insecticides.

REFERENCES

[1] Mound LA, Marullo R. The Thrips of Central and South America: an introduction. Gainsville: Associated Publishers 1996.

[2] Morse JG, Hoddle MS. Invasion biology of Thrips. Annu Rev Entomol 2006; 51: 67-89.

[3] Mituda-Sabado EC, Calilung VJ. Survey, identification and life history of anthurium thrips. Philipp Entomol 2000; 14 (2): 121-9.

[4] Del Bene G, Gargani E. *Chaetanaphothrips orchidii* (Moulton) (Thysanoptera Thripidae), specie nuova per l’Italia. Redia 2001; 74: 119-28.

[5] Wysoki M. Thysanoptera in Israel. Phytoparasitica 1999; 27 (2): 73-4.

[6] Simmonds NW. Bananas. Great Britain: Longman Publishers1959.

[7] EPPO/PQR 46. EPPO’s plant quarantine data retrieval system. vers.4.6. 2007.

http://www.eppo.org./quarantine/Pest_Risk/PRA_documents.htm (accessed 2007).

[8] Childers CC. Flower thrips: *Frankliniella bispinosa* (Morgan), *F. kelliae* Sakimura (Thysanoptera: Thripidae) and post bloom fruit drop disease are economic pest on citrus. In: Proc Flo State Hort Soc 1999; 112: pp. 88-95.

[9] Abeles FB, Morgan PW, Saltveit ME. Ethylene in plant biology. 2nd ed. San Diego: Akademic Press 1992.

[10] Avila Y, Stavisky J, Hague S, Funderburk J, Reitz S, Momol T. Evaluation of *Frankliniella bispinosa* (Thysanoptera:Thripidae) as a vector of Tomato Spotted Wilt Virus in pepper. Fla Entomol 2006; 89 (2): 204-7.

[11] Del Bene G, Gargani E. Contributo alla conoscenze di *Frankliniella occidentalis* (Perg.) (Thysanoptera Thripidae). Redia 1989; 72 (2): 403-20.

[12] Navarro C, Pastor MT, Ferragut F, Garcia Marí F. Trips (Thysanoptera) asociados a parcelas de citricos en la Comunidad Valenciana: abundancia, evolucion estacional y distribucion espacial. Bol San Veg Plagas 2008; 34: 53-64.

[13] Teksam I, Tunc I. An analysis of Thysanoptera associated with citrus flowers in Antalya, Turkey: composition, distribution,abundance and pest status of species. App Entomol Zool 2009; 44 (3): 455-64.

[14] Bournier A. Les thrips. Biologie-Importance agronomique. Paris: Inra 1983.

[15] Marullo R. Host-plant ranges and pest potential: habits of some thrips species in areas of southern Italy. B Insectology 2009; 63: 30-5.

[16] Moritz G, Mound LA, Morris DC, Goldarazena A. Pest thrips of the world – Visual and molecular identification of pest thrips. Brisbane: CBIT 2004.

[17] Marullo R. Conoscere i Tisanotteri. Bologna Edagricole 2003.

[18] Mound LA, Monteiro RC. A review of the Genus *Heliothrips* (Thysanoptera: Thripidae), with a new sister-species of the greenhouse thrips from south eastern Brazil. J New York Entomol Soc 1997; 105 (3-4): 154-60.

[19] Baker G, Jackman D, Keller M, MacGregor A, Purvis S. Development of an Integrated Pest Management system for thrips in Citrus. HAL Final Report CT97007 2002.

[20] Marullo R. In: Atti XIX Congr Naz It Entom, Catania,10-15 Giugno 2002; pp. 595-99.

[21] Webster KW, Cooper P, Mound LA. Studies on Kelly's citrus thrips, *Pezothrips kellyanus* (Bagnall) (Thysanoptera. Thripidae): sex attractants, host associations and country of origin. Austral J Entomol 2006; 45: 67-74.

[22] Smith I.M. In : Smith I.M., Mc Namara D.G., Scott P.R., Holderness M, Eds. Quarantine pests for Europe. Wallingford: CAB International 1996.

[23] Parrella MP, Lewis T. In: Lewis T, Ed. Thrips as crop pests. Wallingford: CAB International 1997; pp. 595-614.

[24] Viggiani G. Lotta biologica ed integrata nella difesa fitosanitaria. Vol. II. Napoli: Liguori 1997.

[25] Conti F, Tumminelli R, Amico C *et al.* In: Marullo R, Mound LA, Eds. Thrips and Tospoviruses. Proceedings of the 7th International Symposium on Thysanoptera. Reggio Calabria, Italy, July 2-7 2001; ANIC, Università degli Studi Mediterranea di Reggio Calabria 2002; pp. 207-10.

[26] Conti, F., Tumminelli, R., Fisicaro R *et al.* An IPM system for new citrus thrips in Italy. Bull IOBC WPRS 2003; 26(6): 203-8.

[27] Vassiliou VA. Chemical control of *Pezothrips kellyanus* (Thysanoptera:Thripidae) in citrus plantations in Cyprus. Crop Prot 2007; 26: 1579-84.

[28] Parker BL, Skinner M. In: Lewis T, Ed. Thrips as crop pests. Wallingford: CAB International 1997; pp. 615-38.

CHAPTER 8

Cicadellidae

Hüseyin Başpinar[1*], Nedim Uygun[2] and Alfonso Hermoso De Mendoza[3]

[1]Department of Plant Protection, Adnan Menderes University, Agricultural Faculty, Aydın, Turkey; [2]Department of Plant Protection, Çukurova University, Agricultural Faculty, Adana, Turkey and [3] Institut Valencià d'Investigacions Agràries, Carretera de Nàquera km 5, 46113 Montcada (València), Spain

Abstract: The large hemipteran family Cicadellidae, or leafhoppers, has more than 2300 species described in 338 genera in the Palearctic region. Many are world-wide pests of crops. Sixty-one leafhopper species were collected in citrus orchards in the Mediterranean Region, but many have low economic importance. *Asymmetrasca decedens* (Paoli), *Empoasca alsiosa* Ribaut, and *E. decipiens* Paoli are common and devastating cicadellids. They live on vegetables, vineyards and cotton in summer, moving to citrus later on, when other host-plants are no longer available. On the other hand, members of the *Circulifer haematoceps* (Mulsant and Rey) complex are less common than the three aforementioned species, but are capable of transmitting *Spiroplasma citri* Saglio *et al.*, the causal agent of citrus stubborn disease. Many parasitoids and predators of cicadellids have been identified. *Gonatopus lunatus* Klug on *Circulifer haematoceps* and *Aphelopus* sp. on *Asymmetrasca decedens* and *Empoasca decipiens* were effective parasitoids (Hymenoptera: Dryinidae). *Chrysoperla carnea* (Stephen) (Neuroptera: Chrysopidae), *Deraeocoris pallens* (Rt.) (Hemiptera: Miridae), *Nabis ferus* (L.) (Hemiptera: Nabidae), *Geocoris* sp. (Hemiptera: Geocoridae), *Paederus* f.c. *kalalovae* Curtis (Coleoptera: Staphylinidae) are common predators feeding on cicadellids in citrus orchards.

Keywords: Citriculture, Mediterranean Region, citrus leafhoppers, bio-ecology, damage, control.

1. INTRODUCTION

The family Cicadellidae, or leafhopers, is very large, with more than 2300 species described in 338 genera in the Palearctic region [1]. Many are world-wide crop pests. They are small jumping insects that rarely exceed ½ inch in length and may be only a few millimeters long [2]. They vary in form, color and size and are often marked by beautiful color patterns. Their nymphs can walk sideways on the leaves when slightly disturbed.

Members of the Cicadellidae are very similar to hoppers of related groups, which hinders their identification. Thus it might be useful to distinguish this family from its close relatives.

Leafhoppers have one or more rows of small spines on the tibiae of the hind leg (Fig. **1A**), whereas cercopids have one or two stout spines and usually a series or circlet of spines at apex (Fig. **1B**). Delphacids bear a broad movable apical spur on their hind tibiae (Fig. **1C**).

2. FIELD KEY TO SOME FAMILIES OF THE CICADOIDEA (Modified from [2])

1. Hind tibiae with one or more rows of spines; hind coxae transverse (Fig. **1A**) ....**Cicadellidae**

-. The spination of the hind tibiae is different.. ..2

2. Hind tibiae with 1 or 2 stout spines and usually a series of circlet of spines at apex; hind coxae short and conical (Fig. **1B**)...**Cercopidae**

-. Hind tibiae with a broad movable apical spur (Fig. **1C**)...**Delphacidae**

***Address correspondence to Hüseyin Başpinar:** Department of Plant Protection, Faculty of Agriculture, Adnan Menderes University, 09100 Aydın, Turkey; Tel: ++ 90 256 772 70 23; E-mail: hbaspinar@adu.edu.tr

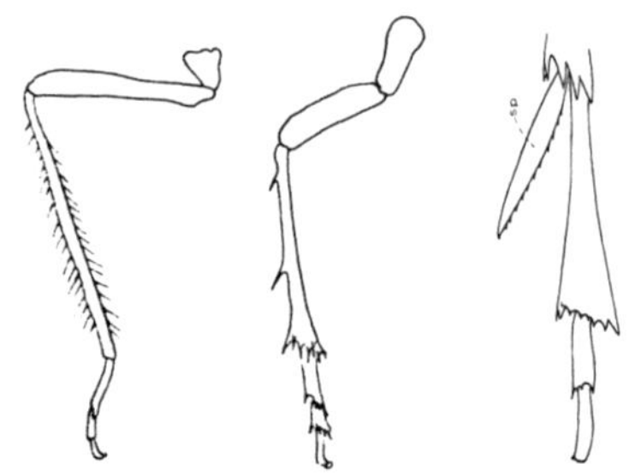

Fig. (1). Spination on the hind leg: A, Cicadellidae; B, Cercopidae; C, Delphacidae [2].

The life cycle of leafhoppers consists of seven stages; egg, five nymphal instars and adults. Along the Mediterranean coast many species may overwinter as adults.

Leafhoppers embed their eggs into plant tissues with their ovipositor. They feed by sucking excessive amounts of sap [3], thereby reducing or destroying the plant's chlorophyll. Leafhoppers may cause indirect damage to citrus by transmitting plant pathogens. More than 150 cicadellid species have been reported to be disease vectors [4] and they are assigned to seven subfamilies, the majority being in the subfamily Deltocephalinae [3].

Sixty-one leafhopper species were collected in citrus orchards [5, 6]. Leafhoppers are very active insects that in the orchards feed more on weeds than on citrus. For instance, *Asymmetrasca decedens* (Paoli), *Empoasca alsiosa* Ribaut, and *Empoasca decipiens* Paoli feed on citrus as well as on weeds in the orchards, which helps them to build up large populations on the trees. The other species found on weeds in citrus orchards are not economically important and are either very rare or never found on the trees. *Circulifer haematoceps* (Mulsant and Rey), however, even when occurring only in very low numbers on the trees, is an important pest because it transmits citrus stubborn disease (caused by *Spiroplasma citri* Saglio *et al.*), which causes severe economic losses. Thus *E. decipiens, E. alsiosa, A. decedens* and *C. haematoceps* are treated as cicadellid pests of citrus.

Usually it is difficult to identify leafhoppers to species level only by a visual examination in the field, because many species, even of different genera, are similar in size and coloration. Examples are *Asymmetrasca decedens* and *Empoasca decipiens,* which occur together the year around on weeds and on citrus trees. Morphologically they are almost indistinguishable from each other, requiring a microscopic examination of their male genitalia. The following key for the identification of the four pest species is thus based on the male genitalia, as observed from microscopic preparations.

3. MICROSCOPIC KEY TO THE CITRUS CICADELLIDAE

1. The aedeagus is symmetric..2

-. The aedeagus is asymmetric, with an small appendix dorsally at apex (Fig. **2**).................. ....***Asymmetrasca decedens***

2. The aedeagus is bilaterally ramified and circled (Fig. **6**) ....***Circulifer haematoceps***

-. The aedeagus is simple and not ramified 3

3. The apophyse of the pygophore lobe straight and gradually tapering in lateral view (Fig. **4**)......... ....***Empoasca alsiosa***

-. The apophyse of the pygophore lobe are in a bended position and tapering immediately at apex in lateral view (Fig. **5**)..***Empoasca decipiens***

4. CITRUS CICADELLID SPECIES

Asymmetrasca decedens (Paoli)

Diagnostic characters: This species resembles *Empoasca decipiens.* The body is 3.5-4.1 mm long, with whitish spots on the anterior of the pronotum. There are usually two dark spots in the vertex and a whitish band (cicatrice) that crosses its middle. The aedeagus is asymmetric (Fig. **2D**); the pygophore lobe apophyse and anal block are as in Fig. **2**.

Life history: *Asymmetrasca decedens* is very widely distributed in the Mediterranean region [7]. It is a polyphagous pest that attacks many crops, such as cotton, summer vegetables, citrus as well as many weeds. Winter is passed in the adult stage but not in hibernation, surviving in very low populations along the mild Mediterranean coasts where the pest can find food. The leafhopper has 4-5 annual generations in the Mediterranean coasts of Turkey, but the number of generations may vary in cooler or warmer area in the Mediterranean Region.

The population fluctuations of *A. decedens* and *E. decipiens* were monitored on citrus, maize and cotton fields [8]. *Asymmetrasca decedens* was the most common species on citrus, cotton, and maize, averaging 83% of the combined populations, whereas *E. decipiens* was of minor importance, with 17%.

The populations of both species peaked on citrus in November and remained at high levels till January. On citrus *A. decedens* made up 89% of the leafhopper population, *E. decipiens* only 11%, whereas it was 45% and 55%, respectively, on the weeds in citrus orchards. The first damage in the season was observed on Washington Navel oranges, later on the Hamlin variety, and then on mandarins. Little damage was observed on grapefruits and lemons [8]. Only very few nymphs of both leafhopper species were noticed on citrus during the samplings. This suggests that the high leafhopper population on citrus was due not to multiplication on these trees, but most likely to immigration from the surrounding summer host crops, which are being cultivated in a big scale in that area.

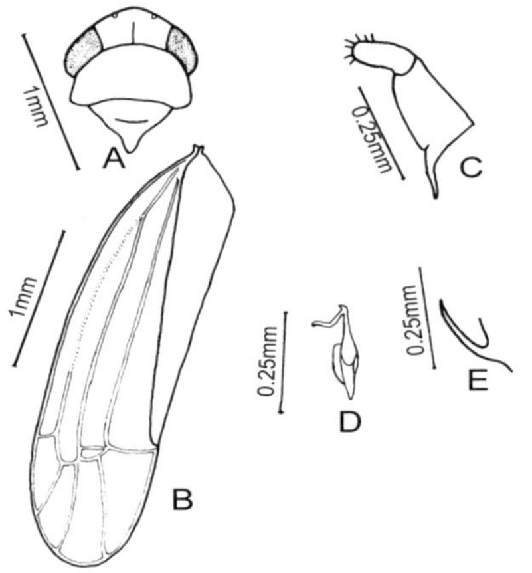

Fig. (2). *Asymmetrasca decedens* (Paoli): A, head; B, forewing; C, anal block (lateral view); D, aedeagus (dorsal view); E, pygophore lobe apophyse (lateral view).

Economic importance: *Asymmetrasca decedens* individulas occur in mixed populations with *E. decipiens* in Turkey, and with *Empoasca alsiosa* and *E. decipiens* in Spain. *Asymmetrasca decedens*, *E. decipiens* and *Cicadulina chinai* Ghauri were reported in Egyptian citrus orchards [9]. The leafhopper feeds on maturing citrus fruits by puncturing the rind cells, thus causing yellowish to brown, rounded scars (Fig. **3**), thereby reducing the market value of fruits [8]. Serious damage caused by *A. decedens* and *E. decipiens* was recently observed on citrus in the East Mediterranean region of Turkey.

Management: Control measurements should be taken to suppress *A. decedens* populations in fields where summer hosts are grown, so that pest populations that immigrate to citrus orchards will not attain high levels in the autumn. Effective weed control may help to reduce pest numbers. In Turkey, if damage is 1 patch of 1 cm diameter per 250 fruits, it is recommended to spray a lime solution (4 kg pure lime/100 l water) under high pressure in October, in order to cover the tree canopy [10].

Fig. (3). Damage caused by *Asymmetrasca decedens* (Paoli) and by *Empoasca decipiens* Paoli on citrus fruit (left), and an adult of *Asymmetrasca decedens* (Paoli) on the fruit (right).

4.2. Empoasca alsiosa Ribaut

Diagnostic characters: This species resembles *E. decipiens*; its body is 3.2-3.7 mm in length. The pygophore lobe apophyse is straight and resembles a netting needle in lateral view (Fig. **4D**).

Life history: In the Mediterranean Region *E. alsiosa* occurs in France, Italy and Spain [5, 11, 12]. Its biology is not well known, but is believed to be similar to related *Empoasca* species, such as *E. decipiens*. No specific habitat was observed for this species [11], which is also found in Italian vineyards [12]. *Empoasca alsiosa* lives on vegetables, in vineyards and on cotton in summer, moving to citrus later when other host-plants are no longer available [13]. This leafhopper was the most abundant cicadellid on the citrus tree canopy in Spain; *A. decedens*, *E. decipiens* and *E. solani* also occurred there, but in smaller numbers [5]. The populations of *E. alsiosa* on citrus in Spain had three annual peaks, in spring, summer (the highest) and in autumn.

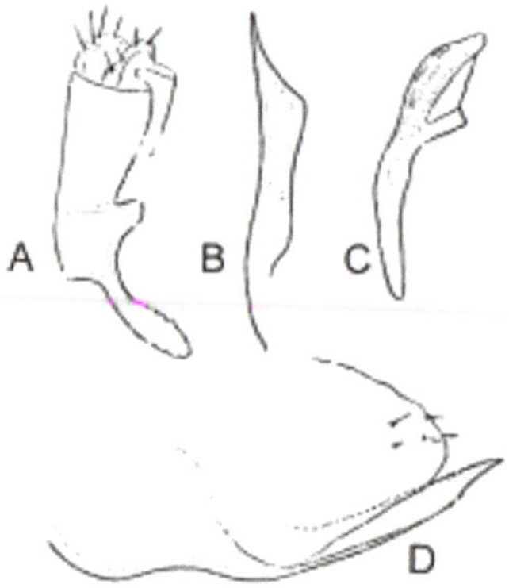

Fig. (4). *Empoasca alsiosa* Ribaut: A, anal block in lateral view; B, pygophore lobe apophyse (ventral view); C. aedeagus (lateral view); D, pygophore lobe apophysis (lateral view) [11].

Economic importance: The damage caused by *E. alsiosa* to citrus fruits is similar to that caused by other *Empoasca* spp. [14]. On citrus foliage *E. alsiosa* was dominant within a population mixed with Typhlocybinae species such as *E. solani*, *A. decedens* and others [5].

Management: Control measurements should be taken to suppress *E. alsiosa* populations in fields where summer hosts are grown, in order that pest populations that subsequently migrate to citrus orchards will not, in autumn, attain high levels. Effective weed control may reduce pest numbers.

4.3. Empoasca decipiens Paoli

Diagnostic characters: The green body is 3.2-4.1 mm in length, rarely with whitish spots on the head. Male pygophore lobe apophyse and aedeagus as in Fig. **5**.

Life history: *Empoasca decipiens* is very widely distributed in Mediterranean countries, and its life cycle is similar to that of *A. decedens* [7]. It is widely distributed in Spanish citrus orchards [15]. According to Raupach *et al.* [16]

the duration of the egg stage was around 11 days, and the newly hatched nymphs developed to adults in approximately 15 days on broad beans at 25°C. It was estimated that 126 nymphs hatched from the eggs laid by a single female.

Economic importance: *Empoasca decipiens* occurs in mixed populations with the other related species, and their combined damage has been discussed under *A. decedens*.

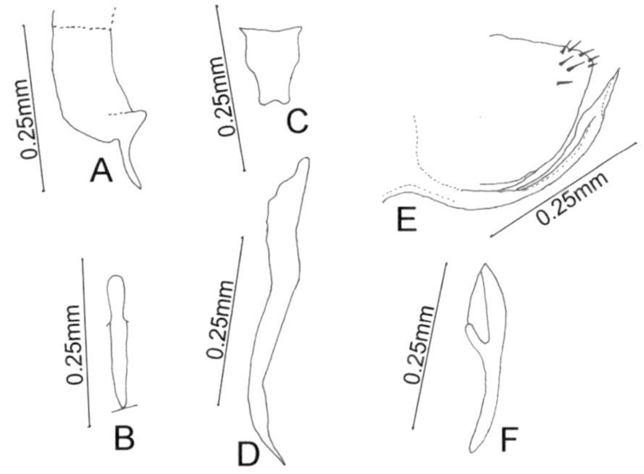

Fig. (5). *Empoasca decipiens* Paoli: A, anal block (lateral view); B, aedeagus (dorsal view); C, connective; D, stylus; E, pygophore lobe apophyse; F, aedeagus (lateral view).

Management: The recommended control measurements are like those against *A. decedens*.

4.4. Circulifer haematoceps (Mulsant and Rey) Complex

Many studies were conducted in order to resolve the taxonomy of the genus *Circulifer* [14, 15, 17], because the *Circulifer* spp. of the *haematoceps* complex strongly differ in distribution and host plant range in Israel. Several species could thus be in the *C. haematoceps* complex [18]. Three distinct species complexes of the genus *Circulifer*, namely those of *Circulifer haematoceps* (Mulsant and Rey), *Circulifer opacipennis* (Lethierry) and of *Circulifer tenellus* (Baker), were determined in Turkey [17]. Due to the fact that the structure of the male genitalia provides the best diagnostic features for identifying these species, microscopic preparations were used for their determination. The species name was applied according to variations of the plate and style in their genitalia (Fig. **6 E, F**).

Diagnostic characters: The yellowish green body is 3.3-4.1 mm in length and smooth without any spots. The pronotum is generally dull grey, and longer than the vertex. The abdomen and thorax are yellowish or greenish in color dorsally and ventrally. The wings are without any spots and the veins are light. The posterior margin of the female pregenital sternite is incised and dark red. The male aedeagus is as in Fig. **6**.

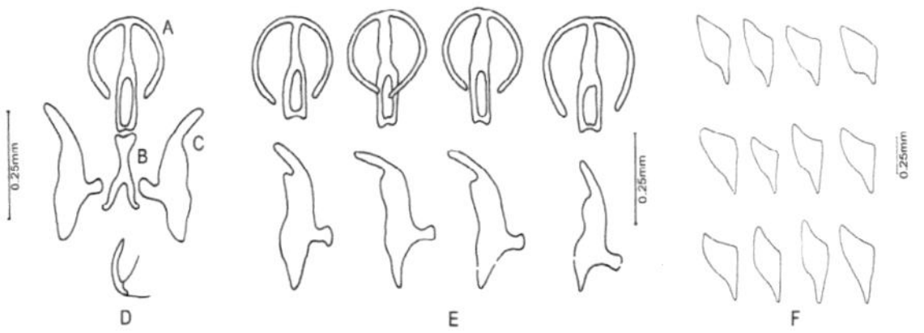

Fig. (6). *Circulifer haematoceps* (Mulsant and Rey): A, aedeagus; B, connective; C, stylus; D, right pygophore lobe apophyse (lateral view); E, variation in styli (second row); F, variation in plates.

Life history: Members of the *C. haematoceps* complex are very abundant and widely distributed in the Mediterranean region [19]. They occur in the subtropical zone along the Mediterranean coast as well as in the sub-Mediterranean climatic zone in the south west and the continental climatic zone of eastern Turkey. They overwinter in the adult stage, becoming active at the beginning of May. The oviposition period may continue for a few weeks and approximately 50-70 eggs were laid by a single female. Nymphs have 5 instars. In Turkey members of the *C. haematoceps* have one-three annual generations [1].

Members of the *C. haematoceps* complex infest a wide range of host plants, monocotyledons as well as dicotyledons, including sesame, cotton, maize, sugar beet and others [17]. The largest number of *C. haematoceps* were found on sesame, a common crop that is cultivated near citrus orchards. By the end of summer, after the sesame harvest, the leafhoppers migrate to other areas, including citrus orchards, for food and shelter. In such orchards they were collected on weeds, such as *Amaranthus* spp., *Chenopodium* sp., *Cyperus* spp., *Portulaca oleracea* Linnaeus, *Prosopis stephaniana* Kunth, *Raphanus raphanistrum* Linnaeus, and *Sorghum halepense* (Linnaeus) Pers. [18]. Sesame as a common crop and *S. halepense* as a common weed in citrus orchards harbor the causative agent of Citrus Stubborn Disease (CSD), namely *Spiroplasma citri* [20, 21]. The plants that serve as hosts for both leafhoppers and CSD are very important in the epidemiology of the disease.

Economic importance: The damage caused by members of the *C. haematoceps* complex consists mainly of vectoring CSD, which results in heavy quantitative and qualitative losses to citrus production.

Management: The populations of members of the *C. haematoceps* complex never attain high levels in citrus orchards. They do not regularly feed on citrus but on the weeds growing in the orchards. Weed control there could thus reduce vector populations as well as disease incidence. Chemicals may be applied in citrus nurseries and in young orchards, especially if adjacent to sesame fields.

5. NATURAL ENEMIES OF CICADELLIDAE

Many parasitoids and predators of cicadellids have been identified. *Gonatopus lunatus* Klug on *Circulifer haematoceps* and *Aphelopus* sp. on *Asymmetrasca decedens* and *Empoasca decipiens* were common natural enemies (Hymenoptera: Dryinidae). *Chrysoperla carnea* (Stephen) (Neuroptera: Chrysopidae), *Deraeocoris pallens* (Rt.) (Hemiptera: Miridae), *Nabis ferus* (L.) (Hemiptera: Nabidae), *Geocoris* sp. (Hemiptera: Geocoridae), *Paederus* f.c. *kalalovae* Curtis (Coleoptera: Staphylinidae) were common predators feeding on *C. haematoceps*, *A. decedens* and *E. decipiens* in citrus orchards. By the end of the season, in September 1991, an overall 70% parasitization rate on *Circulifer* spp. was observed on sugar beet in Ereğli, Central Anatolia [22]. In laboratory studies *G. lunatus* preferred the fourth and fifth nymphal stages of *C. haematoceps* for parasitization, whereas younger nymphs were fed on. In Spain, 27% of *Empoasca alsiosa* adults were parasitized by Dryinidae in a Navel orange orchard in Alberic (Valencia) during 1976 [5].

REFERENCES

[1] Lodos N. Entomology of Turkey II. General, Applied and Faunistic. Bornova-İzmir: Aegean University, Agricultural Faculty Press No: 429, 1982.

[2] Borror DJ, DeLong DM. An Introduction to the Study of Insects. 3rd ed. New York 1970.

[3] Oman PW. The Nearctic Leafhoppers (Homoptera: Cicadellidae): A Generic Classification and Checklist. Memoris of The Entomological Society of Washington, No: 3. Washington, D.C.: The Entomological Society of Washington 1949.

[4] Nielson MW. Leafhopper Systematics. In Nault LR, Rodrigez JG, Eds. The Leafhopers and Planthoppers. New York, Wiley-Interscience Publication 1985; pp. 11-39.

[5] Hermoso de Mendoza A, Medina V. Estudio inicial sobre cicadélidos (Homoptera, Cicadellidae) en los huertos de agrios del País Valenciano. Anales INIA. Serie Protección Vegetal 1979; 10: 43-68.

[6] Başpınar H, Uygun N. Studies on the determination of Cicadellidae species in citrus orchards in East Mediterranean Region of Turkey, monitoring their population changes by different sampling methods, and food plants, and relationship between Stubborn disease and its vector cicadellid. Turk J Agricul Forest 1994; 18: 9-20.

[7] Lodos N, Kalkandelen A. Preliminary list of Auchenorrhyncha with notes on distribution and importance of species in Turkey. XII: Family Cicadellidae, Typhlocybinae: Empoascini. Turk J Plant Prot 1983; 7: 153-65.

[8] Başpınar H, Uygun N. The population Dynamics and damage of *Asymmetrasca decedens* (Paoli) and *Empoasca decipiens* Paoli (Homoptera: Cicadellidae) on citrus in the East Mediterranean Region of Turkey. In: Proceedings of the 2nd Entomological Congress of Turkey, 1992: Adana, Turkey: Turkish Ent Society 1992; pp. 533-40.

[9] Aly H, Desuky WMH, Attia AA, Youssef AAA. Ecological studies on certain homopterous insects infesting some citrus trees in Egypt. Egypt J Agricul Res 2003; 81: 531-50.

[10] Anonymous. Plant Protection Technical Instructions. Directorate General of Agricultural Researches. Ministry of Agriculture and Rural Affairs of Turkey. Ankara 2008 (in Turkish).

[11] Ribaut, H. Faune De France, Homopteres Auchenorhynques. I (Typhlocybidae). Rue De Tournon: Paul Lechevalier et Fils 1936.

[12] Mazzoni V, Cosci F, Lucchi A, Santini L. Occurrence of leafhoppers (Auchenorrhyncha, Cicadellidae) in three vineyards of the Pisa district. Integrated Control in Viticulture, IOBC/WPRS Bulletin 2001; 24: 267-71.

[13] Del Rivero JM. La roseta de los agrios. Boletín de Patología Vegetaly Entomología Agrícola 1954; 20: 193-210.

[14] Oman PW, Knight WJ, Nielson MW. Leafhopper (Cicadellidae). A bibliography, generic check-list and index to the world literature 1956-1985. Wallingford, UK: CAB International Institute of Entomology 1990.

[15] Kersting U. Untersuchungen über Wirtspflanzenspektrum und Insektenvektoren sowie Nachweismethoden des Erregers der Zitrusstauche, *Spiroplasma citri* Saglio *et al.*, im Zitrusbaugebiet der Südtürkei. Dissertation der Universitaet Bonn 1991, FRG, 97 pp.

[16] Raupach K, Borgemeister C, Hommes M, *et al.* Effect of temperature and host plants on the bionomics of *Empoasca decipiens* (Homoptera: Cicadellidae). Crop Prot 2002; 21: 113-9.

[17] Başpınar H, Kersting U, Şengonca Ç, Uygun, N. Studies on taxonomy, distribution and host plants Turkish species of *Circulifer* Zakhvatkin (Homoptera: Cicadellidae). Turk J Entomol 1993; 17: 129-40.

[18] Klein M, Raccah B. Seperation of two leafhopper populations of the *Circulifer haematoceps* complex on different host plants in Israel. Phytoparasitica 1991; 19: 153-5.

[19] Lodos N, Kalkandelen A. Preliminary list of Auchenorrhyncha with notes on distribution and importance of species in Turkey. XVII. Family-Cicadellidae: Deltocephalinae: Grypotini, Goniagnathini and Opsiini (Part I). Turk J Plant Prot 1983; 9: 79-90.

[20] Kersting U, Başpınar H, Çınar A *et al.* New findings on the epidemiology of *Spiroplasma citri* in the Eastern Mediterranean region of Turkey. In: Proceedings of the 12th IOCV; 1993: Riverside, USA: 1993; pp. 336-41.

[21] Kersting U, Başpınar H. Occurence and distribution of *Spiroplasma citri* and sesame phyllody in sesame in Southern Anatolia. J Turk Phytopathol 1997; 26: 1-9.

[22] Başpınar H, Kersting U, Ugun N. Natural enemies of Cicadellidae species in East Mediterranean Region. In Proceedings of the 3rd Biological Control Congress of Turkey, 1994: İzmir Turkey: Turkish Ent Society 1992; pp. 365-74. 1994.

CHAPTER 9

Aphididae

Nedim Uygun[1*], Alfonso Hermoso de Mendoza[2] and Hüseyin Başpınar[3]

[1]*Department of Plant Protection, Çukurova University, Agricultural Faculty, 01100 Adana, Turkey;* [2]*Institut Valencià d'Investigacions Agràries (IVIA), Carretera de Nàquera km 5, 46113 Montcada (València), Spain and* [3]*Department of Plant Protection, Adnan Menderes University, Agricultural Faculty, 09100 Aydın, Turkey*

Abstract: The Aphididae, or aphids, is a large family of nearly 4000 small and soft-bodied insects in the superfamily Aphidoidea. More than 25 aphid species have been found in citrus orchards throughout the world, of which some are of economic importance and others probably occur there sporadically. The aphids infesting citrus in the Mediterranean Region reach their maximal numbers in spring, attaining another, lower peak in autumn, and sometimes a third, of minor importance, during summer. Some of the citrus aphids, like *Aphis craccivora* Koch, *A. fabae* Scopoli, *Macrosiphum euphorbiae* (Thomas), and *Myzus persicae* (Sulzer), never reach high population levels due to the activity of their natural enemies. The others, *Aphis gossypii* Glover, *A. spiraecola* Patch, *Toxoptera aurantii* (Boyer de Fonscolombe) *and T. citricida* (Kirkaldy), are serious pests, either because they occur in very large numbers or due to their ability to transmit many virus diseases, including citrus tristeza virus (CTV). Their management is mostly by their many natural enemies. These include parasitoids of the subfamily Aphidiinae (Hymenoptera, Braconidae) and many predators of the families Coccinellidae (Coleoptera), Syrphidae (Diptera); Cecidomyiidae, Chamaemyiidae (Diptera); Anthocoridae, Lygaeidae, Miridae, and Nabidae (Hemiptera) and Chrysopidae (Neuroptera).

Keywords: Citriculture, Mediterranean Region, citrus aphids, bio-ecology, damage, control.

1. INTRODUCTION

The Aphididae, or aphids, is a large, cosmopolitan family of nearly 4,000 species in the superfamily Aphidoidea (**Hemiptera**). They are small, soft-bodied insects that suck out plant juices. Their mouth parts resemble an elongated beak that arises from the hind part of the head. The external part is the labium, which encloses the four piercing stylets that consist of two mandibles and two maxillae. When held together tightly by labium, the stylets form an elongated filament with two inner channels. One, the salivary channel, injects saliva that dissolves the sap after piercing through the host-plant tissue. The other channel is used to suck the sap [1, 2]. This feeding system enables aphids (and many other Hemiptera) to transmit plant viruses (or other pathogens). These are acquired when aphids suck from a diseased plant and are transmitted upon injecting the pathogen-containing saliva into healthy plants.

Aphids excrete honeydew from their anus, which is processed excessive plant sap that contains sugars and waste materials. The sugars in the honeydew provide a substrate for some fungi, like *Aspergillus niger* Tiegh., which cover the leaf surfaces with their dark mycelia, reducing respiration and photosynthesis.

Aphids possess certain unique morphological and biological characteristics. A winged parthenogenetic female (wings removed on the left in order to see some details) is shown in Fig. **1**, in order to indicate the main characters of aphids. Morphological structures exclusive to aphids are the siphunculi or cornicles (S in Fig **1**), which occur in most species. Their, structure is constant within each species, but varies in form, size and colour between species. The siphunculi have different functions, like defence, by squirting a spray of haemolymph that is rich in wax, which upon solidifying might glue the jaws of a predator. Another structure unique to aphids is the cauda (C), located at the posterior of the abdomen, whose morphological characteristics are peculiar for each species. The pterostigma (P), a dark spot located in the fore wings of aphids, and the number of branches (1^a, 2^a) in the vein media (M) of this wing are also typical for these insects. Finally, the form of the front (F), and the number, colour, proportions and characteristics of the antennal segments (1 to 6), mainly of the last, which consists of a base (B) and a *processus terminalis* (PT) are useful for the determination of aphid species.

***Address correspondence to Nedim Uygun:** Department of Plant Protection, Çukurova University, Agricultural Faculty, 01100 Adana, Turkey; Tel: +90 3223386754; E-mail: nuygun@cu.edu.tr

Vincenzo Vacante and Uri Gerson (Eds)

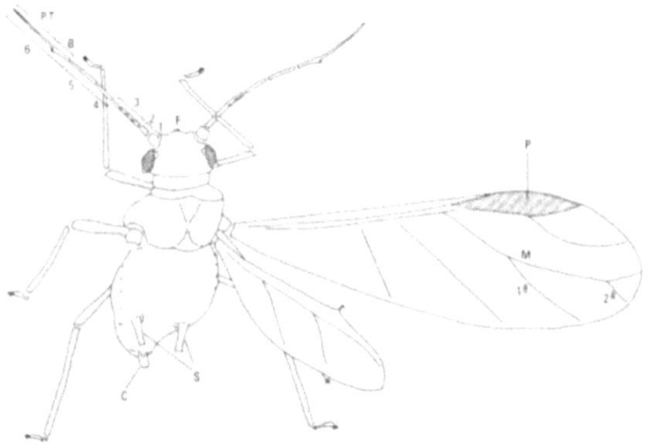

Fig. (1). A winged (alate) aphid (wings removed on left).

Aphids have complicated life cycles, often alternating between amphigenic (females and males) generations and parthenogenetic ones (only females), with different apterous (apterae) and winged (alatae) adult forms and different nymphal instars necessary to produce adults [3]. Most species overwinter in the egg stage. Nymphs appear in the spring and become adult apterous females that reproduce parthenogenetically, giving birth to living nymphs. Winged females (alatae) appear after several generations in order to migrate to other plants of the same species (a monoecious cycle) or to other host plants (a heteroecious cycle). In the latter case the plant of origin is called the primary host and the other plant is the secondary host. On these hosts the alatae continue to reproduce by parthenogenesis or viviparity. By autumn female sexuparae appear and produce winged females and males that return to the primary host, whereon they copulate and the females lay the wintering eggs. This complete cycle is typical for holocyclic aphid species. In many other aphids, known as anholocyclic species, an amphigenic generation does not occur because they reproduce parthenogeneticly throughout the year.

In the Mediterranean Region the aphid populations that infest citrus usually peak in spring, with another, generally lower peak in autumn, and sometimes with another, minor population increase during summer [4].

More than 25 aphid species have been recorded from citrus world-wide [5], but many are probably only chance visitors, and others are of low economic importance. The most abundant species in the Mediterranean Region are in the subfamily Aphidinae of the family Aphididae. They consist of *Aphis spiraecola patch*, *Aphis gossypii* Glover and, in some areas, *Toxoptera aurantii* (Boyer de Fonscolombe). Locally *Myzus persicae* (Sulzer), *Aphis craccivora* Koch, *Aphis fabae* Scopoli and *Macrosiphum euphorbiae* (Thomas) can also be important. In addition, *Toxoptera citricida* (Kirkaldy), the most efficient vector of Citrus Tristeza Virus (CTV) has recently invaded the western Mediterranean countries [6].

Two identification keys for adult aphids (parthenogenetic females: apterous or alatae, as the case may be) of all nine species are presented below. The first is a field key, based mainly on color, whereas the other key uses microscopic structures for the accurate identification of aphids in the laboratory.

2. FIELD KEY TO CITRUS APHID SPECIES (apterous or alatae adults are needed, depending on the species)

1. Apterous dark.. .2

-. Apterous pale (green, yellow or pink).. 6

2. Alatae with dark pterostigma; forewing media once-branched (Fig. **6**) .... ***Toxoptera aurantii***

-. Alatae with pale pterostigma; forewing media branched twice... 3

3. Alatae with antennal segment III dark, segment IV partially pale (Fig. **7**) .... ***Toxoptera citricida***

-. Alatae with uniformly colored antennae...4

4. Apterae shiny black (Fig. **2a**)..***Aphis craccivora***

-. Apterae otherwise, not shiny black.. 5

5. Apterae glossy black (frequently with whitish wax markings) (Fig. **3a**)..............................***Aphis fabae***

-. Apterae variably colored (almost black to almost white, but usually dark, light or dirty green) (Fig. **4a**) ....***Aphis gossypii***

6. Cauda ensiform; siphunculi pale and long (Fig. **8b**)......................................***Macrosiphum euphorbiae***

-. Cauda triangular or finger-like..... 7

7. Siphunculi pale (Fig. **9b**).................. ....***Myzus persicae***

-. Siphunculi black..................... 8

8. Cauda black apterae uniformly green (Fig. **5a**).......... ....***Aphis spiraecola***

-. Cauda pale; apterae variably coloured (if green, then uneven) (Fig. **4a**) ....***Aphis gossypii***

3. MICROSCOPIC KEY TO CITRUS APHID SPECIES (for apterous and alatae adults)

1. Siphunculi pale; frons with a groove.... 2

-. Siphunculi dark; frons without groove... 3

2. Siphunculi with reticulated apex (Fig. **8b**) ... .***Macrosiphum euphorbiae***

-. Siphunculi without reticulations (Fig. **9b**).. ....***Myzus persicae***

3. Abdomen with stridulatory apparatus.. .4

-. Abdomen without stridulatory apparatus..5

4. Cauda usually with less than 20 setae... ....***Toxoptera aurantii***

-. Cauda usually with more than 20 setae.......... ....***Toxoptera citricida***

5. Cauda pale (Fig. **4b**)........ ....***Aphis gossypii***

-. Cauda dark..6

For apterous aphids only

6. Colour in life green...***Aphis spiraecola***

-. Colour in life black.............. 7

7. Abdomen with an extensive black patch........ ....***Aphis craccivora***

-. Abdomen without patch..***Aphis fabae***

For alatae aphids

6. Abdomen without stripes (Fig. **5b**)....... ....***Aphis spiraecola***

-. Abdomen with several stripes... 7

7. Abdomen with a long stripe connecting the bases of the siphunculi (Fig. **2b**) ....***Aphis craccivora***

-. Abdomen with a short stripe that does not connect the bases of the siphunculi (Fig. **3b**).....***Aphis fabae***

4. CITRUS APHID SPECIES

The basic sources for the following text are in references [7-12].

4.1. *Aphis craccivora* Koch

Diagnostic characters: The live parthenogenetic apterous females are black with a characteristic metallic shine (Fig. **2a**), due to an extensive sclerotized patch on the dorsal abdomen, best seen in microscopic preparations. Parthenogenetic alatae females lack that patch, but several transversal stripes are evident on the abdomen in microscopic preparations, as is the stripe between the bases of the siphunculi (Fig. **2b**). The cauda usually bears 7 setae.

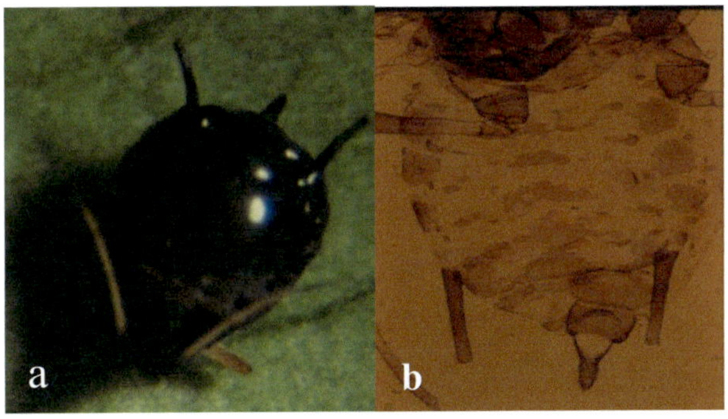

Fig. (2). *Aphis craccivora* Koch: a, apterous adult; b, abdomen of cleared alata.

Life history: *Aphis craccivora* is usually anholocyclic, although sexual forms have occasionally been seen. It lives on many host plants, especially legumes.

Economic importance: Due to its polyphagous nature, *A. craccivora* is a pest of many cultivated plants wherever it occurs. Although its origin appears to be in the Palearctic, the current distribution of this pest is almost world-wide. On citrus trees *A. craccivora* is a minor pest, probably because these plants are not its favourite hosts. *Aphis craccivora* has been reported as a vector of more than 30 viruses of cultivated plants. It is known to transmit CTV only in India (13), but not in any other country (14).

4.2. *Aphis fabae* Scopoli

Diagnostic characters: Live apterous parthenogenetic females with a glossy black body (Fig. **3a**), often with white waxy markings. In microscopic preparations (Fig. **3b**) alatae show several transversal stripes on the abdomen (like *A. craccivora*), but the stripe between the siphunculi of *A. fabae* does not connect their bases, and the cauda has more than seven setae.

Life history: *Aphis fabae* is mostly holocyclic heteroecious, whose primary host is *Euonymus europaeus* L. and many plants serve as secondary hosts. It is exclusively anholocyclic on several secondary hosts.

Fig. (3). *Aphis fabae* Scopoli: a, apterous adult; b, abdomen of cleared alata.

Economic importance: *Aphis fabae* is present in most of the Northern hemisphere, including North America, as well as in Africa. It is polyphagous, infesting many crop plants, but is usually not very harmful to citrus. *Aphis fabae* transmits several viruses affecting different crops, but has not been found to be a vector of any citrus viruses [14].

4.3. *Aphis gossypii* (Glover)

Diagnostic characters: The colour of the parthenogenetic apterous females is very variable (Fig. **4a**), from almost black to almost white, although mainly dark green or unevenly green. Cauda pale or dark, according to the colour of the body; the siphunculi are always dark. In cleared preparations parthenogenetic alatae females show a variable number of abdominal stripes (Fig. **4b**). The stripe between the siphunculi is always present, without connecting them. The cauda is pale and usually with not more than 7 setae.

Life history: *Aphis gossypii* is generally anholocyclic on numerous hosts that belong to many plant orders, but some cases of holocyclic development are also known. The aphid performs better on grapefruit than on sweet orange, mandarin or lemon [15].

Economic importance: *Aphis gossypii* infests many crops in many parts of the world. On citrus it was of secondary importance until the middle of 1980's, when the aphid developed resistance to certain insecticides [16]. In consequence it became one of the predominant aphid species on citrus, mainly in the Mediterranean Region [17, 18].

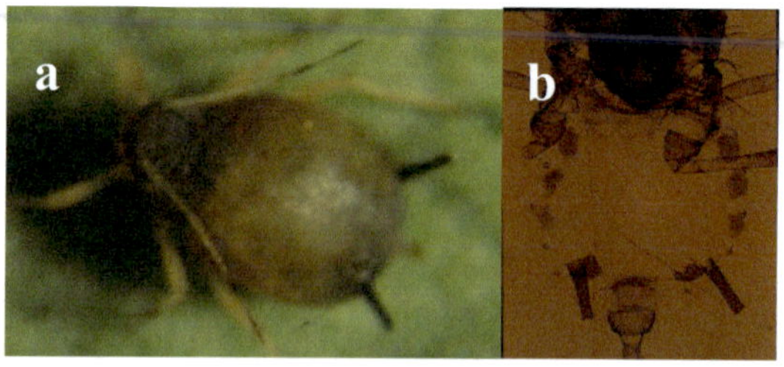

Fig. (4). *Aphis gossypii* Glover: a, apterous adult; b, abdomen of cleared alata.

The main economic risk represented by *A. gossypii* is due to its capacity to transmit more than 50 plant viruses. These include CTV [19] as well as the citrus vein enation virus [20]. As a vector of CTV the aphid is almost as efficient as *Toxoptera citricida*, especially in the absence of the latter. This was observed when *A. gossypii* was mainly responsible for the epidemic of CTV that ruined large parts of the Spanish citrus industry starting from 1957, and which forced growers to substitute the traditional rootstock, sour orange, by new rootstocks tolerant to CTV.

4.4. *Aphis spiraecola* Patch

Diagnostic characters: Parthenogenetic apterous females are typically green, with dark cauda and siphunculi (Fig. **5a**). In cleared specimens parthenogenetic winged females do not show stripes in the abdomen between or above the siphunculi, and the cauda carries more than seven setae (Fig. **5b**).

Life history: *Aphis spiraecola*, which is of Oriental origin and had become naturalized in North America, has spread in the 20th century throughout most of the world, where it is anholocyclic on numerous plants. In its areas of origin the aphid is holocyclic. The developmental of *A. spiraecola* on Washington navel oranges in the laboratory required 12.1 days at 15°C and 5.8 days at 25°C [21].

Economic importance: Although a pest of many crops, on citrus *A. spireacola* causes the most intensive damage, deforming the leaves on which it feeds. The aphid is more destructive than *A. gossypii* to the mandarin CV Fremont in the laboratory [18]. Although the pest was introduced into the Mediterranean Region many years ago, it began to infest citrus only from about 1960. *Aphis spireacola* then displaced the previously dominant species, *Toxoptera aurantii* [22], in many orchards, probably due to having fewer natural enemies. This aphid is known to vector several plant viruses and, in citrus, it transmits CTV [23, 24]. Despite not being an efficient CTV vector, its large populations on citrus indicate that this aphid has a role in the spread of the disease.

4.5. *Toxoptera aurantii* (Boyer de Fonscolombe)

Diagnostic characters: Parthenogenetic apterous females are dark brown or black, their antennae with white and black bands. Cleared parthenogenetic winged females bear a stridulatory organ on the abdomen, which is characteristic of the genus. In addition, the cauda is usually with less than 20 setae. The most characteristic features of this species are the black pterostigma and the single branch in the media of the forewing; all other citrus aphids bear a pale pterostigma and their media are two-branched (Fig. **6**).

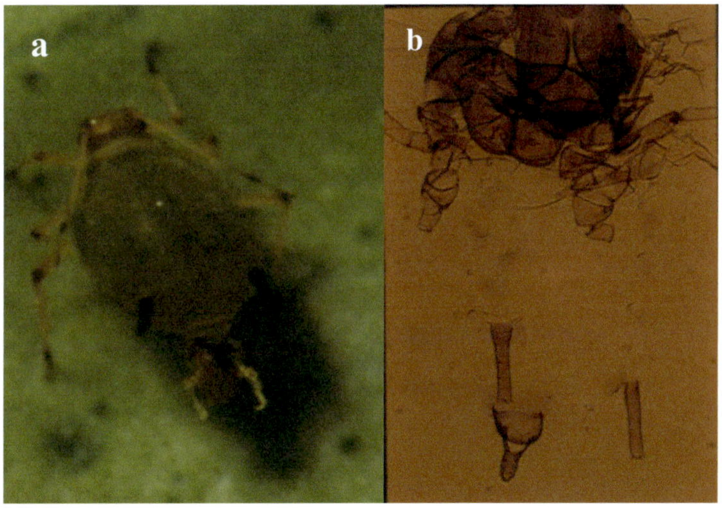

Fig. (5). *Aphis spiraecola* Patch: a, apterous adult; b, abdomen of cleared alata.

Life history: *Toxoptera aurantii* is anholocyclic throughout its distribution, occurring in all tropical and subtropical regions of the world, infesting a large number of plants.

Economic importance: *Toxoptera aurantii* attacks several crops, but mainly citrus. In the Mediterranean Region it was termed "the orange aphid" par excellence, because it was the most abundant aphid on citrus before the invasion of *A. spiraecola*; in certain areas it is still very important.

This aphid transmits several viruses of cultivated plants, citrus included. Like the previous species, it is a vector of CTV but with low efficiency; it can also transmit citrus vein enation virus [24].

4.6. *Toxoptera citricida* (Kirkaldy)

Diagnostic characters: Parthenogenetic apterous females dark (Fig. **7**), their antennae without the white and black bands seen in *T. aurantii*. The cauda is usually with more than 20 setae. Parthenogenetic winged females are identifiable by their third antennal segment that is entirely dark (Fig. **7**), whereas the fourth segment is partially pale. No other alate citrus aphid has similar antennae, including the apterae of *T. citricida.*

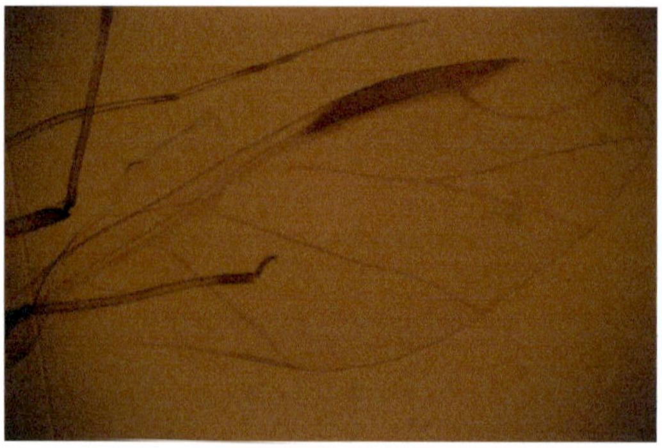

Fig. (6). Forewing of *Toxoptera aurantii* (Boyer de Fonscolombe).

Life history: *Toxoptera citricida* occurs throughout Oceania, eastern and southern Asia, Africa south of the Sahara, and Central and South America; recently it was also found in Mexico and Florida. Except for isolated cases in Japan, the aphid is always anholocyclic. After being detected in Madeira in 1994 [25], by 2002 the aphid was discovered in the north of Spain and later, in 2003, in continental Portugal [6]. At the moment the pest continues to spread to the coastal area of the northwest quadrant of the Iberian Peninsula, corresponding to the Atlantic slope. It has not yet been detected in the Mediterranean coast where the main Spanish citrus areas are located [26]. Although *T. citricida* was sporadically reported on different hosts, it lives almost exclusively on plants of the family Rutaceae, especially on *Citrus* spp.

Fig. (7). Antenna of alata of *Toxoptera citricida* (Kirkaldy).

Economic importance: The importance of *T. citricida* on citrus is due mostly to its capacity to transmit several viruses, such as citrus vein enation virus [27] and, mainly, CTV [28]. This aphid is the most efficient vector of CTV [29]. It also transmits some very aggressive CTV races that induce direct damage to the bark of certain citrus varieties, whether or not grafted on rootstocks tolerant to the normal races of CTV. The recent invasion of *T. citricida* into the Mediterranean Region thus represents a great hazard to the citrus industry, especially where trees are grafted on the traditional sour orange rootstock, which is very sensitive to CTV. In addition *T. citricida* causes citrus leaf rolling.

4.7. *Macrosiphum euphorbiae* (Thomas)

Diagnostic characters: This species is the largest aphid infesting citrus in the Mediterranean Region. The parthenogenetic apterous females are usually green (Fig. **8a**), although sometimes yellow or pink, with a pale, long cauda and siphunculi. The reticulation on the apex of the siphunculi, which is absent in all other citrus aphids, can be seen in cleared specimens of the apterae or alatae (Fig. **8b**). The front has a closed groove with convergent sides. Alatae with a large dark patch on the abdomen (Fig. 8**b**).

Life history: *Macrosiphum euphorbiae* is of world-wide distribution. In North America it is holocyclic, but mostly anholocyclic in the rest of the world. The aphid lives on many host plants.

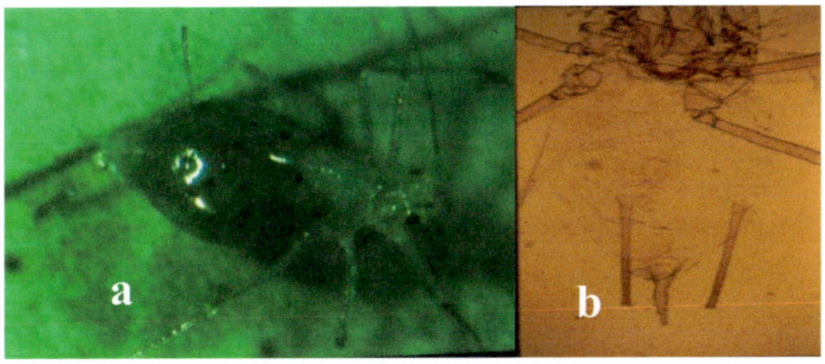

Fig. (8). *Macrosiphum euphorbiae* (Thomas): a, apterous adult; b, abdomen of cleared alata.

Economic importance: *Macrosiphum euphorbiae* attacks several crop plants, especially of the family Solanaceae, particularly potatoes. On citrus it occurs only sporadically. This aphid vectors over 40 plant viruses, but none of citrus.

4.8. *Myzus persicae* (Sulzer)

Diagnostic characters: Parthenogenetic apterous females green, yellow or pink, with pale siphunculi and cauda (Fig. **9a**). Siphunculi without reticulation, and the front with a closed groove of convergent sides. Alatae with an extensive dark patch on the abdomen (Fig. **9b**).

Life history: *Myzus persicae* occurs everywhere. Its primary hosts are *Prunus* spp. (especially peach tree), on which it is holocyclic. A heteroecious aphid, it has many secondary host plants. In warm areas it can be anholocyclic permanently.

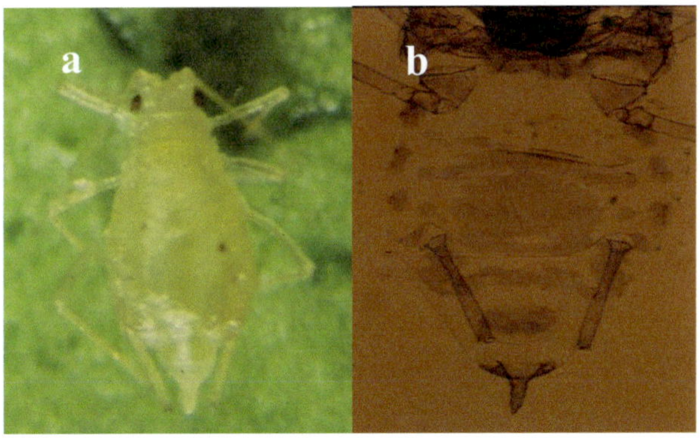

Fig. (9). *Myzus persicae* (Sulzer): a. apterous adult; b, abdomen of cleared alata.

Economic importance: *Myzus persicae* attacks a number of crops. On citrus it is usually not very abundant, but the ability of this aphid to develop resistance to certain insecticides may at times elevate it to pest status. It transmits more than 100 plant viruses, but only in India does this aphid transmit CTV to citrus [13]. *Myzus persicae* also transmits the citrus vein enation virus [30].

5. INTERVENTION THRESHOLDS OF CITRUS APHIDS

Several pesticides are recommended against citrus aphids, but their efficacy varies according to circumstances. For example, the Spanish Services of Vegetal Protection currently recommends the following: chlorpyriphos, pimetrozina, pirimicarb and, only in young orchards, acetamiprid where Integrated Pest Management (IPM) is practiced [31].

Intervention thresholds of pesticides to be applied against citrus aphids have been established, using different indices within an IPM program. The oft-used index is the percentage of infested shoots in 0.25 - 2.0 m rings located on the outside of the tree canopy [32, 33]. Other, more simplified indices for field use have also been suggested, such as the number of infested shoots per ring, the index of infestation and the percentage of occupied rings [34], and the number of aphids captured weekly in a yellow water trap [35]. Whatever the index used, practitioners should bear in mind that these thresholds are only approximations, resulting in much variability in the thresholds established by different authors.

In order to eliminate this variability and accurately decide on the intervention thresholds against citrus aphids, a function was calculated that relates clementine yield losses to aphid (*A. gossypii* and *A. spiraecola*) density [36, 37]. The relevant intervention thresholds [38], which include the Economic Injury Level (EIL) and the Economic Threshold (ET) can then be calculated from that function. With these thresholds, the Environmental Economic Injury Level (EEIL) and the Environmental Economic Threshold (EET) may also be obtained, where the ecological cost due to pesticide application is considered [39]. To calculate these thresholds, besides the function yield loss, several parameters should be known [40]: pesticide cost per hectare (C), fruit price per kilogram (V), the pesticide efficacy, with values between 0 and 1 (K), and yield per hectare of a minimal pest level orchard (P_0).

The EIL can be obtained for each aphid species and expressed as number of aphids per m^2 of canopy, as:

EIL (for *A. gossypii*):

$$\frac{218333 \cdot C + 11130.6 \cdot VP_0K}{57.08 \cdot VP_0K - 100 \cdot C}$$

EIL (*for A. spiraecola*):

$$\frac{879666 \cdot C + 23804 \cdot VP_0K}{108.2 \cdot VP_0K - 100 \cdot C}$$

Upon calculating with the values of the variable parameters currently used in the Mediterranean Region, an EIL of 271 aphids/m^2 is obtained for *A. gossypii* and 370 aphids/m^2 for *A. spiraecola*. Starting with these data, corresponding values of ET, EEIL, and finally EET may be calculated. The EET, transformed to the more usual index and adapted to be similar for both aphid species, is approximately 25% of infested shoots: when this aphid density is reached, a pesticide treatment is required within 24 hours. This is the threshold currently recommended in Spain in citrus orchards under IPM [39].

6. MANAGEMENT OF CITRUS APHIDS

Some of the aphid species that infest citrus, *e.g. A. craccivora*, *A. fabae*, *M. euphorbiae* and *M. persicae,* are controlled by their natural enemies and never reach high population levels. Others, like *A. gossypii, A. spiraecola, T. aurantii* and *T. citricida* are very injurious because they attain very high population levels and/or transmit virus diseases, including CTV. Aphids attacking citrus, including the main virus vector, *T. citricida*, have many natural enemies. In Spain they include members of the families Braconidae (Aphidiinae) and Aphelinidae as parasitoids, along with predators of the families Anthocoridae, Cecidomyiidae, Coccinellidae, Chamaemyiidae Chrysopidae, Lygaeidae, Miridae, Nabidae and Syrphidae.

Some authors had stated that these natural enemies are very common and could control citrus aphids [26, 41]. It was also stated that aphid populations usually declined naturally after periods of khamsin (hot and dry winds that blow mostly in the spring) in Israel [42].

In laboratory trials *Lysiphlebus testaceipes* (Cresson) was a more effective parasitoid of *Aphis gosspii* than *L. confusus* Tremblay and Eady and *L. fabarum* (Marshal). *Aphis fabae,* reared on *Vicia faba* Linnaeus, was a suitable host for *L. testaceipes* in a mass production [43].

Trioxys angelicae (Haliday) was the most common parasitoid in Turkey, its attack rate on citrus aphids coming to 56.4%. It was followed by *Aphidius colemani* Viereck (11.9%), *L. fabarum* (9.6%), *L. confusus* (5.8%), and *Aphidius matricariae* (Haliday) (5.0%) [43].

In conclusion, it appears that in Turkey there is a rich and widespread complex of natural enemies, which could successfully control aphids in mature citrus orchards [41]. The aphids could however become a problem from time to time in nurseries or in young orchards, for which specific aphidicides are recommended.

REFERENCES

[1] Borror DJ, DeLong DM. An Introduction to the Study of Insects. 3rd ed. New York 1970.

[2] Miyazaki M. In: Minks AK, Harrewijn P, Eds. Aphids. Their Biology, Natural Enemies and Control (A). Amsterdam: Elsevier 1987; pp. 1-25.

[3] Miyazaki M. In: Minks AK, Harrewijn P, Eds. Aphids. Their Biology, Natural Enemies and Control (A). Amsterdam: Elsevier 1987; pp. 27-50.

[4] Hermoso de Mendoza A, Michelena JM, González P. Evolución global y por colonias de los pulgones (Hemiptera: Aphididae) y sus enemigos naturales en los clementinos valencianos. VI Congreso Nacional de Entomología Aplicada, Palma de Mallorca 2009.

[5] Hermoso de Mendoza A. Pulgones de cítricos y transmisión de virosis. Phytoma España 1994; 58: 40-6.

[6] Ilharco FA, Sousa-Silva CR, Álvarez A. First report on *Toxoptera citricidus* (Kirkaldy) in Spain and continental Portugal (Homoptera, Aphidoidea). Agron Lusit 2005; 51: 19-21.

[7] Stroyan HL. La identificación de los afídidos que viven sobre Citrus. Bol Pat veg Ent agr 1961; 25: 71-100.

[8] Holman J. Los áfidos de Cuba. La Habana: Editorial Organismos 1974.

[9] Nieto JM, Díaz TE, Mier MP. Catálogo de los pulgones (Homoptera Aphidoidea) de España y de sus plantas hospedadoras. Leon: Universidad de León 1984.

[10] Nieto JM, Mier MP, García F, Pérez N. Hemiptera Aphididae III. Madrid: Museo Nacional de Ciencias naturales, CSIC 2005.

[11] Blackman RL, Eastop VF. Aphids on the World's Crops. Chichester: John Wiley & sons 1985.

[12] Belliure B, Pérez P, Marcos MA, Michelena JM, Hermoso de Mendoza A. In: Jaca JA, Urbaneja A, Eds. Control biológico de plagas agrícolas. Phytoma España 2009; pp. 209-38.

[13] Varma PM, Rao DG, Capoor SD. Transmission of tristeza virus by *Aphis craccivora* (Koch) and *Dactynotus jaceae* (L.). Indian J Entomol 1965; 27: 67-71.

[14] Hermoso de Mendoza A, Ballester-Olmos JF, Pina JA. Transmission of citrus tristeza virus by aphids (Homoptera, Aphididae) in Spain. In: Garnsey SM, Timmer LW, Dodds JA, Eds. Proceedings 9th Conf IOCV; 1984: Riverside; pp. 23-7.

[15] Satar S, Kersting U, Uygun N. Effect of different citrus host plants and temperatures on development rate and fecundity of apterous *Aphis gossypii* Glover (Homoptera: Aphididae). Turk J Entomol 1998; 22: 187-97.

[16] Melià A, Blasco J. Resistencia de *Aphis frangulae gossypii* Glover (Homoptera: Aphididae) a insecticidas en el cultivo de los cítricos. Bol Sanid Veg Plagas 1990; 16: 189-93.

[17] Hermoso de Mendoza A, Moreno P. 1989. Cambios cuantitativos en la fauna afídica de los cítricos valencianos. Bol Sanid Veg Plagas 1989; 15: 139-42.

[18] Yumruktepe R, Uygun N. The aphids (Homoptera: Aphididae) determined in citrus orchards in East Mediterranean Region of Turkey and their natural enemies. Proceedings of the 3rd Turkish Congress of Biological Control. Turk. Entomol Assoc. Publ. 1994; **7:** 1-12 (in Turkish).

[19] Dickson RS, Flock RA, Johnson MM. Insect transmission of citrus quick decline. J Econ Entomol 1951; 44: 172-6.

[20] Laird EF, Weathers LG. *Aphis gossypii*, a vector of citrus vein-enation virus. Plant Dis Rep 1961; 5: 877.

[21] Satar S, Uygun N. Life cycle of *Aphis spiraecola* Patch (Homoptera: Aphididae) in East Mediterranean Region of Turkey and its development on some important host plants. Bull IOBC/WPRS 2008; 38: 216-24.

[22] Hermoso de Mendoza A, Fuertes C, Serra J. Proporciones relativas y gráficas de vuelo de pulgones (Homoptera, Aphidinea) en los cítricos españoles. Inv Agraria Prod y Prot Veg 1986; 1: 393-08.

[23] Norman P, Grant TJ. Preliminary studies of aphid transmission of tristeza virus in Florida. The Citrus Industry 1954; 35: 10-2.

[24] Manjunath KL. Studies on vein enation virus disease of citrus in South India. Indian J Plant Pathol 1987; 5: 121-5.

[25] Fernandes J, Cruz de Boelpaepe MO. Programa de prospecção de organismos nocivos em citrinos. Sub-programa pragas. 1, *Toxoptera citricidus*. Lisboa: Centro nacional de Protecção da Produção Agrícola 1994.

[26] Hermoso de Mendoza A, Álvarez A., Michelena JM, González P, Cambra M. *Toxoptera citricida* (Kirkaldy) (Hemiptera, Aphididae) and its natural enemies in Spain. Bull IOBC/WPRS 2008; 38: 225-32.

[27] McClean APD. Citrus vein-enation virus. South Afr J Sci 1954; 50: 147-51.

[28] Meneghini M. Sobre a naturaleza e transmissibilidade da doença "Tristeza" dos Citrus. O Biológico 1946; 12: 285-7.

[29] Yokomi RK, Lastra R, Stoetzel B *et al.* Establishment of the brown citrus aphid *Toxoptera citricida* (Kirkaldy) (Homoptera: Aphididae) in Central America and the Caribbean basin and its transmission of citrus tristeza virus. J Econ Entomol 1994; 87: 1078-85.

[30] Wallace JM, Drake RJ. Citrus vein enation. In: Citrus Virus Diseases. Berkeley: Univ. California Press 1959; pp. 163-5.

[31] Conselleria d'Agricultura, Pesca i Alimentacio. Normes per a la produccio integrada en citricos en l'ambit de la Comunitat Valenciana. Diari Oficial de la Comunitat Valenciana, 6266 (13-5-2010): 19661-81.

[32] Ortu S, Prota R. In: Cavalloro R, Prota R, Eds. Validità dei metodi di campionamento e delle relative soglie di intervento per il controllo dei principali fitofagi della arancicoltura. Standardisation de metodologies bio-techniques sur la lutte intégrée en agrumiculture. S. Giuliano, Corse 1980; pp. 35-52.

[33] Cavalloro R, Prota R. Integrated control in citrus orchards: sampling methodology and threshold for intervention against the principal phytophagous pests. Commission of the European Communities, Brussels-Luxembourg 1983.

[34] Ripollés JL, Marsà M, Martínez M. Desarrollo de un programa de control integrado de las plagas de los cítricos en las comarcas del Baix Ebre-Montsià. Levante Agrícola 1995; 34: 232-48.

[35] Melià A. Muestreo de poblaciones y actividad de vuelo de *Aphis frangulae gossypii* Glover (Homoptera, Aphididae) y otros pulgones sobre cítricos en Castellón. Bol San Veg Plagas 1995; 21: 601-10.

[36] Hermoso de Mendoza A, Belliure B, Carbonell E, Real V. Economic thresholds for *Aphis gossypii* (Hemiptera: Aphididae) on citrus clementina. J Econ Entomol 2001; 94: 439-44.

[37] Hermoso de Mendoza A, Arouni R, Belliure B, Carbonell EA, Pérez-Panadés J. Intervention thresholds for *Aphis spiraecola* (Hemiptera: Aphididae) on Citrus clementina. J Econ Entomol 2006; 99: 1273-9.

[38] Stern VM, Smith RF, Van Den Bosch R, Hagen KS. The integrated control concept. Hilgardia 1959; 29: 81-101.

[39] Higley LG, Wintersteen WK. In: Higley LG, Pedigo LP, Eds. Thresholds for Integrated Pest Management. Lincoln: University of Nebraska Press 1996; pp. 249-74.

[40] Higley LG, Pedigo LP. In: Higley LG, Pedigo LP, Eds. Thresholds for Integrated Pest Management. Lincoln: University of Nebraska Press 1996; pp. 9-21.

[41] Uygun N, Satar S. The current situation of citrus pests and their control methods in Turkey. Bull IOBC/WPRS 2008; 38: 2-9.

[42] Swirski E, Amitai S. Annotated list of aphids (Aphidoidea) in Israel. Israel J Entomol 1999; 33: 1-120.

[43] Satar S, Uygun N, Demirhan G, Karacaoğlu M. Potention of *Lysiphlebus confusus* Tremblay and Eady, *Lysphlebus fabarum* (Marshall) and *Lysiphlebus testaceipes* (Cresson) (Hymenoptera: Braconidae: Aphidiinae), parasitoids on *Aphis gossypii* (Glover) (Homoptera: Aphididae) in biological control. The Scientific and Technological Research Council of Turkey. Agriculture, Forestry and Veterinary Grant Group, Project No: 105-O-581, Adana 2009 (Final Report in Turkish).

CHAPTER 10

Endemic and Emerging Vector-Borne Mediterranean Citrus Diseases and their Epidemiological Consequences

Moshe Bar-Joseph[1] and Antonino Catara[2*]

[1]*Formerly, the S. Tolkowsky laboratory, ARO, the Volcani Center, Bet Dagan, Israel and* [2]*Department of Phytosanitary Sciences and Technologies, University of Catania, 100 via S. Sofia, Catania 95123 Italy*

Abstract: The chapter presents a brief historical account on citrus cultivation in the Mediterranean basin and on major epidemics that affected citriculture in the past. The citrus industries mostly survived these serious drawbacks by changing about 150 years ago their horticultural practices from non-grafted to grafting trees on the sour orange rootstock which produced a stionic combination highly resistant to fungal root-rots. Basing the industry on the sour orange eventually turned into the road map for the emergence of a subsequent major global pandemic caused by the aphid borne *Citrus tristeza virus* (CTV). This led growers to shift to new stionic combinations based on a variety of rootstocks that provided not only tolerance to the disease along with adaption to a variety of edaphic conditions. Adaptability is an ongoing process as recently experienced in certain parts of Brazil, where the Rangpur lime rootstock which for almost half a century has been successfully providing both drought and CTV tolerance is succumbing to a recent localized epidemic caused by citrus sudden death, a new vector borne disease caused by an infectious agent that is still not sufficiently characterized. Rootstock replacement is however not an universal remedy for all citrus mainly because some threatening diseases such as greening /huanglobing (HLB), *Citrus variegated chlorosis* (CVC), witches broom and stem pitting isolates of CTV damage directly the scion part and thus not helped by rootstock-replacement. The chapter focuses on three vector-borne pathogenic agents that cause three important citrus diseases, stubborn, CTV and citrus greening/HLB respectively representing three types of disease problems (i) endemic, (ii) emerging and (iii) threatening of concern for Mediterranean growers. Some possible ways to minimize the damaging impacts of these diseases on the future of citrus production in the Mediterranean Region are discussed and some measures validated by past experience with endemic disease are used to obtain an estimate on the potential damage of two disease problems threatening the Mediterranean citrus industries, the spread of *Toxoptera citricidus*, the most effective vector of CTV and the possible spread and emergence of greening/HLB epidemic in the area. Information on other vector born citrus diseases as citrus variegated chlorosis, citrus sudden death, citrus leprosis, Omani lime witches broom disease, in relation to their potential threats to Mediterranean citriculture is provided. Essentially the message is simple and follows the slogan "better safe than sorry" as a clear indication that the Mediterranean citrus industries must place high priorities both on national and regional certification and R&D programs to prevent the spread of these potentially devastating diseases.

Keywords: Citriculture, Mediterranean Region, citrus viruses, damage, control.

1. INTRODUCTION

The spring time smell of citrus blossom and gold colored fruits hanging on long rows of green citrus trees continue to be among the most recognized scenes of Mediterranean visitors. Although citrus is cultivated in the region for at least two millennia, it was mainly the Portuguese[1] introduction of sweet oranges that turned citrus in this region into such an important horticultural commodity. Ever since the fresh fruit markets of the Mediterranean and throughout Europe enjoy the winter harvests and delicacy of these highly tasty and flavoured fruits.

Interestingly, however, despite the long tradition of cultivation and popularity of citrus fruits within this region, as indicated by the association of the names of some the most famous citrus varieties with coastal places along the

***Address correspondence to Antonino Catara:** Science and Technology Park of Sicily, Z.I. Blocco Palma I, stradale V. Lancia 57 Catania 95121 Italy; Tel: +39095292390; E-mail: acatara@pstsicilia.org

[1] Indeed in many countries oranges became since then named as Portucales.

Vincenzo Vacante and Uri Gerson (Eds)

Mediterranean, *e.g.* oranges like the Valencia, Jaffa, Ovale Calabrese and Maltese and lemons like Lisbon and Femminello di Siracusa, the progenitors of most cultivated citrus produced in the area were originally domesticated elsewhere and imported from India and China and later also from Africa, Australia, and other secondary citrus growing countries (*e.g.*, the navel oranges, from California where these citrus varieties were adapted after importation from Bahia, Brazil).

Initial imports of citrus to the Mediterranean region were mainly through seed sources and only at later stages citrus was also introduced as budwood and as rooted plants. The rather common availability of the "apomyxis" trait in many *Citrus* spp., further contributed to the widely accepted practice of cultivating citrus trees from seed sources. This practice was abandoned only about 150 years ago, with the emergence of the root rot pandemic (to be discussed below).

Citrus cultivation involved two main changes compared with the traditional horticultural practices of this region. Firstly, unlike most of the native and traditional cultivated fruit crops such as olives, grapevines, pomegranates, figs and carobs that were mostly cultivated without irrigation (except at the first few year's following planting) and the continued productivity of matured groves was dependent solely on the Mediterranean climate winter rainfall. This natural water supply regime was however insufficient for cultivating citrus trees and citriculture in the Mediterranean region was always dependent on locating adequate amounts of high quality water and in considerable investments in irrigation schemes and arrangements to provide sufficient amount of water during the dry summer months. In addition, orange consumption as fresh fruits differed from the most common and important Mediterranean fruits trees listed above, the fruits of which were rarely consumed freshly and were mostly processed first for long storage, to be turned into subsistence products to support growers' needs in an area where rains are often erratic and water supply scarce. Thus the main products of traditional Mediterranean fruit tree growers were not consumed fresh but first turned into oil, wine and concentrated honey, like syrups made of grapevines, dates and carobs and as dried dates, raisins carobs, and figs.

Early interest in cultivation of citrus in the region most probably started with adoption of Etrog citrons as a religious ornament. Other early uses of citrus fruits, including sour oranges and lemons, were mainly as spices and as food additives. The introduction of the sweet and tasty oranges revolutionized and advanced fruit production in the Mediterranean areas, when during the last century citrus became among the most important cash crops of this region.

Naturally, the considerable expansion of citriculture in this region, as in other geo-graphic areas, was also associated with numerous problems, including some serious and devastating diseases. As a result research on a variety of local citrus plant protection problems had been intensified and certification and prevention schemes were successfully introduced to prevent the spread of damaging diseases. However, there is a need of continuous alertness because new disease problems emerge constantly and, with recent advances in transportation and commerce, the possibility of preventing their global spread becomes more and more difficult.

The present chapter deals mainly with three insect-borne citrus diseases that differ considerably in almost every aspect. They are caused by different groups of pathogenic agents, they occur in different geographical origins, they are differently dispersed by vectors and they also differ widely in their potential threats to citrus production throughout the Mediterranean basin. The specific nature of the components of the epidemiological triangles of each of these vector-borne diseases will be discussed and their significance will then be used for estimating their potential damage to citrus production throughout the Region, if left uncontrolled.

2. CITRUS CULTIVATION AND CITRUS VIRUS AND VIRUS LIKE DISEASES IN THE MEDITERRANEAN BASIN

The history of citrus cultivation and its special role in the cultures of ancient civilizations has been excellently recorded by Tolkowsky [1]. Etrog citrons (*Citrus medica*) were apparently the first *Citrus* sp. to reach the East Mediterranean shores. Probably brought over from Persia and Medea where they attracted the formidable appreciation of early famous Greek encyclopedians. Etrogs were adapted for at least two millennia as a sacramental fruit for the Jewish tabernacle holiday, eventually leading to its scattered cultivation throughout the different parts of the Mediterranean basin. Sour oranges (*C. aurantium*) and lemons (*C. lemon*) followed, while there is still an unsettled debate whether sweet oranges (*C. sinensis*) were present or commonly grown in the Region before the arrival of these highly favoured fruits to the shores of Portugal with the Vasco de Gama expedition. The considerable economic success of citrus cultivation and the excellent productivity of citrus trees and groves in areas far away from the center of these species' origin encouraged growers in Mediterranean countries with suitable climatic conditions and water resources to extend citrus cultivation continuously for the last two centuries.

Grafting, although known from antiquity, had little or any importance in the early stages of citrus cultivation in this Region and most trees and groves were maintained on their own roots. Early spread of *Citrus* spp. was mainly with fruits and seeds; hence several of the most important citrus disease agents commonly found in China, India and other Far East countries were not carried along in these seed sources. The absence of seed transmission of phloem limited pathogens, such as *Citrus tristeza virus* (CTV) and probably also of the dieback/greening/ huanglongbing agents, prevented the spread of these serious diseases along with plants produced from the introduced seed sources to new cultivation areas in the Mediterranean Region.

However, this only temporarily helped to escape from old diseases, because later increased interest in adapting new citrus varieties led explorers and rare-plant-hunters, along with eager collectors of unique plant species and varieties, to search and introduce not only seed sources, but also budwood infected with some dangerous plant disease agents for vegetative propagation.

Generally, citrus like other newly adopted crops, was often found to produce excellently, both in quality and yields in many areas distant to their centers of origin. However, as also experienced with some other newly imported crops, citrus plants often became exposed and succumbed to new pathogens endemically present in new cultivation areas. Thus, for example, the absence of both *Spiroplasma citri* and its efficient leafhopper vector *Circullifer tenellus* from the citrus growing countries in the Far East, where citrus had originated, and their presence on the Eastern Mediterranean natural vegetation, is a clear case where only after citrus cultivation began in some parts of the Mediterranean Region had the casual agent of the stubborn disease become adapted to become a citrus pathogen.

With the development of technologies that allow the shipment of rooted plants over long distances, were catastrophic damages experienced all through the Mediterranean shores by the root rot or crown or collar rot, or gummosis, caused by *Phytophthora* spp. Reported in 1832 in the Madeira Island, the root rot epidemic of citrus trees swept through the Mediterranean Region, killing all trees grown from seedlings and eventually turned into the road map of the tristeza pandemic. With the decline in the use of seedlings planted in groves, the sour orange plant which was found to be tolerant to *Phytophthora* infections was embraced as a rootstock for new citrus plantations, in the Mediterranean and also in other regions where citrus cultivation was then continuously expanding.

The first historical report of sour orange decline, which later became known as the tristeza disease, was apparently noticed in Australia and documented by the late Vivian Fraser. New plantings of oranges grafted on sour (Seville) orange were initially (1843 to 1851) doing well, however following several imports of rooted citrus trees from China, Japan and other Far Eastern regions, it was observed that the trees became sick. The thick dark green foliage of healthy groves became yellow by 1865 and the trees were dying. This led to the following declaration at a Fruit Growers Conference in Sydney [2] "*never use Seville orange stocks as they have proved a complete failure*". A similar catastrophic phenomenon was noticed about 70 years later all through South America's citrus groves. Alarming reports all indicated the emergence of serious decline and death of sweet orange trees grafted on sour orange rootstocks. As it often happens, the mysterious natures of these apparent incompatibility diseases were often associated with serious debates on their possible causes.

That tristeza epidemic was in many cases a "man-made" disease-problem was noted by the pioneering work of JM Wallace [4], which suggested that the South American CTV outbreak was started with the introduction of infected, but symptomless trees grafted on tolerant rootstocks from South Africa. Unintentional introduction of tristeza infected trees and budwood was also the practice in other areas including the Mediterranean.

During the late 1920s' and 1930s' citrus cultivation in the Mediterranean Region expanded considerably, and with expansion growers and citrus specialists often introduced budwood of exotic citrus varieties mostly from California, Florida and the Far East. The origins of some of these CTV infected varieties were imports from China (*e.g.* the Meyer lemon) and Japan (Kumquats). Unfortunately, although neither of these became of commercial significance, they were popular for home gardens, and as a result CTV became widely dispersed geographically. In addition, several Mediterranean countries had also imported thousands of tristeza-infected Satsuma plants from Japan when that the tristeza was still unnoticed there as a disease problem.

The year 1946 was a turning point in global tristeza research, with the first demonstration by Fawcett and Wallace of Quick Decline (QD) transmission from infected to healthy orange/sour orange trees by grafting [5] and the Meneghini [6] report of transmission with the Brown Citrus Aphid (BCA) *Toxoptera citricidus*. Hughes and Lister [7] shortly afterwards showed that the disease causing serious dieback of acid limes trees (*C. aurantifolia*) in Gold Coast (now Ghana) was transmissible both grafting and by the BCA and that infected lime-plant seedlings were showing conspicuous vein clearing symptoms.

It was subsequently found that budwood from plants with CTV symptoms in Brazil and Argentina, QD from California, stem pitting of grapefruit in South Africa and sour orange incompatibility problem in South Africa and Australia, were all responding with similar symptoms when graft inoculated on sour lime seedlings. The regular and fast (1-3 months or less) reaction of sour lime plants from different geographic localities to inoculations with a range of tristeza isolates made lime seedlings universal indicators of CTV and considerably advanced tristeza research. Using the new indicator it became clear that incidences of introduction of tristeza-infected plants were also the case throughout the Mediterranean citrus growing areas.

The most serious outbreak of a tristeza epidemic was noticed since the 1960s' in Spain, where it was estimated that several million trees grafted on sour orange had succumbed to the disease [8].

A different tristeza-infection scenario was noticed since the late 1960s' in Israel. The first outbreak of the epidemic involved the collapse of orange trees grafted on sour orange. To prevent the spread of the disease the local citrus industry made a serious eradication effort, which seemed to be effective for about fifteen years. However, by 1986 it was discontinued, mainly because it was realized that infection rates had reached a level that could not be controlled by the allocated resources, and by observations that the majority of the CTV infections detected were caused by tristeza isolates that were not harmful to the most sensitive citrus combination, sweet on sour orange trees.

Indeed, almost 25 years since these eradication efforts were given up, there were only a few minor CTV decline problems and growers prefer planting trees grafted on the sour orange rootstock. It should however be emphasized that the coexistence of tristeza with trees grafted on sour orange might not last, once the more efficient BCA vector will became established in the Mediterranean basin.

Other virus and virus-like pathogens of Mediterranean citrus. As mentioned, this chapter will deal mainly with three important vector-borne citrus diseases. However, there are several other citrus diseases of considerable importance for regional certification and sanitation needs, either with unknown vectors or with still uncharacterized etiological agents, listed below:

a. Viroids (CVds), such as citrus exocortis, cachexia and xyloporosis and a few CVds with less defined symptoms. It is important to note that viroids that infect citrus are neither known to be transmitted by vectors nor to be seed-borne in citrus plants. They are mechanically transmitted by grafting and on pruning tools, but the incidence of mechanical transmission in citrus groves is normally of little economic consequence. Means of preventing mechanical transmission are of considerable significance in budwood certification programs, where even if a single mother plant is infected, the diseased material will later be extensively propagated.

b. Viruses such as citrus psorosis, mostly affecting sweet oranges, mandarins and grapefruits, and crinkle leaf and infectious variegation, all quite common once on lemon trees.

c. Graft transmissible agents that remain to be characterized, such as concave gum, impietratura and cristacortis.

These diseases were rather common in old Mediterranean groves, yet had almost disappeared from new citrus plantings in the Region following the use of propagation material that was freed of these diseases, either by using budwood from nucellars or through pathogen elimination by shoot tip grafting.

d. Citrus yellow vein clearing, not a destructive virus disease, first described in lemon and sour orange, may affect sweet orange and other citrus and non-citrus hosts. Originally described in Pakistan [9] and associated with a filamentous virus, has later been reported in India and Turkey, where the vectoring of the virus by *Aphis craccivora* and *A. spiraecola* has been ascertained, and more recently in China.

e. *Citrus vein enation - woody gall virus* (CVE -WGV) is an aphid-borne virus affecting plants grafted on rough lemon or volkameriana lemon. It is widely distributed, but causes little economic damage.

f. Citrus leaf blotch is a disease that shows chlorotic blotching, stem pitting and vein clearing on indicator plants, and bud union crease in commercial orchards of different geographic areas (e.g Spain and Italy), affecting citrus varieties grafted on citrange or trifoliate rootstocks. The causal agent is the *Citrus leaf blotch virus* (CLBV), the type member of genus *Citrivirus*, family *Flexiviridae.* The virus is the only citrus virus which has been proved to be seed transmitted.

g. A new vector-borne disease, the virus-like citrus chlorotic dwarf, recently recognized in Turkey, is transmitted by the bayberry whitefly (*Parabemisia myricae*). Previously believed to be a crinkle leaf type disease (Korkmaz *et al.,* 1994), it causes crinkled and puckered areas in lemon, tangelo and orange trees. It recently became epidemic especially in Kutdiken lemon (a common local variety) and Minneola tangelo. The symptoms are similar to those induced by citrus infectious variegation: the infected plants show smaller leaf size. Olive latent virus-l has been isolated from CCD-infected Kutdiken lemons and Mexican limes.

Adapting to living with vector-borne viruses and virus-like disease agents. Vector-borne agents such tristeza, stubborn, greening and others cannot be eliminated once they become established in a citrus growing area by any of the currently used certification schemes. Hence there is a continuous need to develop vector spread preventive means and to locate new citrus sources that show either tolerance or resistance to the invading pathogenic agents. Biotechnological methods offer considerable potential to develop transgenic citrus with traits that will allow plants to produce well despite infections. However, as there are no proven reports of resistance to any of the vector-borne citrus pathogens, this subject will not be discussed in great detail in this chapter.

Significant progress was achieved mainly by practicing adaptability, allowing citrus growers to escape disease problems like CTV long before the viral nature of a pathogenic disease was realized. This was amply exemplified by the success of Japanese growers to produce citrus in CTV infested areas by cultivating Satsuma and other mandarins on trifoliate orange (*Poncirus trifoliata*), which not only offered tristeza but also cold tolerance. Australians and South African growers, realizing that trees on sour orange were declining soon after planting, were using new tolerant rough lemon (*C. jambhiri*) and mandarin rootstocks, which allowed their industries to produce excellent crops despite infections. Similarly, forced by the death of trees grafted on sour orange, Brazilians growers found the Rangpur lime (*C. limonia*) rootstock to be both tristeza and drought tolerant.

However, not all citrus diseases that are vector-borne can be managed by replacing the rootstock. The recent spread of citrus greening/HLB through South, Central and Northern America and the presence of diseased trees and vectors at rather short distances from the Mediterranean citrus growing areas should be considered as constant serious threats to citrus producers in the Region.

3. ENDEMIC, EMERGING AND THREATENING VECTOR BORNE CITRUS DISEASES OF THE MEDITERRANEAN REGION

Citrus stubborn spiroplasma

Historical Background. A disease named "little leaf" was first reported in Israel by Reichert and Perlberger [10]. The symptoms of diseased young trees included stunting and bushiness. The leaves were showing yellow discoloration and/or mottling which caused the authors to define the problem as a "yellows" disease. Interestingly, they considered little leaf to be a xeromorphic disease problem that resulted from the unusual severe hot spells (Chamsins) that occurred in the area shortly after planting.

A similar disease, named stubborn because of the recurrence of little leaf-like symptoms in diseased trees, despite being their top- grafted with budwood of normal healthy trees, was observed about two decades earlier in California. The Californian stubborn disease was considered to be caused by a virus until the 1960s', when the seminal finding of MLO in a number of yellows diseases was reported by the Japanese group of Doi *et al.* [11]. These developments prompted efforts by Californian and French researchers that led to the thin section Electron Microscopical (EM) finding of mollicutes in sieve tubes of stubborn-infected trees. The stubborn associated cells were cultured axenically and found both in culture and in thin sections to have a spiral like morphology, thus eventually named *Spiroplasma citri* [12]. The completion of Koch's postulates for *S. citri* clearly proved their pathogenic nature in citrus and other hosts [13, 14].

The optimum temperature for *S. citri* cell growth in culture is 32ºC and this preference for high temperature coincides with the *S. citri* pathogenicity in infected trees. Stubborn disease symptoms are indeed most frequently expressed under warm conditions, whereas even some highly susceptible hosts remain symptomless in cool weather.

Stubborn is rarely a lethal citrus disease, but it has a marked economic effect on productivity, especially when infection occurs soon after planting. Such trees are severely stunted, produce poorly, mainly unmarketable fruits. Fruits are often lopsided and do not attain the orange colour at the stylar-end as they mature. The seeds of deformed fruits are frequently brownish and aborted and the seed coats are often loaded with *S. citri.*

Typically infected leaves are cupped and unnaturally thick, with variable chlorotic patterns which at late summer and autumn show mottle and veinal chlorosis. These symptoms closely resemble those associated with greening/HLB disease and could also be confused with severe girdling effects.

Spiroplasma citri is a phloem-restricted pathogenic mollicute transmitted in a persistent and circulative manner by phloem feeding leafhoppers. Its cells lack a true cell wall, are culturable, motile, helically filamentous. Several plasmids were reported and cultured cells were used to raise antibodies that however have only limited value for detecting the pathogen in citrus, mainly because the concentration of *S. citri* cells, even in symptomatic citrus, is often below the threshold of detection by ELISA. Leafhopper vectors such as *Circulifer tenellus* contain high titres of the pathogen when tested individually by ELISA. It should however be noted that not all leafhoppers that carry *S. citri* are also able to transmit the pathogen to citrus plants.

Host range and geographical distribution. Sweet oranges, grapefruits and tangelos are highly sensitive to natural infections. Most mandarins and lemons will only rarely show symptoms, unless they are top grafted on infected rootstocks.

Spiroplasma citri is also naturally infecting periwinkle (*Catharanthus roseus*) and some wild and cultivated crucifers. Recently it was reported to infect and cause severe witches'-brooms of carrots. In addition several leguminous plants show severe stunting when infected experimentally. The disease infects citrus groves in Eastern Mediterranean countries (Turkey, Syria, Lebanon, Cyprus, Israel and Egypt) and also in Morocco, but apparently it is of little significance in Italy and Spain. Except for the Mediterranean, the disease is known as a serious problem only in California, where it probably became established following its introduction along with the *Circulifer* sp. leafhopper vector in shipments of beet roots carried by immigrants from the Mediterranean Region.

Management and control of stubborn infected trees. The disease remains an enigmatic control problem mainly because of its unusual air-borne spread by vectors that primarily dwell on wild plants. Furthermore, due to the rather erratic spread of the vector into young citrus groves, disease incidence fluctuates considerably, varying between years from almost 0 to >50% infections. The realization that stubborn is caused by a mollicute prompted considerable efforts of control, applying a variety of measures such as white wash sprays and preventive covering with nets. None of these were effective in preventing stubborn spread to citrus, mostly due to experimental difficulties with the erratic natural infection

Similarly, although *S. citri* in culture media are highly sensitive to antibiotics, such as tetracycline and its derivatives, symptom remission could be achieved by injection of antibiotics into infected trees. Antibiotic toxicity was observed in treated trees, probably suggesting that the failure of these treatments was not due to insufficient adsorption of these compounds (Raccah & Bar-Joseph, unpublished). In the absence of disease control the removal of symptomatic trees is recommended, especially when two or more infected trees cluster together. It is interesting to note that under certain conditions trees that were infected with stubborn recovered spontaneously after 12-14 years. However due the poor productivity of infected trees during the years before such recovery, the continued maintenance of poorly performing infected trees will not be justified economically.

Citrus tristeza closterovirus

Historical background. See above in the citrus history part.

CTV virions and defective RNAs. CTV is the largest known plant RNA virus with a single positive-stranded genome RNA (gRNA) of c. 19.3kb, bipolarly encapsidated by two different coat proteins a major (CP) and a minor (CPm) of 25 and 27 kDa, respectively. The CP encapsidates the major part of the virions, whereas about 3% of the 5' of gRNA is encapsidated by CPm. The sequences of the Florida isolates T36 the Israeli VT strains and several other isolates from the Mediterranean Region were determined. The CTV genome (Fig. **1**) encodes 12 ORFs, potentially translated and processed into at least 17 products, which include replication-associated proteins, the homologue of the HSP70 proteins, the two coat proteins, three RNA silencing suppressors and a few translational products with unknown functions. The CTV-infected plants contain a nested set of 3'-coterminal single- and double stranded (ds) subgenomic RNAs (sgRNAs) and two different sizes of single stranded 5' sgRNAs. In addition a large number of different size and different architecture defective (d)-RNAs mostly composed of the two termini of the CTV genome are found in plants infected with different CTV isolates.

The reactions of citrus to CTV infections. Under natural conditions CTV infections are mostly limited to *Citrus* sp. in which the result could range from negligible to catastro-phic, depending on the host genotype, the stionic combination and severity of the prevailing CTV isolate(s). The most dramatic effect is quick decline of trees grafted on sour orange. However, dieback of sour lime and the stem pitting decline of grapefruits and oranges also often result in considerable economic damage. In addition infection of sour orange, lemon and grapefruit seedlings with certain aggressive CTV isolates will cause a severe Seedling Yellows (SY) reaction. However, in most cases these SY symptoms are only observed on experimentally grafted young plants. Occasionally SY symptoms will be manifested when certain rootstocks infected with CTV-SY isolates are top grafted lemon or grapefruit buds.

The virus infects experimentally several citrus relatives such as *Aeglopsis*, *Afraegle*, *Fortunella* and *Pamburus* and also some *Passiflora* spp. Molecular studies on CTV has greatly benefited with the finding that CTV infects *Nicotiana benthamiana* protoplasts [15].

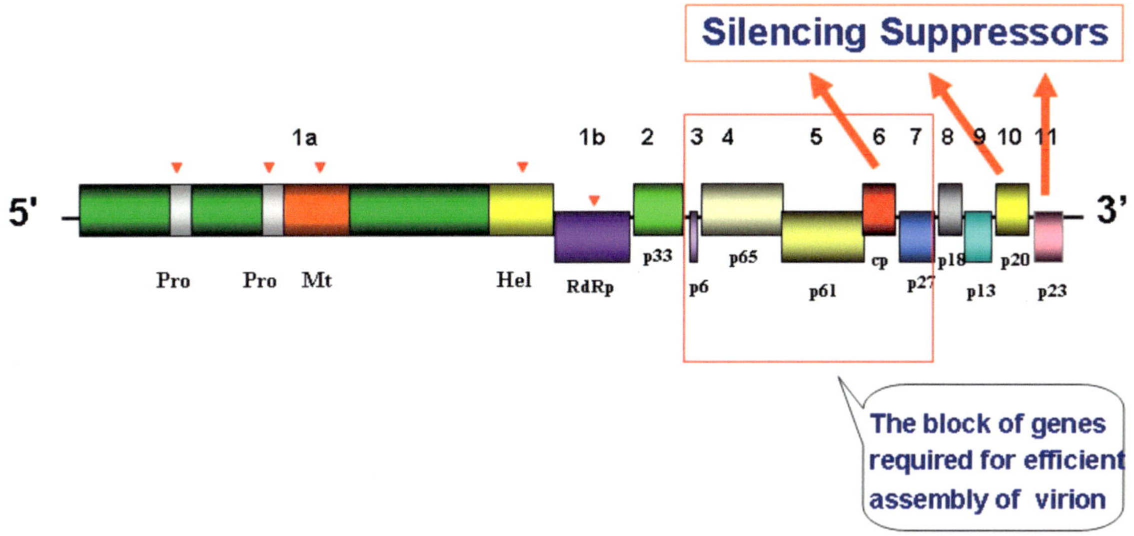

Fig. (1). CTV genome organization and gene functions.

The vectors that transmit CTV. The disease is naturally transmitted in a semi-persistent manner by several aphids, the most efficient of which is *Toxoptera citricidus*. The absence of this aphid from the Mediterranean Region was probably among the main reasons for the rather slow spread and damage of CTV to the Mediterranean citrus production. However, as a warning notice it should be noted that *T. citricidus* was recently reported in some parts of Portugal. This unfortunate new situation needs to be given the most serious attention by the Mediterranean citrus growers, especially for those that continue planting on the sour orange rootstocks. The melon or cotton aphid (*Aphis gossypii*) is less common on citrus and is less effective in transmitting the disease, as compared with BCA. However, certain CTV strains are highly transmissible by this aphid and incidences of natural infections that reached 80-90% within about five years were noticed (M Bar-Joseph unpublished). Other aphids, including *A. citricola* and *T. aurantii*, are far less efficient vectors than *A. gossypii* and despite their abundance on Mediterranean citrus trees; their contribution to natural CTV spread in the area seems to be limited. This is exemplified by observations during 1971-1973, shortly after *A. spiraecola* became widely spread in Israel. The incidences of the natural spread of CTV when this aphid reached explosive populations was actually low as compared with later years, when the populations of *A. spiraecola* were considerably reduced (M. Bar-Joseph, unpublished).

Transmission rates by *A gossypii* are strain- dependent [16]; the VT isolate reached >40% transmission compared to only 5-6% transmissibility of other isolates. Raccah *et al.* [17] showed that CTV isolates in a single plant could be segregated by *A. gossypii* into a number of individual isolates with different transmission rates. Interestingly, the Meyer lemon isolate was found almost non-transmissible by *A. gossypii*.

A long lag period, often of 30 years, was observed between imports of CTV-infected propagation material from areas where it is primarily vectored by BCA, and the appearance of natural transmission by *A. gossypii* in areas where this aphid is absent [18]. Moreno *et al.* [19] monitored the spread of CTV in a heavily infested citrus area in Spain and reported that after four years 98.7% of Newhall navel trees were infected, whereas after eight years only 17.8% of Marsh grapefruit trees became diseased. Similar variations in infections rates of these two citrus hosts were also observed in Israel, where Bar-Joseph and Nitzany [20] described an unusually destructive CTV isolate that was killing sweet orange and Minneola trees very rapidly and yet had only affected few of the similar-age start Ruby grapefruits.

The status of CTV in Italy was reviewed by Davino and Catara [21]. The disease is now however far more spread then at that time and considerable efforts are being exerted for its containment [22].

The most common detection methods of CTV infected trees. The outcome of CTV infections will depend mostly on the genotypes of the host and CTV isolates. Trees grafted on sour orange when infected with severe isolates will collapse within a few years post infection, usually during late autumn. The collapse event is mostly sudden as indicated by the fact that fruits on such trees develop normally up to the defoliation stage and will remain on trees, but will not reach proper maturation. The collapse is often followed by partial recovery of a short upward spring growth. However, trees remain unproductive and often die within the next few years. Some CTV isolates, spread by *T. citricidus* in the Far East, South Africa and Brazil, cause stem pitting and stunting of grapefruits and sweet oranges. Stem Pitting (SP) is associated with decline, small fruit sizes and with considerably reduced production. It is thus considered to be the most serious CTV problem, especially since, unlike the QD problem; it is uncontrollable by changing the rootstocks.

In the Mediterranean area the presence of SP isolates has been reported in Spain and in other countries.

For most practical needs, CTV is both conveniently and reliably diagnosed by enzyme immunoassay (ELISA). This technique allows the handling of large amounts of samples in a short time, a critical issue for budwood certification needs. Other molecular technologies, *e.g.* Northern hybridization with non radioactive cDNA or cRNA probes, SSCP and different PCR procedures are generally either more sensitive than ELISA, or allow strain specific identification. However, defining the biological characteristics of a new CTV isolate, based on these molecular tools, needs further elucidation.

For strain specific diagnostic purposes, there were a number of attempts to use immunoassay and molecular techniques. However biological characterization on a battery of indicator plants remains the only reliable technique. Three main groups of citrus hosts were commonly used including sour limes, which show vein clearing symptoms by most CTV isolates; sour orange, lemon and grapefruit seedlings for the SY reaction and grapefruit and sweet orange seedlings for detecting SP. The vein clearing symptoms of CTV infections will appear within a short period

(1-2 months), SY on sour orange is manifested within less than three months, whereas SP in orange and grapefruit could require up to a year to determine (Fig. **2**).

Fig. (2). Symptoms of tristeza decline on Valencia sweet orange grafted on sour (top) and stem pitting on sweet orange (left) and grapefruit (right)

Management and control of CTV infections. For long periods CTV management has worldwide been initially based on eradicating the virus by the elimination of infected trees. However the results achieved in Israel, California, Spain and other countries have convinced researchers of the need to change this strategy to the selective removal of severe, economically important, CTV strains.

The main measure applied to control an emerging CTV epidemic is by using tristeza-tolerant rootstocks, mostly hybrids of trifoliate, as citranges (*P. trifoliata* x *C. sinensis*), citrumelo (*P. trifoliata* x *C. grandis*). But there are two main difficulties with the practical approach of replacing the sour orange rootstock. The first has to do with the excellent adaptability of sour orange rootstock to calcareous and high pH soils. That is most evident when comparing this adaptability with trifoliate and trifoliate hybrids, which show chlorosis and poor performance in heavy water-logged and calcareous soils that are most common in many of the new plantings in the Mediterranean Region. And second, because most Mediterranean citrus fruits are aimed at fresh fruit markets, problems with the quality of fresh fruit from tristeza-tolerant rootstocks, like rough lemon and volkameriana lemon (*C. volkameriana*), prevents Mediterranean growers from turning their backs to the sour orange rootstocks.

Will this situation remain unchanged? Probably only up to the stage that the more efficient *T. citricidus* will show up in the Region. This is a highly possible scenario which, as indicated above, remains a serious threat for Mediterranean citriculture.

In areas where SP isolates were damaging sweet oranges (Brazil) or grapefruits (South Africa), the SP problem was successfully controlled by using mild strain cross protection, a phenomenon in which a primary viral infection prevents a secondary infection with the same or a closely related virus strain. Although almost 80 years have passed since the cross protection phenomenon was first noticed, its molecular basis remains to be elucidated. Thus, for example, tests in Brazil to locate mild CTV isolates demonstrated that protection is only conferred between certain isolates, while many other mild CTV isolates did not offer protection [26]. Hybridization studies carried out in the late eighties with two CTV isolates, VT and HT, which could be differentiated by using specific hybridization probes [27], suggested that neither of these CTVs isolates prevented the infection of the other.

A recent paper from Dawson's laboratory [25], using a GFP-labelled CTV clone, elegantly demonstrated that cross protection occurred only between isolates of the same strain and not between isolates of different strains. When isolates of the same strain were used for sequential plant inoculation, the primary infection provided complete exclusion of the challenge isolate, whereas isolates from heterogonous strains appeared to have no effect on replication, movement or systemic infection by the challenge virus.

Much research is being conducted in many laboratories in order to develop new protection strategies against tristeza decline, intended to reduce damage in those areas where calcareous or saline soils cause poor performance of decline-tolerant rootstocks.

Citrus greening disease/ huanglongbing (HLB)

Management and control of HLB infections. Citrus greening or huanglongbing (HLB) is considered as one of the most destructive diseases of citrus. Although the disease is still absent from the Mediterranean basin, its recent spread and economic damages to citrus industries in South (Brazil, 2004) Central (Cuba and Mexico, 2004) and North (Florida, 2005) America warrant considerable attention by those involved in the Mediterranean citrus industries to this seriously threatening disease. It is naturally spread by citrus psyllids and is associated with low titre phloem-limited infections of three closely related bacteria that so far have frustrated all attempts of their axenic culturing. Comprehensive reviews have been published [27].

The complete circular *Candidatus Liberibacter asiaticus* genome from an infected psylla [28] and from the phloem of an infected citrus tree was elucidated, and the resulting genomic sequences were found to be essentially identical. Phylogenically *Ca. L. asiaticus* (CLas) belongs to the family Rhizobiaceae, order Rhizobiales, class Alphaprotobacteria (Fig. **4**). Its cells have an average 36.5% GC content and gene annotation revealed that 4.5% of the bacterial genes are involved in cell motility, 8.0% in active transport that is probably also contributing to their virulence. *Ca. L. asiaticus* appears to have a limited ability for aerobic respiration and is likely auxotrophic for at least five amino acids. CLas has all the genes of a type I secretion system that is needed for both multidrug efflux and toxin effector secretion. They lack, like other intracellular pathogens, the type III and type IV secretion systems and the extracellular degradative enzymes, found in typical free-living or plant colonizing bacteria [28].

The citrus plants sensitive to greening/HLB infections. The disease is restricted to *Citrus* spp. and to some related plants: *Severinia buxifolia*, *Limonia acidissima Vepris lanceolata*, and to the ornamental orange jasmine (*Murraya paniculata*). Other non-citrus hosts, such as *Catharantus roseus*, could be infected by dodder transmission. Symptoms include yellow shoots, twig dieback, leaf drop, and leaves with blotchy yellow/green coloration similar to the symptoms of stubborn and of zinc nutritional deficiency. Veins are often enlarged and corky, with excessive fruit drop, small and misshapen fruits that remain green at the stylar-end after maturity (Fig. **3**). Fruits are deformed with aborted seeds, necrotic vascular bundles in the fruit pith center, and the fruit juice is mostly of poor quality. In some cases silver spots are left on the fruit skin after pressing with a finger. However, this symptom although simple to be followed should not be considered as universal for all incidences of greening /HLB infections. As with stubborn, the symptoms of greening /HLB are far more pronounced during late summer to autumn, and new young flush will often not show symptoms. The time from infection to the appearance of symptoms is variable, fluctuating during the time of year, environmental conditions, tree age, host species, cultivar and horticultural conditions. Generally symptoms on younger trees are observed faster and easier.

Fig. (3). Greening/HLB symptoms: poorly coloured and lopsided fruits (normal on the right) and blotchy mottle - affected leaves.

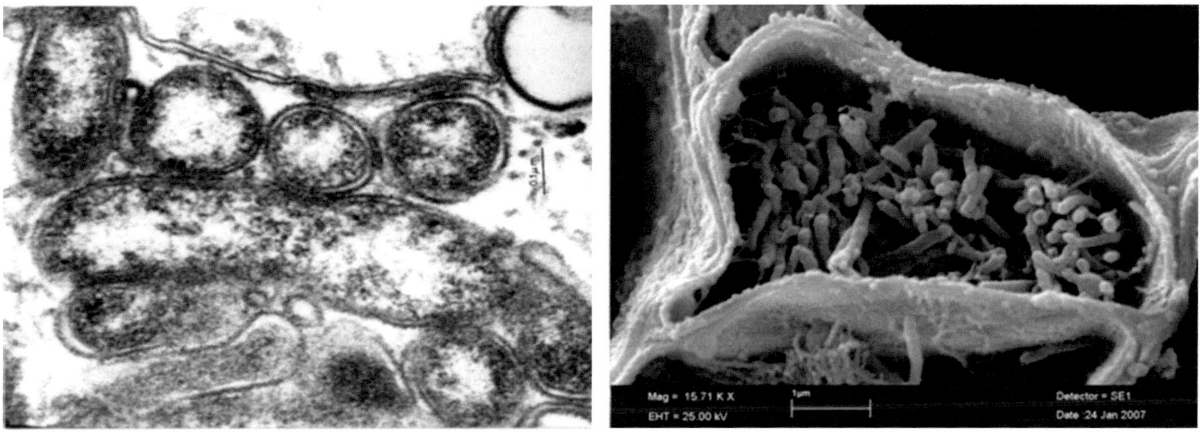

Fig. (4). Cells of *Candidatus Liberibacter*, causal agent of HLB in sieve tubes (courtesy of E.W. Kitajima).

So far no distinguishable differences could be observed among the three different *Ca. Liberibacter* spp. (*Ca. L. asiaticus, africanus,* and *americanus*).

Vectors and epidemiology of greening/HLB infections. The CLas and CLam pathogens are transmitted by the citrus psyllid *Diaphorina citri*. CLaf, transmitted by a different psyllid species, *Trioza erytreae*, is found at elevations greater than 700m and is less heat tolerant than CLas; CLam resembles CLaf. Psyllids will acquire the bacterium during their nymphal stages and transmission continues throughout their lives. Eggs that hatch in 2-4 days are laid on newly emerging leaves and the five nymphal instars of *D. citri* complete their development in 11-15 days. The entire life cycle takes 15-47 days, depending on temperatures. Adult *D. citri* may live several months and during their lifetime each female will be laying up to 800-1200 eggs.

Viruliferous *D. citri* appear to disperse over long distances. Recent epidemiological analyses of infected groves in Florida demonstrated higher incidences of HLB infections at the edges of citrus groves. Edge effect could have been the result of an invading vector that has a high tendency to land on the first-met citrus tree in a plantation, or alternately edge effect could result from the preference of psyllid vectors to land initially in a spacey environment. However, unlike in the case of stubborn epidemiology where leafhoppers will not spread the disease from infected to non infected mature trees, greening/HLB will spread continuously once *D. citri* populations become established, either in young or in mature orchards.

4. EPIDEMIOLOGICAL TRIANGLES OF STUBBORN, TRISTEZA AND GREENING/HLB, AND THE ESTIMATED PRESSURES OF THESE DISEASES ON CITRUS SURVIVAL IN THE MEDITERRANEAN REGION

Disease incidence is the outcome of interactions amongst the main factors that define the epidemiology triangle: the host plant, the vector, and the environmental conditions.

The following discussion summarizes some of the main differences between the three citrus diseases and provides a rough estimate of past, present and future impacts, of the stubborn, CTV and greening/HLB diseases, respectively, on the Mediterranean citriculture.

Temperature and pathogen populations in infected trees. High titres of CTV are mostly at moderate temperature of ca. 25°C, isolates with higher temperature optima are also known. *Spiroplasma citri* cultures show a rather strict optimal growth at 32°C, whereas the greening /HLB bacteria differ considerably, with CLaf restricted to low temperatures, the CLas to far higher temperatures.

Clustering and edge effect. The clustering of CTV is common both for *T. citricidus* and *A. gossypii* infections. There is an apparent clustering of stubborn infected trees; this is however mostly the result of the unfavourable nature of citrus hosts to the feeding and breeding of the stubborn leafhopper-vectors. The observed clustering thus differs from the common tree-to-tree transmission, and it occurs because leafhopper carrying the pathogen propagatively will not settle on the first plant visited, but will continue and inoculate other nearby plants. A definite edge effect was noted for greening /HLB infections in Florida, with far higher infection rates at the edges of mature citrus groves.

Tree-to-tree spread. There are considerable differences between the three pathogens as to secondary spread from the initially infected trees. Tree-to-tree spread of CTV with *A. gossypii* and with *T. citricidus* was frequently noticed, indicating the continuity of infection along the row in an orchard where the disease is spread by *A. gossypii*. A similar situation is expected for greening /HLB, where citrus psyllids feeding and breeding on young growth will regularly transmit the *Ca. Liberibacter* sp. from infected to non infected tree parts and/or from infected to non-infected trees. In contrast the situation is different for stubborn, because the leafhopper vectors do not breed on citrus and infections thus take place with inoculative vectors that had acquired the pathogen mostly from wild plants. That is probably why most of infections are restricted to young plantations with apparently no secondary spread of the disease from infected citrus trees. The question however remains why infections are not observed on trees located on the spacey edges of mature groves.

Sectorial infection. CTV virions, like the agents of stubborn and greening/HLB, are phloem restricted. There are however considerable differences in the systemic nature of these pathogens. Depending on stionic combination, CTV will either spread systemically very efficiently, whereas in other combinations and also with some highly virulent isolates trees decline prior to the virus systematization throughout the canopy of infected trees. The common explanation of this phenomena, which is expected to have beneficial effects as it will limit, on one hand, disease spread to neighbouring trees while also imposing considerable diagnostic problems for routine CTV detection is as follows. Aphid infection of the external canopy of a sweet on sour orange by such isolates does not lead to horizontal spread of the virus, which instead first spreads basipetally, toward the rootstock, where it evokes a hypersensitive reaction leading to rapid collapse of this sensitive stionic combination.

Greening (HLB) is also often limited to few tree parts; secondary infections will however eventually lead to the systemic spread of the pathogen throughout most tree parts.

Systemic spread of stubborn pathogen depends on the timing of infection; young plants will show rather even distribution, whereas other trees infected at some later stage will only become infected on one or more branches. Sectorial infection of stubborn could be stable for long periods, but may be manifested once the trees are severely pruned, thus causing the systemic spread of *S. citri* to other none-infected parts. Similar situations could also occur when stems of partially stubborn infected trees are top grafted.

Symptomless infection. This possibility is rather common for CTV infections mainly because many citrus species or stionic combinations will be tolerant and not show apparent symptoms of the disease, despite the multiplication and accumulation of the virus in infected tissues.

Stubborn infections will normally lead to clear symptoms in sensitive citrus plants; reports from some cooler areas suggest that the disease may be present in a symptomless phase in trees growing in certain Mediterranean areas. It is interesting to note that typical stubborn symptoms, even in the East Mediterranean Region, are mainly manifested in leaves that develop at high temperatures and are almost absent from the early cooler season flushes during spring.

The Greening /HLB agent is causing severe symptoms at temperature conditions similar to stubborn, and most citrus varieties will eventually display typical disease symptoms.

Diagnosis based on visual inspections. For many practical purposes early and rapid diagnosis of infections of either of these epidemic diseases is of critical importance. It is rather easy to diagnose CTV isolates by inducing SP or detecting CTV on filed samples of sour lime or citrus macrophylla leaves; the majority of the citrus genotypes are however tolerant to CTV infections and the diagnosis must therefore be based on detecting the pathogenic agent rather than on visual inspections.

The situation is different in stubborn cases, and diseased trees are conveniently recognized by expert eyes during the late summer and autumn months.

The methods and means of detecting greening (HLB) will depend mostly on the stage of disease prevalence. In regions with established infections, visual inspection is a convenient mean of early diagnosis. It can be conducted either from the ground or with crews on elevated platforms that are run through the groves. Although the detection rates of trained personnel are high, some PCR and preferably qPCR backing of the visual inspection should be followed.

Possibilities of accurate diagnosis using laboratory techniques. The detection of CTV by ELISA is probably the most common and most feasible technology for routine CTV testing. The highly conserved CTV coat protein allows the use of a single antibody as a useful reagent for CTV detection by ELISA. Strain specific monoclonal Abs are commercially available, however their use is probably only needed to validate infections by certain severe isolates and from certain geographical areas, such as Florida, where distinct differences were found among the common severe and other less severe CTV isolates.

Other methods, including Northern hybridization and various PCR applications could also be used for CTV detection, but their practical application is limited mainly because ELISA is sufficiently effective for diagnostic purposes and can be carried out on a much wider scale, for most practical certification needs.

Stubborn diagnosis can be done by PCR, using ribosomal spacer specific primers. However the uneven distribution of the pathogen in citrus trees turns even the most sensitive molecular techniques useful only for validating the presence of *S. citri* in citrus tissues showing suspected symptoms.

The situation is rather similar for greening /HLB detection by PCR or by qPCR.

Detection in vectors. The vectors of CTV transmit the disease semi-persistently and only harbour it for about a day following feeding on infected plants. The leafhopper and psyllids vectors of stubborn and greening/HLB could harbour the specific pathogenic agents *S. citri* and *Liberibacter*, respectively. ELISA assays were highly effective in detecting *S. citri* in leafhoppers and PCR could provide important information on the presence of *Liberibacter* in the psyllid vectors. It should be remembered that *S. citri* could occur in leafhoppers that are not natural vectors of this pathogenic agent. Some leafhoppers can thus probably acquire the agent, but are lack the ability for its transmission. In conclusion, it is apparent that only the detection of *Ca. Liberobacter* in psyllids provides an important tool for preventing the greening disease.

Estimated disease pressures. Our calculations of estimated Disease Pressure (DP) are based on arbitrary values given to the main components of the epidemiological triangles the Plant (P), Pathogen (Pat) and the Vector (V) (**Fig. 5**).

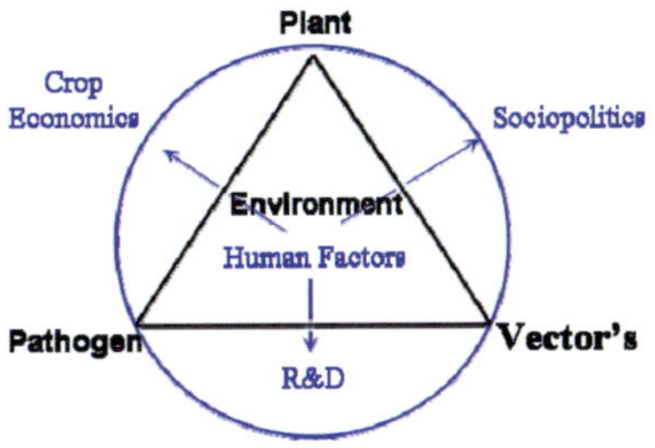

Fig. (5). The estimated Disease Pressures (DP) of three citrus pathogens, based on differences of their epidemiological triangles.

Stubborn. The DP of stubborn is calculated by P x Pat x V, estimated as follows:

a. P=0.25, since only some of the citrus varieties are sensitive, and the sensitivity of the others is limited to young trees.

b. Pat= 0.5 the pathogen shows limited systemic infection and it is active only at a rather limited temperature range.

c. V= 0.25 because of the considerable large variation both in seasonal and especially annual presence of the vector in young citrus groves.

The estimated DP value of stubborn is about 0.031 (3.1%).

Citrus tristeza virus. Because the DP of CTV is dependent on the transmission efficiency of the vectors, the triangle values may be variable, as follows:

a. P=0.5 Because only some stionic combinations are sensitive.

b. Pat=0.25 to 0.75. Because of the considerable variation in the aggressiveness of CTV isolates towards sensitive hosts and combinations.

c. V=0.25 to 0.75. Due to differences in vector abundance and in transmission efficiency between *A. gossypii* and *T. citricidus*.

The DP of CTV ranges considerably from just 3.1 to 9.3% for *A. gossypii* and from 9.3 to 28% for *T. citricidus*, respectively

Greening/HLB. The following calculations are mainly based on CLas infections that are vectored by *D. citri*:

a. P= 0.5. Most citrus trees are sensitive, but there are major differences in the intensity of systemic spread and its effect on productivity between young and mature citrus groves.

b. Pat = 0.5.

c. V = 0.5 to 1.0. The two different values for V reflect expected changes in vector populations. V is expected to be high soon after the vector will invade the Mediterranean Region and will most probably slow down at some later stage, due the build-up of local and imported natural enemies. In addition, it is expected that the breeding grounds on citrus trees in the Mediterranean zone, with long dry summers and cold winters will be less favourable for *D. citri* increase, as compared with faster tree growth and longer flushing periods in the hot and wet climates of Asia and the Americas.

Therefore, the DP of greening HLB transmitted by *Diaphorina citri* is expected to range between 12.5 to 56.2 %.

The actual disease losses and potential threats. Although calculations of disease impacts are difficult to asses for future problems, in the above calculations we included two diseases, stubborn, which is endemic in Israel and CTV that is naturally transmitted by *A. gossypii*. Unfortunately, there are no recent official figures about trees damaged by either of these diseases. However, it is the common experience of one of the authors (MB-J) and of several citrus extension specialists, that the prevalence of stubborn and CTV diseased trees in Israeli citrus groves, where these diseases have been spread naturally for the last 80 and 25 years, respectively, closely fit the DP estimates.

These confirmatory results point to the serious threats posed to the Mediterranean citrus production by the entrance and spread of *T. citricidus* with a DP of 38%, almost six times higher than the present spread by *A. gossypii*, and with an even far higher potential for an epidemic (DP up to 75%) soon after the region will experience the emergence of a *D. citri*-vectored greening/HLB epidemic.

5. OTHER VECTOR-BORNE CITRUS DISEASES AND THEIR POTENTIAL THREATS TO MEDITERRANEAN CITRICULTURE

Citrus variegated chlorosis (CVC)

This disease has been causing considerable damage in Brazil since 1987. The causal agent is *Xylella fastidiosa*, a xylem-limited bacterium (Fig. **7**), spread by at least 11 species of sharpshooter leafhoppers that feed on xylem vessels. CVC causes important diseases of several plants beside citrus, including grape (Pierce's disease), almond (Almond leaf scorch) and oleander (Oleander leaf scorch).

Most citrus trees are highly sensitive to CVC infections, but disease severity in lemons, limes and some mandarins is limited and Rangpur lime and pummelo (*C. grandis*) are tolerant. Symptoms include severe chlorosis between the veins, flowering out of season and the infected trees are less vigorous. Symptoms become more apparent with disease progress, often starting as sectorial and eventually spreading to the whole tree (Fig. **6**). Although the genome of this pathogen has been known for almost a decade, there is still almost no cure, and the only way of handling the disease is by eliminating inoculum sources. Field diagnosis of CVC is difficult because symptoms are often confused with those associated with other diseases and deficiencies. Molecular tests based on PCR are available and should be included in the Mediterranean certification schemes.

Fig. (6). Citrus variegated chlorosis symptoms on leaves and fruits (courtesy of E.W. Kitajima).

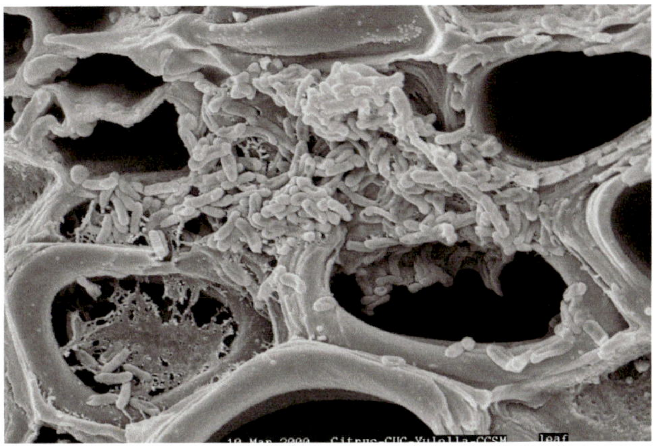

Fig. (7). Scanning electron microscopy of citrus vessels with *Xylella fastidiosa* (courtesy of E.W. Kitajima).

Citrus sudden death. A disease of rather recent prevalence in certain parts of Brazil (Minas Gerais and Sao Paulo States), it affects mainly trees grafted on Rangpur lime and, less severely, on volkameriana and rough lemons. The causal agent is still not completely proved; early suggestions included a specific isolate of CTV, whereas others associated the disease with a *Marafivirus,* named *Citrus sudden death associated virus* (CSDaV). Symptoms include pale green discoloration of the tree canopy, followed by decline and tree death. The most typical symptom is observed in the Rangpur lime and volkameriana rootstocks, which express a yellow stain; this symptom however is normally observed in declining trees and it is difficult to reproduce on indicators. The disease can be controlled by inarching trees prior to their natural infection or even in early stages with tolerant rootstocks, like the Cleopatra mandarin. No vector of the disease agent has so far been reported.

Interestingly, the spatial distribution of CSD resembles the epidemiological patterns observed for CTV; this had lead to theories on a possible association between their pathogenic agents. The recent wide use of volkameriana rootstocks in certain Mediterranean areas makes CSD a possible threat. However, with few exceptions, there is a great difference between the climates of the regions where the disease has so far spread, namely Brazil and the Mediterranean.

Citrus leprosis virus. A highly important citrus virus disease transmitted by several species of *Brevipalpus* mites in a persistent circulative manner. The most important is *B. phoenicis*, widely spread in many countries, including in the Mediterranean Region. The disease is causing chlorotic localized lesions which turn to black spots on leaves, fruits, and twigs of orange and mandarins (Fig. **8**) and also premature leaf and fruit drop.

Fig. (8). Symptoms of leprosis on citrus leaves, fruits and twigs (courtesy of E.W. Kitajima).

Two different types of leprosis viruses exist: the most common (>99% of incidence) occurs in the cytoplasm, whereas the nuclear type is almost rare (Fig. **9**).

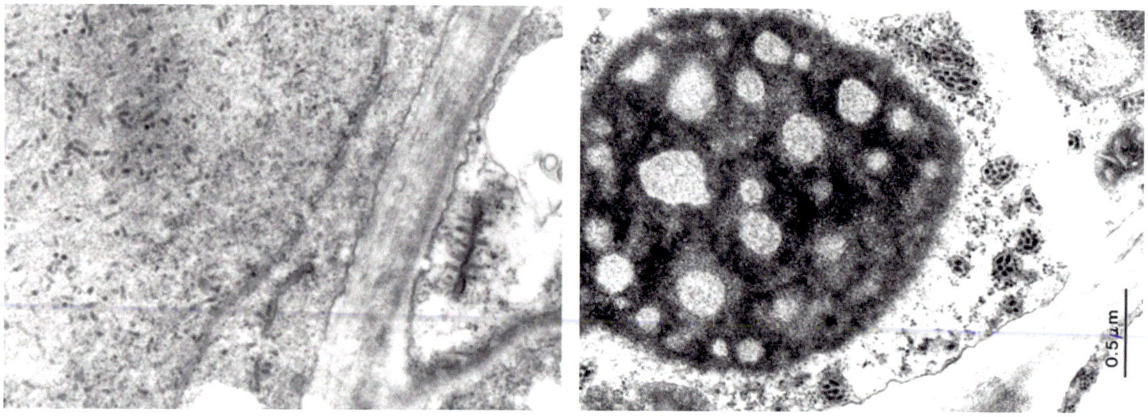

Fig. (9). Bacilliform-like virions and viroplasm of Citrus leprosis virus-C in parenchymal cells of sweet orange (courtesy of E.W. Kitajima).

The disease is widely spread in Brazil and other areas of South America, where control is based mainly on acaricidal treatments, and on pruning the affected parts to reduce the inoculum sources. The high cost of acaricides treatments makes the disease even more economically important.

Omani lime witches broom disease. A serious witches' broom disease affecting acid limes, sweet lime and citron trees was initially reported from Oman and recently from the United Arab Emirates [29], Iran and India. Symptoms include witches' brooms, proliferating twigs with small, pale leaves followed by dieback and tree death in about 3-5 years. In graft transmission tests disease symptoms are intensively manifested at high temperatures. The disease was graft transmitted to Meyer lemon and several rootstocks, including rough lemon, Rangpur lime, *P. trifoliata*, Troyer citrange (*P. trifoliata* x *C. sinensis*), *C. macrophylla*, *C. ichangensis*, *C. hystrix* and *C. karna*.

Most commercial sweet orange, grapefruit or mandarins and pummelo cultivars are not affected. Early electron microscopical observations associated a phytoplasma with diseased trees and also with the leafhopper (*Hishimonus phycitis*) that is common on citrus in Oman, UAE, and Iran; however the vector potential needs to be experimentally confirmed. Molecular characterization of the ribosomal spacer genes placed *Candidatus phytoplasma aurantifolia* into the 16SrII group, which contains several Phytoplasma diseases affecting a wide range of annual and other cultivated plants from Oman, Pakistan, Africa and Central America. Other Phytoplasmas were recently associated with trees showing greening/HLB-like symptoms in Brazil and the Far East.

Citrus tatter leaf virus (CTLV). Citrus tatter leaf is an economically important disease that causes bud-union in trees grafted on trifoliate or hybrids rootstocks. The causal agent is a mechanically transmitted virus (*Citrus tatter leaf virus*, CTLV), which is also transmitted by seeds.

The virus is closely related to *Apple stem grooving virus* and a *Lily latent virus*. While natural spread in citrus remains unconfirmed, its wide distribution in many Asian countries makes this virus a matter of concern, because most current citrus rootstocks are genetically vulnerable to it.

Finally, we have to mention Satsuma dwarf, spread by an unknown soil-borne vector, and a severe form of psorosis (Citrus necrotic ringspot), which is spreading in Argentina.

6. CONCLUSIONS

The forecasting and assessment of the epidemiological consequences of emerging plant diseases, both as to their expected prevalence and especially for their economical impacts, remains a rather difficult task, often leading to rather erratic conclusions. Indeed, summarizing some forecasts made in the past, prior or even during the early stages of recently occurring epidemics of citrus diseasees, makes one wonder how do most trees, groves and industries continue being productive after such gloomy forecasts. Furthermore, summarizing the losses often attributed to each of the less dramatic disease and pest problems of Citrus, or of other crop plants, one obtains total losses of a couple of hundreds percent, which again are not in accordance with the reality of continuous increases in citrus productivity in many areas. In the present chapter we made a rather naïve attempt to analyse the expected losses of the three vector-borne citrus diseases dealt with in this chapter.

Considering rather approximate estimates as values of the three main epidemiological triangles, components of the DP values that could also be correlated with expected crop loss for infections, were stubborn (3.1%), CTV (3.1 to 9.3% or 9.3 to 28%, according to the prevalent vector) and greening/HLB (12.5% for *D.citri* and 56.2% for *T. erythreae*).

The outcome of *T. citricidus* vector-spread CTV and the greening epidemics fortunately remain unknown. The prevalence of Stubborn disease, which has been spreading without any control measures in the Eastern Mediterranean for at least 80 years, and the CTV epidemics and losses that were experienced during the last 50 and 40 years in Spain and Israel, respectively, seem to be in close accordance with the DP values calculated for these diseases.

The question remains, will the dangerous greening/HLB disease spread into the Mediterranean. And if it does, what will the outcome be of such a doom epidemic in countries where most of production is aimed for the fresh fruit market, where wholesalers are already imposing numerous restrictions on pesticide use, which is probably at this stage the only effective means of controlling the rapid spread of this dreadful pathogen by its effective vector.

The question of how to deal with the dangerous prospect of the emergence of a greening/HLB epidemic in the Mediterranean citrus growing areas remains a challenging problem for everyone concerned with this economically important industry. Tables **1** and **2** lists some of the essential steps to be taken prior and during the early stages of such a threatening situation. There are two main points to remember; the first that solving the problem is apparently a difficult task for any of the Mediterranean producers, and yet some locally important varieties may eventually be neglected once the disease emerges. An alternative measure could be to develop resistant varieties by transgenic means, once the disease will turn epidemic in the Mediterranean area.

Thus what remains is buying time. The recent unprecedented mobilization of the US citrus industry and academy to develop new technologies to control the disease will most probably provide, within ten years or less, some possible remedies for controlling the disease. Meanwhile it is up to the Mediterranean citrus community to make every effort to prevent and or delay the emergence of this disease, before the remedy turns into a reality.

Table 1. Guidance for possible action steps to be taken prior and following the emergence of greening/HLB epidemic

Phase 1- Prior to the disease detection

1. It is a devastating disease, different from most other problems of citrus.
2. Do not panic.
3. Do not give up easily.
4. Familiarize with the disease in advance.
5. Establish a dedicated greening diagnostic laboratory, with protocols that should be run bi-weekly or monthly, using mock samples with proper controls.
6. Establish & Follow - natural traps on a geographical grid of citrus psyllids. These should be regularly scouted by 20-50 people present in citrus groves.
7. Advance a program of transgenic R& D to allow adapting novel technologies to your own particular citrus varieties.
8. Learn the lessons of past eradication efforts and do not over-rely on committees as problem solvers; all you need are just a few dedicated professionals.
9. Start a carefully balanced public relation program, providing the growers, the public and especially the Agriculture and Finance Ministry officials with an in-depth understanding of the significance of this serious disease problem.
10. Do not over-do this, because that could work against you when/if the problem will really turn catastrophic.

Phase 2- Once the disease was detected

1. Do not panic.
2. Stop immediately pruning and topping.
3. Immediately map the geographical distribution of infections.
4. Consider the prospect of eradication only if the incidence is extremely low!
5. Introduce differential control programs for young and mature groves.
6. Consider differential control programs for small holders and large groves, especially those that are in massive blocks of non-urban areas.
7. Start an immediate program of psylla control at the regional level.
8. Remove Murraya ornamentals and force, by law, the removal of abandoned and non-profitable citrus groves, especially when close to the major groves.
9. Consider an embargo on new planting for at least the first year, to allow and help nurseries to maintain plants for longer periods under controlled conditions.
10. Intensify R&D.

REFERENCES

[1] Tolkowsky, Hesperides S. A history of the culture and use of citrus fruits. John Bales, Sons & Curnow; Ltd. 1938.

[2] Bitancourt AA. A podridão das radicelas dos citrus na provincial de Corrientes, Argentina. O Biológico 1940; 6: 356-364.

[3] Zeman V. Una enfermedad nueva en los naranjales de Corrientes. Physis 1931; 19: 410-11.

[4] Wallace JM. Tristeza disease of citrus, with special reference to its situation in the United States. FAO Plant Protection Bulletin 1956; 4: 77-87.

[5] Fawcett HS, Wallace JM. Evidence of virus nature of citrus quick decline. California Citrograph 1946; 32: 88-9.

[6] Meneghini M. Sobre a naturaleza e trans- missibilidade da doença "Tristeza" dos Citrus. O Biologico 1946; 12: 285-7.

[7] Hughes WA, Lister CA. Lime disease in the Gold Coast. Nature 2010; 164: 880.

[8] Moreno P, Ambros S., Albiach-Martì MR, *et al.* Citrus tristeza virus: a pathogen that changed the course of the citrus industry. Mol. Plant. Path 2008; 9 (2): 251-68.

[9] Catara A, Azzaro A, Davino M, Polizzi G. Yellow vein clearing of Lemon in Pakistan. In Moreno P, da Graca JV, Timmer LW, Ed. Proceedings of the 12th Conference of the International Organization of Citrus Virologists 1992; Riverside, California; International Organization of Citrus Virologists, pp. 364-7.

[10] Reichert I, Perlberger J. Little leaf disease of citrus trees and its causes. Hadar 1931; 4: 193-4.

[11] Doi Y, Teranaka M, Yora K, Asuyama H. Mycoplasma or P.L.T. group-like microorganisms found in the phloem elements of plant infected with mulberry dwarf, potato witches'-broom, aster yellows or paulownia witches'-broom. Annals of Phytopathological Society of Japan 1967; 33: 259-66.

[12] Saglio P, Lhospital M, Lafleche D, *et al. Spiroplasma citri* gen. and sp.n: a mycoplasma-like organism associated with "stubborn" disease of citrus. Int J Syst Bacterial 1974; 23: 191-04.

[13] Daniels MJ. Mechanisms of spiroplasma pathogenicity. Annu Rev Phytopathol 1983; 21: 29-43.

[14] Markham PG, Townsend R, Bar-Joseph M, *et al.* Spiroplasmas are the causal agents of citrus little leaf disease. Ann Appl Biol 1974; 78: 49-57.

[15] Satyanarayana T, Gowda S, Boyko VP, *et al.* An engineered closterovirus RNA replicon and analysis of heterologous terminal sequences for replication. Proc Natl Acad Sci USA 1999; 96: 7433-38.

[16] Bar-Joseph M. Loebenstein G. Effect of strain, source plant and temperature on the transmissibility of Citrus tristeza virus by the melon aphid. Phytopathology 1973; 63: 716-20.

[17] Raccah B, Loebenstein G, Bar-Joseph M, *et al.* Transmission of citrus tristeza virus by aphids prevalent on citrus, and operation of the tristeza suppression programme in Israel. In Calavan EC, Ed. Proceedings of the 7th Conference International Organization Citrus Virologists 1975, Riverside, University of California, 1991, pp 47-9.

[18] Bar-Joseph M. A possible cause for natural spread of citrus tristeza virus after a prolonged lag in Israel. Phytopathology 1978; 68: 1110-11.

[19] Moreno P, Piquer J, Pina JA, *et al.* Spread of Citrus Tristeza virus in a heavily infested citrus area in Spain. In Timmer LW, Garnsey SM, Navarro L, Ed. Proceedings of the 10th Conference of the International Organization of Citrus Virologists, 1986; Riverside, California; International Organization of Citrus Virologists, pp. 71-6.

[20] Bar-Joseph M, Nitzany Y The spread and distribution of Citrus tristeza virus isolates in sour orange seedlings. In: Brlansky RH, Lee RF, Timmer LW, Proc. 11th Conf. International Organization of Citrus Virologists. 1989; Riverside, California; International Organization of Citrus Virologists, pp. 162-5.

[21] Davino M, Catara A. La tristezza degli agrumi. Informatore Fitopatologico. 1986; 36: 9-18.

[22] Catara A, Barbagallo S, Saponari M. Il caso "tristeza" degli agrumi. Globalizzazione e difesa delle colture. I Georgofili, quaderni 2007, Firenze.

[23] Costa AS, Muller GW. Tristeza control by cross protection: A U.S.-Brasil cooperative success. Plant Dis 1980; 64: 538-541.

[24] Rosner A, Lee RF, Bar-Joseph M. Differential hybridization with cloned cDNA sequences for detecting a specific isolate of Citrus tristeza virus. Phytopathology 76: 820-824. Plant Pathol 1986; 42: 35.

[25] Folimonova SY, Robertson CJ, Shilts T *et al.* Infection with strains of Citrus tristeza virus does not exclude superinfection by other strains of the virus. J Virol 2010; 84: 1314-25.

[26] Bovè JM. Huanglongbing: A destructive, newly-emerging, century-old disease of citrus. Plant Pathol J 2006; 88: 7-37.

[27] Committee on the Strategic Planning for the Florida Citrus Industry: Addressing Citrus Greening Disease (Huanglongbing). National Research Council 2010. http://www.nap.edu/catalog/12880.html.

[28] Duan Y, Zhou L, Hall DG *et al.* Complete Genome sequence of citrus huanglongbing bacterium, '*Candidatus* Liberibacter asiaticus' obtained through metagenomics. Mol Plant Microbe In 2009; 22: 1011-20.

[29] Garnier M, Zreik L, Bovè JM. Witches' broom, a lethal mycoplasmal disease of lime trees in the Sultanate of Oman and the United. Arab Emirates. Plant Dis 1991; 75: 546-51.

CHAPTER 11

Aleyrodidae

Yael Argov[1], Nedim Uygun[2], Francesco Porcelli[3*] and Hüseyin Baspinar[4]

[1]Israel Cohen Institute for Biological Control, Plant Production and Marketing Board, Citrus Division, POB 80 Bet Dagan, 50250, Israel; [2]Department of Plant Protection, Çukurova University, Agricultural Faculty, 01100 Adana, Turkey; [3]Dipartimento di Biologia e Chimica Agroforestale ed Ambientale (DiBCA), Sez. Entomologia e Zoologia, Facoltà di Agraria, Università degli Studi di Bari, Via Amendola 165/a, 70126 Bari, Italy and [4]Department of Plant Protection, Adnan Menderes University, Agricultural Faculty, 09100 Aydın, Turkey

Abstract: The adults of the Aleyrodidae (whiteflies) are small winged insects usually yellow-white in colour; some species bear grey marks on the wings or are darker, even brown-black. Immatures may be misidentified as aphids or scale insects, but the "vasiform orifice" will help in separating whiteflies from other groups in slide mounted and red-stained puparia. All instars secrete wax, in the shape of powder, curls, threads or as vitreous layers. They are called "whiteflies" due to their general whitish appearance. They are mostly bisexual, but several species or "strains" show both arrhenotokous and thelytokous parthenogenesis, usually in relation to insemination. The post-embryonic development is neometabolic, consisting of four larval instars, pupae (immotile and not-feeding) and adults. Whiteflies share with other Sternorrhyncha the piercing-sucking mouthparts and the specialized "filter chamber" mid gut. They suck plant sap, killing their host plants in heavy infestations and excreting abundant liquid faeces as honeydew drops. These drops cover infested plants, which then blacken because of colonization by sooty mold fungi. The blackened plants or products are untradeable due to the sooty mold. Whiteflies may also cause physiological changes and transmit viruses. In the Mediterranean Region whiteflies live mostly on woody perennial plants. Natural or classical biological control programmes have successfully controlled current or potential pest species by the introduction of effective natural enemies. *Aleurocanthus spiniferus* Quaintance is currently of major concern to citriculture in the Mediterranean Region because it was recently introduced and no indigenous natural enemies seem to control it. As in the past, unwanted, introduced whiteflies may become key pests of citrus in the Mediterranean Region.

Keywords: Citriculture, Mediterranean Region, citrus whiteflies, bio-ecology, damage, control.

1. INTRODUCTION

The family Aleyrodidae (Hemiptera) is a taxon of nearly 161 genera placed in about 1,556 species [1]. The adults of both sexes are winged, covered with waxy powder and resemble tiny moths. Because of this they are also known as "whiteflies".

Aleyrodidae are small insects, but species up to 10 mm long occur in tropical areas [2]. Immature instars are mostly oval or elongated-oval but some species may be sub-circular, elongated or nearly cordate [3]. The whitefly life cycle consists of an egg, 4 larval instars, pupa and adult. They are neometabolic, having an immotile and unfeeding pupal stage.

Eggs have a short sub-terminal stalk [4] inserted by the ovipositor into the tissue of the host plant. The first instar larvae (crawlers) are minute, with relatively long legs and antennae. They move on the plant surface till they find a suitable site for feeding. Once settled, they remain sessile until reaching the adult stage.

The pupal stage is immotile, encased in the "pupal case" or "puparium" [5], actually the shed cuticle of the fourth stage. The identification of Aleyrodidae is based on the characters of puparium [6, 7], and the "vasiform orifice" helps in discriminating whiteflies from other Hemiptera. Aleyrodidae secrete wax in the shape of powder, curls, threads or vitreous layers by more or less obvious epithelial glands that open in simple or compound pores, mainly on exposed body surfaces.

*Address correspondence to Francesco Porcelli:** Dipartimento di Biologia e Chimica Agroforestale ed Ambientale (DiBCA), Sez. Entomologia e Zoologia, Facoltà di Agraria, Università degli Studi di Bari, Via Amendola, 165/a, 70126 Bari, Italy; Tel: 0039-080 5442880; E-mail: porcelli@agr.uniba.it

Whiteflies damage host plants by sucking phloem sap; heavy infestations can kill the plant. Moreover, due to their specialized "filter chamber" in the mid gut, they excrete large amount of honeydew. This is subsequently colonized by sooty mold fungi, which hinder the development of the host plant. Whiteflies also induce irregular ripening and others physiological changes and can transmit more than 100 viral diseases [8].

In Europe and in the Mediterranean Region the whiteflies fauna consists of 57 native or naturalized species in 26 genera [9, 10], mostly living on woody perennials [4, 5].

Six species of whiteflies are major pests of citrus in the Mediterranean Region, whereas three others occasionally feed on citrus. *Aleurodicus dispersus* Russell is not yet present in the continental area, having been found only in Madeira and the Canary Islands.

2. FIELD KEY TO CITRUS ALEYRODIDAE

1. Puparia black, bordered by white wax, adults bright red; wings grey-bluish with white markings .... ***Aleurocanthus spiniferus***

-. Puparia not black, adult yellowish-white 2

2. Waxy secretion abundant in curls, segments and threads..3

-. Waxy secretion scarce, obscure 6

3. Puparia hidden by a copious woolly wax thick felt on the undersides of leaves; dense, sticky, large and confluent honeydew drops ....***Aleurothrixus floccosus***

-. Puparia not thus hidden even in dense populations..4

4. Puparia oval and prominent on the leaf undersurface because of a waxy marginal palisade...... ....***Trialeurodes vaporariorum***

-. Puparia without a waxy marginal palisade 5

5. Puparia bearing long arched threads and curled wax ribbons, eggs on long spiralling tracks all around adults and nymphs..***Aleurodicus dispersus***

-. Puparia wax-bordered, wax threads cover and are around the nymphs; eggs laid in circles, adults nest in their centre, among the egg-circles..***Paraleyrodes minei***

6. Puparia visible on the underside of leaves 7

-. Puparia almost flattened and transparent, scarcely visible on the underside of leaves. Anterior spiracular and anal furrows are evident in living puparia..***Dialeurodes citri***

7. Puparia white to yellowish, surrounded by a translucent and faint wax fringe ....***Parabemisia mirycae***

-. Puparia spindle-like, not surrounded by a translucent and faint wax fringe, flattened on the underside of leaves, dorsally convex 8

8. Puparia less than 1 mm long, anal setae visible..***Bemisia tabaci***

-. Puparia more than 1 mm long, anal setae not visible..***Bemisia afer***

3. MICROSCOPIC KEY TO CITRUS ALEYRODIDAE

1. Compound pores present..**Aleurodicinae**...2

-. Compound pores absent..**Aleyrodinae**..3

2. Five pairs of compound pores on the dorsum: one on the cephalic area, the others on the abdomen and all are of the same size .... ***Aleurodicus dispersus***

-. Seven pairs of compound pores, one cephalic and six abdominal. Abdominal pores of the first two pairs are smaller than the others....... ....***Paraleyrodes minei***

3. Dorsum with many (> 40) long and acute glandular spines. ....***Aleurocanthus spiniferus***

-. Dorsal disc without spines.......... 4

4. Lingula head distinctly lobulate, vasiform orifice triangular with a lobulate lingual head ... .***Trialeurodes vaporariorum***

-. Lingula head not lobulate.......... 5

5. Vasiform orifice triangular.......... 6

-. Vasiform orifice usually circular or round...8

6. Lingula with two blunt lateral tubercles, 26 short marginal setae and six anterior, posterior and caudal. Simple pores scattered on dorsum. Anterior spiracular furrows are barely visible and caudal furrow is slightly, longitudinally ridged.. ....***Parabemisia mirycae***

-. Lingula without tubercles................ 7

7. Lingula hairy because of microtrichia. Ventral abdominal setae are almost as long as vasiform orifice.. ....***Bemisia tabaci***

-. Ventral abdominal setae are considerably shorter than vasiform orifice.......................... . ***Bemisia afer***

8. Wax pores at the base of the marginal teeth resemble a row of double teeth...................... ....***Aleurothrixus floccosus***

-. No double teeth row, round vasiform orifice with indented inner posterior margin. Anterior spiracular and anal furrows evident. The puparium with one pair of dorsal papillae on prothorax, on metathorax and on the 2nd abdominal segment. Many wax pores scattered all over the pupal case dorsum.... .... ***Dialeurodes citri***

4. CITRUS ALEYRODID SPECIES

4.1. *Aleurocanthus spiniferus* Quaintance

Aleurocanthus spiniferus (Orange Spiny Whitefly, OSW) was recently found in Europe. It was collected in the Apulia Region (South Italy), where it has become acclimatized and is spreading [10]. To date the insect is widespread not only in citrus groves but also in fruit orchards, vineyards and in private gardens, on weeds and ornamental plants [11].

Diagnostic characters: Puparium: ovate, shiny-black, bordered by a fringe of white wax secretions. The female puparium is about 1.25 mm long, while that of males is about 1 mm long (Fig. **1**). OSW adults are distinctive in being bright red with metallic grey-blue wings, which cover most of the body. Light markings on the wings appear to form a across band. The eyes are reddish-brown and the antennae and legs are white with pale-yellow markings. Adult males are smaller than females (Fig. **2**). Many (>40) long and acute glandular spines project from the dorsum. Corbett [12] showed that specimens collected from *Rosa* sp. possess six pairs of submarginal spines posterior to the transverse suture, whereas specimens from *Citrus* spp. bear seven pairs, the fourth and fifth pairs being close together.

Fig. (1). Puparia of OSW on the underside of an orange leaf.

OSW identification requires slide mounting of puparia (Fig. **3**) and reliable knowledge, because of its similarity with *Aleurocanthus woglumi* Ashby, another major citrus pest not yet introduced into the EPPO area [13].

Fig. (2). *Aleurocanthus spiniferus* Quaintance Adult female.

Life history: Adults aggregate in groups on the underside of leaves to mate and lay eggs, which are fixed by a short pedicel in an arched line, consisting of 10-20 elements (Fig. **4**). Adult aggregation may be spotted by observing young leaves against the sun even from a considerable distance. Each female may lay 35 to >100 eggs, which are initially yellowish, becoming dark to brown-black as the embryo develops. Immature leaves are preferred for egg laying; infestation thus starts from the new twigs in spring. Active, black, flattened crawlers emerge and soon insert their mouthparts into the leaves to suck phloem sap. After moulting, their legs atrophy and the nymphs remain fixed to the underside of leaves by their ventral surface. Immature stages often form dense aggregations, of up to several hundred individuals, on a single leaf. Adults fly around briefly if disturbed, landing everywhere and thus dispersing passively. Dispersion may take place even by passing cars or by adhering to humans. The life cycle generally takes 2-4 months, with three to four annual overlapping generations, the number being mainly climate-dependent. OSW development slows down considerably during winter. *Citrus* spp. are the main hosts of economic importance, but in Italy the insect also infests *Vitis* cv, *Parthenocissus tricuspidata* (Sieb. and Zucc.) Planchon, *Hedera helix* Linnaeus, *Fatsia* sp., *Diospyros kaki* Linnaeus, *Laurus nobilis* Linnaeus, *Malva* sp., *Ficus carica* Linnaeus, *Morus alba* Linnaeus, *Punica granatum* Linnaeus, *Hybiscus rosa-sinensis* Linnaeus cv, *Prunus armeniaca* Linnaeus, *Cydonia* cv, *Malus* cv, *Pyrus pyraster* Burgsd., *Pyrus* cv, *Rosa* cv, *Pyracantha coccinea* M.J. Roemer, Plum cv, and *Eryobotrya japonica* (Thunb.) Lindley.

Economic importance: OSW is a well-known pest of fruit trees, mainly of citrus. In Italy *A. spiniferus* excretes copious amounts of sugary honeydew, which coat the leaf and fruit surfaces. Sooty mold fungi then quickly develop and form thick black layers that can be peeled off the leaves. Plant blackening reduces respiration and photosynthesis, resulting in low quality, small and permanently blackish products unsuitable for commerce (Figs **5-7**). This pest is therefore of much concern to citrus growers in the Mediterranean Region today. Other plants, like apples, are heavily infested by OSW, which can cause their death. Ornamentals like *Pyracantha* or *Parthenocissus* may be ruined due to blackening. The pest is spreading on wild plants, which are now its main refuge in Italy. EPPO recently moved the OSW to the A2 list (List of Pests Recommended for Regulation as Quarantine Pests). It is also a quarantine pest for NAPPO and threatens citrus production throughout the Mediterranean Region.

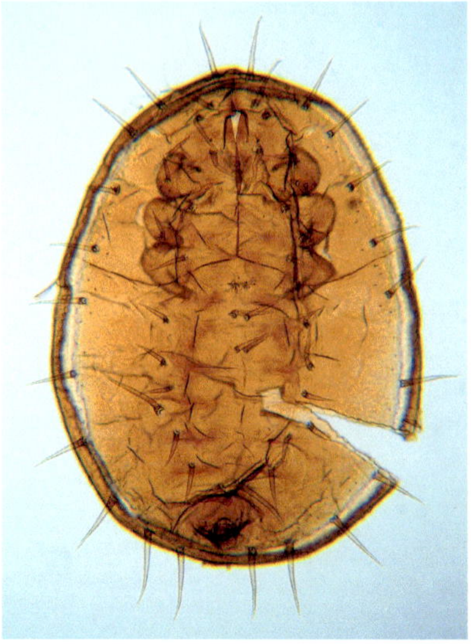

Fig. (3). A slide-mounted puparium of *Aleurocanthus spiniferus* Quaintance. Note the long dorsal glandular spines.

Management: Chemical control is expensive and ineffective, and infestations build-up considerably in spring-summer, starting from neglected orchards and the wild flora. Nor is chemical control an option in citrus orchards, because it disrupts the effective classical biological control of *Aleurothrixus floccosus,* with consequent pest resurgence and pesticide resistance. Infested plants weaken and in order to dispose of blackened plant parts, growers prune the orchards severely, burning the infested cut branches. This measure however results in much re-sprouting of twigs and thus in new pest infestations. As a result the orchard can be severely damaged and even destroyed by the pest. Low impact insecticides and fungicides, like soap salts and oils, could be useful in controlling the crawlers and reducing the sooty mold, until effective classical biological control will be established.

Fig. (4). A reproductive aggregation of OSW adults underside immature orange leaves.

Pesticides should be applied when the pest populations are at a very low level and may be sampled on young, fully extended leaves, due to the oviposition preferences of the pest. A Hymenopteran natural enemy complex was applied against *Aleurocanthus spiniferus* by Silvestri [14], and has proven to be effective in different parts of the world [15 - 17]. *Encarsia smithi* (Silvestri) and *Amitus hesperidum* Silvestri have been successfully used to control the pest in Japan and in Guam [17] and in Ponape, Federated States of Micronesia [18]. The introduction of such useful parasitoids in a IPM program seems to be the only effective measure to control *A. spiniferus.*

4.2. *Aleurodicus dispersus* Russell

The spiralling whitefly (SW) is not yet recorded in continental Europe and Mediterranean countries, but it occurs in the Canary Islands [19 - 21], and more recently became established in Madeira [9] where it is a major pest.

Fig. (5). Sooty mold on OSW infested mandarin leaves.

Fig. (6). Orange blackened by sooty mold due to OSW honeydew.

Both places are politically linked with Mediterranean countries, so the pest is discussed herein due of its quarantine importance.

Fig. (7). Stained, unmarketable fruits from OSW infested plants.

Diagnostic characters: SW adults are almost uniformly white, about 2 mm long. The fourth instar nymph is colourless or yellowish. It secretes wax in long arched threads by its multilocular pores and curled ribbons by the dorsal and marginal pores. The shape of these secretions is singular but ephemeral and may be obscured in dense, overcrowded populations. SW secretes an impressive amount of wax and a thick layer of white matter covers leaves under the surface of infested plants (Fig. **8**). On slides the puparium is 1 mm long and 0.75 mm wide; five pairs of compound pores are on the dorsum: one on the cephalic area, the others on the abdomen [22].

Life history: Each female lays about 21 eggs along a semicircular track marked by white wax, hence the common name of the species (Fig. **9**). Each egg is fixed at right angles to the leaf veins. The eggs hatch after 7-10 days and the crawlers settle near the eggshell, thus maintaining the spiral pattern.

Fig. (8). SW infesting a *Ficus reticulata* (?) urban ornamental tree in Las Palmas de Gran Canaria (photo of May, 2003).

Few larvae move to the leaf margin. Total development time is 22-47 days and adults live about two weeks [23].

Fig. (9). The typical egg laying pattern of the spiraling whitefly on the underside of a leaf of *Strelitzia nicolai* Regel and K. Koch.

Populations of SW may decrease following heavy rains and low temperatures but the pest resurges in warmer and drier weather. Adults suffer at temperature over 35°C and there is much mortality of the young above 40°C. The low temperature limit for SW is around 10°C [24]. This whitefly is highly polyphagous, being common on many different plant families of shrubs and trees, but *Citrus* spp. are among the less susceptible economic host plants known [25].

Economic importance: *Aleurodicus dispersus* damages plants directly by sucking sap, and indirectly by the wax, by the excretion of honeydew and due to the resulting sooty mold. As with *Aleurocanthus spiniferus*, recently introduced and established species are a problem because of their sudden economic impact as they enter a new biogeographical area. Although at this time *A. dispersus* presents only a moderate phytosanitary risk to the tropical and subtropical areas at the limits of its distribution [9], its introduction nevertheless represents a considerable economic risk for citrus and fruit producers in the Mediterranean Region.

Management: The avoidance of unwanted pests is beyond the scope of this chapter. Whiteflies are quite small and cryptic organisms that can pass even the most diligent quarantine inspection services.

However, a careful check of the citrus orchard and the exclusive use of official or certified propagation material are good starting points for a less risky citrus orchard management. Ornamental plants also play a role in the dispersion of such polyphagous pests like SW or *A. spiniferus*, as there are ornamental or garden citrus species that can host these whitefly pests.

Finally we emphasize the role of classical biological control in whitefly IPM, even though some natural enemies are not sufficiently effective in reducing pest populations to below their economic threshold levels. Nevertheless, natural enemies greatly contribute to whitefly IPM.

4.3. *Aleurothrixus floccosus* Maskell

Diagnostic characters: The Woolly Whitefly (WW) pupal case is yellowish, with brown to black individuals being scattered in the population, and the case is usually hidden by copious amounts of greyish white wax threads. The underside of infested leaves is layered by thick woolly wax with dense, sticky, large and confluent honeydew drops (Fig. **10**). The pupal case is less than 1.0 mm long to 0.6 mm wide. The dorsum of the pupal case bears metathoracic setae, 8th abdominal setae, and stout caudal setae. The vasiform orifice is nearly circular, almost filled by the operculum (Fig. **11**). Wax pores occur at the base of the marginal teeth that resemble a double teeth row (Fig. **12**). Adult woolly whiteflies are small, flying insects similar to small white-winged moths, with a yellowish-white abdomen and a light dusting of white powder.

Life history: Each female may lay up to 200 eggs that are inserted into the undersides of fully expanded leaves by a stalk. Eggs are small (<0.3 mm in length), kidney shaped, and are placed in circle or semicircle. The egg laying site is distinctive for the species because of the light dusting of wax that surrounds the egg circles. The eggs hatch in 4-12 days. WW require about 21 day to complete its life cycle and has five to seven broods per year [26-28], with a moderate summer decline. Population outbreaks of *Aleurothrixus floccosus* occur from late May to November, depending on the climate and on host plant sprouting. The pest overwinters mainly as third and fourth instars or as eggs.

Economic importance: WW sucks phloem sap and excretes large amount of dense honeydew droplets that are trapped by the woolly wax layer, thus constituting a thick sticky felt (Fig. **6**). Black sooty mold fungi soon colonize this felt, causing leaves to wilt and drop in heavy infestations, which can result in the blackening of entire trees. Moreover, the viscous honeydew collects dust particles, thus further reducing photosynthesis and gas exchange and resulting in decreased fruit size. Although some of this contamination may be washed off at packaging, harvesting is slowed in infested orchards and workers hesitate or refuse to enter heavily infested orchards. This species was one of the most important citrus pests in the Mediterranean Region until it was controlled by two small Hymenopteran parasitoids, namely *Amitus spiniferus* (Brethes) (Platygastridae) and *Cales noacki* Howard (Aphelinidae).

Fig. (10). A moderate infestation of the woolly whitefly on the underside of a citrus leaf. Note the large confluent drops of thick honeydew.

Management: Chemical control of *A. floccosus* is mostly ineffective, whereas biological control is the best and most effective means of controlling this pest (Fig. **13**). There are several natural enemies that attack the immature stages of WW. Apart from parasitoids, which have proven to be extremely effective and are the key antagonists of the pest, useful arthropods include the mite *Tydeus* spp., lacewings, coccinellid beetles, nabids, *Orius*, and six-spotted thrips. Unfortunately parasitoids are less heat tolerant than WW, which results in local pest outbreaks in late summer. This engenders pesticide applications that eliminate the natural enemies, resulting in pest resurgences. Outbreaks of WW can be contained by the release of suitable parasitoids, by "soft" insecticides and by tree pruning.

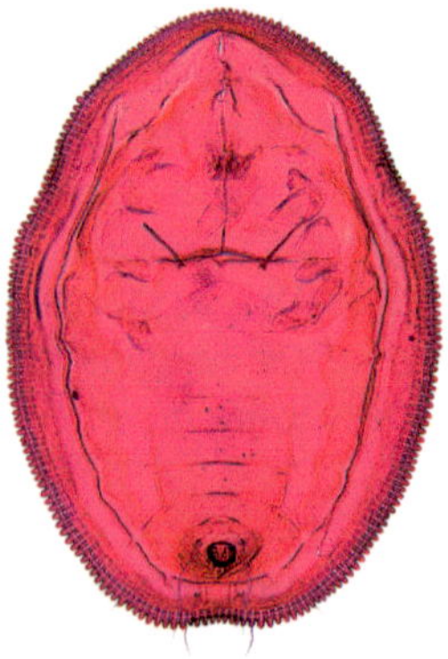

Fig. (11). Slide mounted puparium of *Aleurothrixus floccosus*. The metathoracic dorsal setae are broken (red stain is artificial).

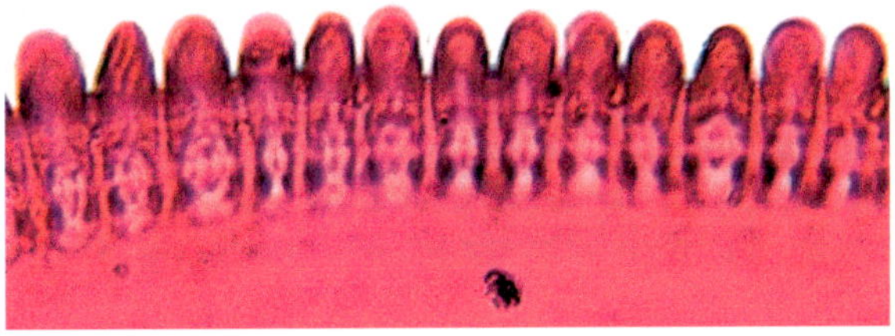

Fig. (12). The row of false double teeth at the border of the WW puparium.

4.4. *Bemisia afer* Priesner and Hosny

Diagnostic characters: The puparium is oval in outline, yellow, flattened, placed on the undersides of leaves, dorsally convex. It is about 1 mm long. The vasiform orifice is triangular in shape. The ventral abdominal setae are short and invisible on living puparia when observed through a dissecting microscope.

Life history: Life cycle of the species is unknown; in general its populations are low because of the natural biological control brought about by parasitoids. *Bemisia afer* is very polyphagous [4, 29].

Economic importance: *Bemisia afer* has no economic importance.

4.5. *Bemisia tabaci* Gennadius

Diagnostic characters: The Sweet Potato Whitefly (SPW) puparium is similar to that of *B. afer*, but smaller, being about 0.75 mm long and bearing long and apparent ventral abdominal setae. The adults are pale yellow and the wings are powdered by white wax. Slide mounted puparia show the wide morphological variability of the species, which is caused by the host plant. The vasiform orifice is triangular in shape; the lingula is devoid of a tubercle but hairy because of microtrichia. The ventral abdominal setae are considerably longer that those of *B. afer,* almost as long as the vasiform orifice.

Fig. (13). Three puparia of the woolly whitefly with holes bored by emerging adult parasitoids. From the fourth (top, right) a whitefly had emerged.

Life history: *Bemisia tabaci* reproduces by parthenogenetic arrhenotoky, unmated females laying haploid eggs from which only male crawlers emerge. Fertilized females lay both haploid (male) and diploid (female) eggs, placed on the leaves' surface [3] [30]. The pest is very polyphagous, infesting over 500 plant species in 63 families [31, 4], including vegetables, field and ornamental crops. Citrus trees are not among its preferred hosts.

Economic importance: SPW causes direct and indirect damage to plants [30]; on citrus it is only a secondary pest.

Management: Several natural enemies regulate SPW populations in citrus orchards. They include the Hymenopterous *Encarsia pergandiella* (Howard), *Encarsia transvena* (Timberlake), *Encarsia nigricephala* (Dozier) and several species of *Eretmocerus*. Effective parasitoids, such as *Eretmocerus eremicus* Rose and Zolnerowich, or lacewings and coccinellid predators are commercially available to control the SPW.

4.6. *Dialeurodes citri* Ashmead

Diagnostic characters: The Citrus Whitefly (CW) adults are pale yellow, 1.4 mm long, coated with white wax. Each wing has a weak area near the apex of the fore wings. The puparia are almost flattened and diaphanous, thus somewhat obscure on the undersides of the leaves (Fig. **14**). Wax secretion is meagre and hyaline, thus apparently absent. The pupal case (Fig. **15**) is elliptic, 1.3 mm long and 0.9 mm wide. The confluences of the anterior spiracular and anal furrows form a whitish Y-like mark on living puparia. The exuvial casts are pale after the emergence of the adults and then resemble whitish spots on the undersides of infested leaves. The puparium has one pair of dorsal papillae on the prothorax, on the metathorax and on the 2nd abdominal segment; in addition, many wax pores are scattered all over the dorsum of the pupal case. The vasiform orifice is round with an indented inner posterior margin and covered by the operculum. Anterior spiracular and anal furrows are evident.

Life history: A female lays about 90-110 eggs on the undersides of citrus leaves and on many other host plants. The inner, shaded tree parts are the preferred oviposition sites, the stalked eggs being inserted into fully extended leaves. The pest also occurs on coffee, gardenia, lilac, myrtle, mock olive (*Notelaea venosa* F. Muell.), pear, *Smilax* sp. and wild olive (*Olea europaea* Linnaeus *oleaster*) [32]. SPW reproduces by parthenogenetic arrhenotoky, unmated females laying haploid eggs from which only male crawlers emerge. Under long day conditions at 25°C, a life cycle requires about 54 days. The adults live for about two weeks. On citrus in the eastern Mediterranean Region the pest usually completes three generation, the first (April-June) in about ten weeks, the second (July-August) in seven weeks and the third, which usually overwinters as egg or as IV larva, lasts until the following spring. A small fraction (5%) of the population develops in the autumn, thus creating a fourth generation, whose adults also emerge in the spring [33]. CW populations peak from May to July. The threshold of development is at 11.3°C.

Fig. (14). CW puparia the undersurface of a tangerine leaf in a mixed infestation with *Lepidosaphes beckii* (Newman) (Diaspididae: Coccoidea).

Economic importance: The citrus whitefly invaded the Mediterranean Region in 1945. Most of its damage is due to copious amounts of honeydew secreted by its nymphs that are colonized by sooty mold fungi (Fig. **16**). These fungi contaminate the fruit, obstruct respiration and disrupt photosynthesis, leading to leaf drop and yield reduction [34]. Heavily infested citrus trees weaken and produce small and flavourless low-quality fruits. The black sooty mold interferes in the ripening process, which is delayed in infested trees.

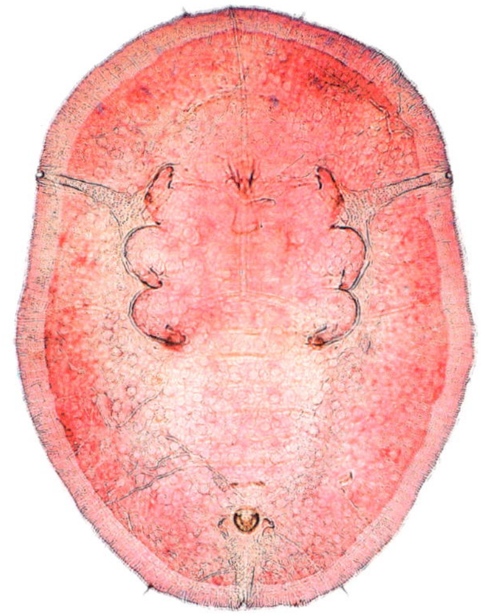

Fig. (15). Slide mounted puparium of *Dialeurodes citri* Ashmead (slide thanks to M.G.M. Jansen, Plant Protection Service, Wageningen, The Netherlands).

Heavily infested leaves become intermittently white-and-black, due to the pest's colouring and the sooty mold. Among citrus hosts it prefers orange and Easy Peeler cultivars.

Fig. (16). Sooty mold blackening a CW-infested tangerine leaf. *Lepidosaphes beckii* (Newman) (Coccoidea Diaspididae) that co-infests the leaf does not excrete honeydew.

Management: The sooty mold may be washed away before marketing, as a commercial management measure [32]. Two natural enemies can be employed to control this pest, both attacking the nymphs: the predator *Clitosthetus arcuatus* (Rossi) (Coccinellidae: Coleoptera) and the chalcid endoparasitoid *Encarsia lahorensis* (Howard) (Hymenoptera), which has been successfully introduced into Italy. Entomopathogenic fungi, like *Aschersonia aleyrodis* Webber and *Aegerita webberi* Fawcett may reduce CW populations, only in tropical moist climates rather than in the Mediterranean Region, where whiteflies are secondary targets of the chemical control of scale insects control.

Two introduced hymenopterous endoparasitoids, namely *E. lahorensis* and an *Encarsia* sp., usually control the pest in Israel [34]. Two predatory coccinellids, *C. arcuatus* and *Serangium montazerii* (Fürsch) also feed on the CW [35]. When chemical control is required, several IGRs (*e.g.* buprofezin that affects molting if applied in early June, or pyriproxifen, which kills developing eggs) should be applied in late April to early May [36, 37]. Oils are useful in removing the sooty mold from the fruits and leaves. Due to the oviposition preferences of the pest, only young, fully extended leaves should be sampled for eggs during its first generation. The recommendation is to sample four leaves per tree, one from the interior of each examined tree, plus three from the periphery, preferably from the northern and southern sides. A good precision level of 0.30 can be achieved by leaves from 11 trees, coming to a total of 44 leaves. For larval sampling (which incorporate sooty mold threshold levels) seven trees (28 leaves) may be taken for the same precision level.

In the second and third generations, one of the four leaves should be young and leaves from nine trees, a total of 36, would suffice to obtain a precision level of 0.25, whereas nine trees (36 leaves) should be sampled for the second and third generations [38]. In the past the CW was a primary citrus pest in the Mediterranean Region, but its importance has considerably decreased during the years [39].

4.7. *Parabemisia myricae* Kuwana

Diagnostic characters: The Japanese Bayberry Whitefly (JBW) lays white eggs that turn black as the embryos develop. Nymphs are white to yellowish, with a faint translucent wax fringe that borders the 4th instar nymph. The puparium is flattened and devoid of prominent setae or spines. It averages 0.85 mm in length and 0.65 mm in width. The adults are smaller than those of WW and CW, about 1.10 mm long, lemon yellow with white wax powered wings. The cuticle of slide-mounted fourth-instar nymphs shows only few details, such as 26 short marginal setae plus six anterior, posterior and caudal setae and scattered dorsal pores. The first abdominal tergum is devoid of setae. The anterior spiracular furrows are barely visible and the caudal furrow is slightly longitudinally ridged. The vasiform orifice is elongate-triangular; the lingula has two blunt lateral tubercles [40].

Life history: JBW is thelytokous, the males being quite rare [41]. Each female produces about 70 eggs that are inserted singly or in semi-circles, mostly into the leaf margins [40] of very young leaves in the "feather" stage [42]. The crawlers usually migrate to the underside of the leaf to settle and feed, but some remain on the upper surface. At 23.3°C a life cycle required 17-23 days. The pest develops throughout the year; its populations surging during the spring growth flush, as the temperatures rise [43]. Among citrus species it developed fastest on lemons, whereas the highest mortality rates of the immature stages was on trifoliate (*Poncirus trifoliata* (Linnaeus) Raf.) [44]. The pest also infests many tropical and subtropical fruit trees.

Economic importance: Injury is due mostly to the copious amounts of honeydew secreted by the juveniles and colonized by sooty mold fungi. These fungi obstruct respiration and hinder photosynthesis, leading to disruptions in growth and to yield reductions, sometimes causing defoliation [41]. There is also a direct feeding damage, because the pest feeds mostly on leaves that develop on the new terminal growth, leading to pit-like leaf distortions and to twig deformations [43].

Management: Natural enemies such as *Encarsia bemisiae* De Santis (Aphelinidae) [45] control *Parabemisia myricae* in Japan. The aphelinid endoparasitoid *Eretmocerus debachi* (Rose and Rosen) [46] completely controls the pest in most Mediterranean countries. In Israel, several parasites and predators have been found on *P. myricae*, including phytoseiid mites [47]. Prospects for biological control thus seem good, should the pest invade other Mediterranean countries. Outdoors JBW is usually under biocontrol, outbreaks occurring only after applications of inappropriate pesticides that disrupt the activities of its natural enemies.

4.8. *Paraleyrodes minei* Iaccarino

Diagnostic characters: The Nesting Whitefly (NW) adults are uniformly pale yellow and lay pale yellow eggs, usually surrounded by fluffy wax. The pupal case is fringed by broad ridged wax ribbons that curve downward and usually cover and surround the case. The pupal case is 0.8 mm long and 0.6 mm wide, with six distinctive pairs of compound wax pores on the dorsal abdominal segments. Seven pairs of compound pores, one cephalic and six abdominal, are on located on the puparium. The abdominal pores of the first two pairs are smaller than the others [48]. Such compound wax pores are diagnostic of the sub-family Aleurodicinae. In the field NW may be misidentified as *Aleurodicus dispersus* (SW) due to its prominent ribbon-like wax matter, as *Metaleurodicus cardini* (Back), which shows a weak ribbon-like wax border, or as *Aleurothrixus floccosus* that secretes abundant thread-like wax.

Life history: This species is arrhenotokous. It lays about 60 eggs/female in rings on the undersides of leaves, protected by the adult or nymphal wax secretions. The common name "nesting whitefly" is due to the behaviour of the adult, as it nests in the centres of the egg-rings. NW is very polyphagous and annually raises up to nine overlapping generations, with continuous development on mandarins. It overwinters as adult in the Middle East [49] as well as in California [50].

Economic importance: NW is not considered an invading species in Syria [51], being only considered as a potential pest [52], but in Lebanon it infests citrus orchards in large numbers [53]. The species is invasive also in California [54]. NW causes direct damage by sucking plant sap and injecting toxic saliva. Sooty mold fungi colonize the large amounts of excreted honeydew, affecting plant metabolism and lowering the quantity and quality of the product.

Management: Chemical control of the NW is expensive and difficult, and no effective natural enemies of this whitefly are known. Pest levels may be lowered by agronomic measures, like pruning, along with mineral oil sprays. Natural enemies, like *Encarsia variegata* Howard [55] are promising, but have not yet been assayed against NW.

4.9. *Trialeurodes vaporariorum* Westwood

Diagnostic characters: The Greenhouse Whitefly (GW) puparium is oval and prominent on the leaf underside because of a waxy marginal palisade. Individuals on hairy leaves bear long thorn-like wax filaments on their dorsum, which may not be seen on glabrous host plants. Adults are faintly yellow, about 1 mm long. The vasiform orifice is triangular with a distinctly lobulate lingula.

Life history: The duration of the development of GW depends on the ambient temperatures. Reproduction is by parthenogenesis, and several generations may overlap on the hosts. The adults aggregate at the tips of shoots, as they prefer young leaves on which they attain high densities, even on unsuitable hosts. When disturbed the adults suddenly fly away. GW is extremely polyphagous, having been recorded from more than 200 plant genera, including many herbaceous and some monocotyledonous plants, and even a cycad [4]. Many more hosts have been recorded since, included *Citrus* spp.

Economic importance: GW is a major pest that reduces plant vigour, sucking sap from leaves and excreting honeydew that is colonized by sooty mold fungi. It may also transmit virus diseases. Nevertheless, it is only a secondary, occasional pest of citrus.

Management: Oil sprays can control WG population outbreaks until the buildup of a local or a temporarily-missing natural enemy complex.

REFERENCES

[1] Martin JH, Mound LA. An annotated check list of the world's whiteflies (Insecta: Hemiptera: Aleyrodidae). Zootaxa 2007; 1492: 1-84.

[2] Martin JH. Giant whiteflies (Sternorrhyncha, Aleyrodidae): a discussion of their taxonomic and evolutionary significance, with the description of a new species of *Udamoselis* Enderlein from Ecuador. Tijdschr entomol 2007; 150: 13-29.

[3] Byrne DN, Bellows TS. Whitefly biology. Ann Rev Entomol 1991; 36: 431-57.

[4] Mound LA, Halsey SH. Whitefly of the World. A Systematic Catalogue of the Aleyrodidae (Homoptera) with Host Plant and Natural Enemy Data. Chichester, New York, Brisbane, Toronto: British Museum (Natural History) and John Wiley and Sons 1978.

[5] Martin JH. An identification guide to the common whitefly pest species of the world (Homoptera, Aleyrodidae). Trop Pest Manage 1987; 33(4): 298-322.

[6] Russell LM. A new genus and four new species of whiteflies from the West Indies (Homoptera, Aleyrodidae). P Entomol Soc Wash 1943; 45(6): 131-41.

[7] Russell LM. A new genus and twelve new species of Neotropical whiteflies (Homoptera: Aleyrodidae). Journal of the Washington Academy of Sciences 1945; 35: 55-65.

[8] Jones DR. Plant viruses transmitted by whiteflies. Eur J Plant Pathol 2003; 109: 195-219.

[9] Martin JH, Mifsud D, Rapisarda C. The whiteflies (Hemiptera: Aleyrodidae) of Europe and the Mediterranean Basin. Bul Entomol Res 2000; 90: 407-48.

[10] Porcelli F. First record of *Aleurocanthus spiniferus* (Homoptera: Aleyrodidae) in Apulia, southern Italy. EPPO Bulletin 2008; 38(3): 518-20.

[11] Nutricato S, D'Onghia AM, Fedele A *et al.* Nuovo Aleirode da quarantena segnalato in Puglia. L'Informatore Agrario 2009; 30: 2-4.

[12] Corbett GH. Malayan Aleurodidae. Journal of the Federated Malay States Museums 1935; 17: 722-852.

[13] Anonymous. "*Aleurocanthus woglumi"* In: Smith IM, McNamara DG, Scott PR, Holderness M, Ed. *Quarantine pests for Europe* 2nd edition CAB INTERNATIONAL. Wallingford, UK: EPPO/CABI 1996.

[14] Silvestri F. Contribuzione alla conoscenza degli Aleurodidae (lnsecta: Hemiptera) viventi su Citrus in Estremo Oriente e dei loro parassiti. Bollettino del Laboratorio di Zoologia Generale e Agraria 1928; 21: 1-60.

[15] Smith HD. Citrus blackfly on the west coast of Mexico and the importation and colonization of *Eretmocerus serius* Silv. for its control. Fitofolio 1945; 4: 67-103.

[16] Quezada JR. Biological control of *Aleurocanthus woglumi* (Homoptera: Aleyrodidae) in El Salvador. Entomophaga 1974; 19: 243-54.

[17] Clausen CP. Introduced parasites and predators of arthropod pests and weeds: a world review VI. Agriculture Handbook No. 480. Washington: USDA 1978.

[18] Muniappan R, Marutani M, Esguerra N. Establishment of *Encarsia smithi* (Silvestri) (Hymenoptera: Aphelinidae) on Pohnpei for control of the orange spiny whitefly, *Aleurocanthus spiniferus* (Quaintance) (Homoptera: Aleyrodidae). Proc Hawaiian Entomol Soc 1992; 31: 243.

[19] Russell LM. A new species of *Aleurodicus* Douglas and two close relatives (Homoptera: Aleyrodidae). Fla Entomol 1965; 48: 47-55.

[20] Martin JH, Hernandez-Suarez E, Carnero A. An introduced new species of *Lecanoideus* (Homoptera: Aleyrodidae) estabilished and causing economic impact on the Canary Islands. J Nat Hist 1997; 31: 1261-72.

[21] Hernandez-Suarez E, Carnero A, Hernandez M, Beitia F, Alonso C. *Lecanoideus floccissimus* (Homoptera: Aleyrodidae): nueva plaga en la Islas Canarias. Phytoma Espana 1997; 91: 35-48.

[22] Aiswariaya KK, Manjunatha M, Mohan I Naik. Biology and Host Range of Spiralling Whitefly. Karnataka Journal of the Agricultural Science 2007; 20(1): 149-52.

[23] Wijesekera GAW, Kudagamage C. Life history and control of 'spiralling' white fly *Aleurodicus dispersus* (Homoptera: Aleyrodidae): fast spreading pest in Sri Lanka. FAO Plant Protection Community South East Asia and Pacific Region, Quarterly Newsletter 1990; 33:22-4.

[24] Cherry R. Temperature tolerance of three whitefly species found in Florida. Environ Entomol 1979; 8: 1150-52.

[25] Cherry R. Host plant preference of the whitefly *Aleurodicus dispersus* Russell. Fla Entomol 1980; 63(2): 222-5.

[26] Vulic M, Beltran JL. Die weibe fliege *Aleurothrixus floccosus*, ein gefahrlicher Schadling der Citruskulturen. Z Pflanzenkr Pflanzenschutz 1977; 84: 202-14.

[27] Onillon JC. Contribution a l'etude de la dynamique des populations d'Hompteres infeodes aux Agrumes. V.2. Possibilités des regulation des populations d'*A. floccosus* Mask. (Homopt. Aleurodidae) sur Agrumes par *Cales noacki* How. (Hymenopt., Aphelinidae). OEPP/EPPO Bullettin 1973; 3: 17-26.

[28] Ortu S, Ibba, I. "*Aleurothrixus floccosus* (Mask.) in Sardinia". Atti XIV Congresso Nazionale Italiano di Entomologia sotto gli auspici dell'Academia Nazionale Italiana di Entomologia, della Societa Entomologica Italiana e della International Union of Biological Sciences. 28 maggio-1 giugno 1985; pp. 607-14.

[29] Anderson PK, Martin JH, Hernandez P, Laguna A. *Bemisia afer* sens. lat. (Homoptera: Aleyrodidae) outbreak in the Americas. Fla Entomol 2001; 84: 317-27.

[30] Berlinger MJ. Host Plant Resistance to *Bemisia tabaci*. Agric Ecosyst Environ 1986; 17: 69-82.

[31] Greathead AH. Host plants. In: Cock MJW Ed. *Bemisia tabaci* - a literature survey on the cotton whitefly with an annotated bibliography. Ascot: CAB International Institute of Biological Control. 1986; pp. 25-7.

[32] Fasulo TR, Weems HV. Citrus Whitefly, *Dialeurodes citri* (Ashmead) (Insecta: Hemiptera: Aleyrodidae). University of Florida IFAS Extension. 2007; Available from: http://edis.ifas.ufl.edu/.

[33] Argov Y, Rössler Y, Voet H, Rosen D. The biology and phenology of the citrus whitefly, *Dialeurodes citri*, on citrus in the coastal plain of Israel. Entomol Exp Appl 1999a; 93: 21-7.

[34] Argov Y, Rössler Y, Voet H, Rosen D. Introducing *Encarsia lahorensis* against Dialeurodes citri in Israel: a case of successful biological control. BioControl 2000; 45: 1-10.

[35] Yigit A. Canhilal R. Establishment and dispersal of *Serangium parcesetosum* Sicard (Coleoptera, Coccinellidae), a predatory beetle of citrus whitefly, *Dialeurodes citri* Ashm. (Homoptera, Aleyrodidae) in the East Mediterranean region of Turkey. Z Pflanzenkr Pflanzenschutz 2005; 112: 268-75.

[36] Peleg, BA. Evaluation of two IGR's, Tiger and Uplord, as selective control agents for the citrus whitefly Dialeurodes citri (Ashmead) (Homoptera, Aleyrodidae). Hassadeh 1990; 70: 720-2.

[37] Bar-Zakay I. Using IGRs in citrus: achievements and problems. Phytoparasitica 1992; 20: 80.

[38] Argov Y, Rössler Y, Voet H, Rosen D. Spatial dispersion and sampling of citrus whitefly, *Dialeurodes citri*, for control decision in a citrus orchard. Agric For Entomol 1999b; 1: 305-18.

[39] Rapisarda C. Gli Aleirodi degli agrumi in Italia. Phytophaga 1985-89; 3: 173-98.

[40] Hamon AB, Nguyen R, Browning H. Bayberry Whitefly, *Parabemisia myricae* (Kuwana) (Insecta: Hemiptera: Aleyrodidae: Aleyrodinae). University of Florida IFAS Extension. 2008; Available from: http://edis.ifas.ufl.edu/

[41] Rose M, DeBach P, Woolley J. Potential new citrus pest: Japanese bayberry whitefly. California Agriculture 1981; 35: 22-4.

[42] Walker GP, Aitken DCG. Oviposition and survival of bayberry whitefly, *Parabemisia myricae* (Homoptera: Aleyrodidae) on lemons as a function of leaf age. Environ Entomol 1985; 14: 254-7.

[43] Swirski E., Wysoki M, Izhar Y. Subtropical Fruits Pests in Israel (In Hebrew). Tel Aviv: Fruit Board of Israel 2002.

[44] Ulusoy MR, Vatansever G, Uygun N. The influence of different host plants on the development of *Parabemisia myricae* (Kuwana) (Homoptera: Aleyrodidae). Z Pflanzenkr Pflanzenschutz 1999; 106: 517-22.

[45] Yasumatsu K, Watanabe C. A Tentative Catalogue of Insect Natural Enemies of Injurious Insects in Japan, Parasite-predator host catalogue Part 1. Fukuoka: Entomological Laboratory, Faculty of Agriculture, Kyushu University 1964.

[46] Swirski E, Blumberg D, Wysoki M, Izhar Y. Biological control of the Japanese bayberry whitefly, *Parabemisia myricae* (Kuwana) (Homoptera: Aleyrodidae), in Israel. Isr J Entomol 1987; 21: 11-8.

[47] Wysoki M, Cohen M. Mites of the family Phytoseiidae as predators of the Japanese bayberry whitefly *Parabemisia myricae*. Agronomie 1983; 3: 823-5.

[48] Iaccarino FM. Descrizione di *Paraleyrodes minei* n. sp. (Homoptera: Aleyrodidae), nuovo aleirodide degli agrumi, in Siria. Bollettino del Laboratorio di Entomologia Agraria 'Filippo Silvestri' 1989; 46: 131-49.

[49] Mohammad A, Abboud R. Study of morphology and biology for mineo-whitefly *Paraleyrodes minei* Iaccarino (Homoptera: Aleyrodidae) in Syria. Arab Journal of Plant Protection 2001; 19(2): 125-30.

[50] Bellows TS Jr, Meisenbacher C, Headrick DH. Field biology of *Paraleyrodes minei* (Homoptera: Aleyrodidae) in southern California. Environ Entomol 1998; 27(2): 277-81.

[51] Uygun N, Ulusoy MR, Karaca Y, Kersting U, Vacante V. Approaches to biological control of *Dialeurodes citri* (Ashmead) in Turkey. Integrated control in citrus fruit crops. Proceedings of the meeting held at Florence, Italy, 29 August 1996. Bulletin OILB SROP1997; 20(7): 52-62.

[52] Ulusoy MR, Uygun N. Two new potential pests in the East Mediterranean Region of Turkey: *Aleurothrixus floccosus* (Maskell) and *Paraleyrodes minei* Iaccarino (Homoptera, Aleyrodidae). Turkiye Entomoloji Dergisi 1996; 20(2): 113-21.

[53] Kfoury L, Abdul Nour H, El Amil R. Les Aleurodes des agrumes au Liban: inventaire raisonne et nouvelles especes introduites (Hemiptera, Sternorrhyncha). Nouv Rev Entomol 2003; 20(4): 345-51.

[54] Anonymous. Nesting whitefly, *Paraleyrodes minei*. The California Plant Pest and Disease Report 1991; 10(1-2): 8-11.

[55] Myartseva SN, Varela-Fuentes SE. *Encarsia variegate* Howard (Hymenoptera: Aphelinidae), a parasitoid of whiteflies *Paraleyrodes* spp. (Hemiptera: Aleyrodidae) in three citrus-producing states of Mexico. Vedalia 2005; 12(1): 23-31

CHAPTER 12

Pseudococcidae and Monophlebidae

Ezio Peri[1*] and Apostolos Kapranas[2]

[1]*Department SENFIMIZO, University of Palermo, Viale delle Scienze, 90128 Palermo, Italy and* [2]*Department of Crop Production, Technological Educational Institute, 41110, Larissa, Greece*

Abstract: The families Pseudococcidae (mealybugs) and Monophlebidae include small scale insects (Coccoidea: Hemiptera) that suck out plant sap. The bodies of young instars and adult females are covered with a wax secretion, whereas adult males carry a pair of small wings. Feeding by these pests can cause premature leaf, flower, and fruit drop, reduce plant vigour and discolour tissues. In addition, they secrete honeydew upon which sooty-mould fungi grow, reducing photosynthesis and the market value of the product. Six species of Pseudococcidae, namely *Planococcus citri* (Risso), *Pseudococcus cryptus* Hempel, *Ps. longispinus* (Targioni Tozzetti), *Ps. calceolariae* (Maskell), *Ps. viburni* (Signoret) and *Nipaecoccus viridis* (Newstead) are economic pests of citrus in the Mediterranean area, along with *Icerya purchasi* Maskell, in the family Monophlebidae.

Keywords: Citriculture, Mediterranean Region, citrus mealybugs, cottony cushion scale, bio-ecology, control.

1. INTRODUCTION

The family Pseudococcidae (Hemiptera: Coccoidea) consists of ca 2,000 described species placed in about 280 genera, including pests of worldwide distribution that affect crops and other cultivated plants [1].

The Pseudococcidae are sexually dimorphic. Adult females are small, 1-4 mm, long, with nymphal characteristics (neoteny), such as a reduced morphology, and lacking wings. The body is oval, covered with a whitish-cottony wax secretion, hence the common name mealybugs. The body has marginal waxy filaments that may be wedge-shaped or spine-like, up to a maximum of 18 pairs. The ovipositing females of several species produce a white cottony ovisac that covers and protects the eggs. Males bear a pair of small wings and caudal two wax filaments.

They live only for about 1-2 days, searching for females to mate; lacking functional mouthparts, they do not feed. Some species are parthenogenetic.

Mealybugs are hemimetabolous insects: females go through three nymphal instars and, except for growing in size, do not change their form. The first instar nymph, known as crawler, is about 0.3 mm long and similar in shape to the adult female, except that its legs and antennae are relatively larger and the body is not covered by wax. The crawlers are mobile, moving along the host plant to find suitable feeding sites. They also disperse passively through the air, thus colonizing new host plants. The subsequent nymphal instars are similar to the adult female but smaller. The male goes through 4 instars: at the end of the second nymphal instar it secretes a loose cottony wax cocoon and moults inside to become a prepupa, a pupa, and finally to emerge as adult.

The family Monophlebidae, originally treated as a subfamily of the family Margarodidae, includes 237 species in 44 genera. They are large-bodied scale insects, commonly called giant scales, being up to 10 mm in length. They bear a well developed anal tube and abdominal spiracles that are usually conspicuous, and dark legs and antennae. In general these insects go through 4 female and 5 male instars. Males however are uncommon for most many species, and when present, the prepupa is quite mobile.

Generally, the species in these two families live in little colonies that appear like small cotton spots, especially when the females produce their ovisacs. They usually colonize all plant parts: roots, stems, branches, leaves and fruits, feeding on plant sap. Their feeding causes reduced plant vigor, discoloration of tissue, leaf and flower drop, and early fruiting. In addition, they secrete honeydew that serves as a substrate for sooty-mould fungi, which reduce photosynthesis and attracts ants that protect these pests from predators and parasitoids.

[*]**Address correspondence to Ezio Peri:** Dipartimento SENFIMIZO, Università degli Studi di Palermo, Viale delle Scienze, 90128 Palermo, Italy; Tel, +3909123896009; E-mail: e.peri@unipa.it

Vincenzo Vacante and Uri Gerson (Eds)

More than 60 Pseudococcidae have been reported from citrus, usually as occasional or minor pests; however, under certain agronomic and/or climatic conditions, some mealybugs may become major pests. In the Mediterranean Basin the pests include the following six species: *Planococcus citri* (Risso), *Pseudococcus longispinus* (Targioni-Tozzetti), *Pseudococcus calceolariae* (Maskell), Pseudococcus cryptus Hempel, *Pseudococcus viburni* (Signoret) and *Nipaecoccus viridis* (Newstead), whereas in the Monophlebidae, only *Icerya purchasi* Maskell is considered a key pest of citrus.

2. FIELD KEY TO CITRUS PSEUDOCOCCIDAE AND MONOPHLEBIDAE

1. Mature females with a smooth ovisac or without ovisac (**Pseudococcidae**).. ..2

-. Mature females with a flutted ovisac (**Monophlebidae**)..***Icerya purchasi***

2. Mature female with oval body, covered by mealy wax, with 17 or 18 pairs of lateral wax filaments................. 3

- Mature female with round or broadly oval body, covered by thick white or pale-yellow wax, without bare areas on dorsum; with 5 or 6 pairs of lateral wax filaments......... ....***Nipaecoccus viridis***

3. Mature female with oval body, covered by mealy wax i with 17 pairs of lateral wax filaments ………..……........… .... …4

-. Mature female with pink body, covered by white wax, body segmentation visible i with 18 pairs of small lateral waxy filaments, 17 of equal length, and one longer, in the anal region ……...................... ....…..***Planococcus citri***

4. Mature female covered by white wax, with longitudinal dorsal lines.. 5

-. Mature female covered by white wax, without dorsal longitudinal lines....... 6

5. Mature female yellowish or greyish, covered by white wax through which the body color may be seen and with three longitudinal dorsal lines i with 17 lateral pairs of waxy filaments, 15 long, up to half body width, another, in the anal region, as long as half body length and caudal filaments that are longer than the body......... .... ….***Pseudococcus longispinus***

-. Mature female red; unevenly covered by white wax i with 17 pairs of lateral waxy filaments, 15 relatively short, one slightly longer in the anal region, and caudal filaments longer, about half the length of the body................................. …………… ....***Pseudococcus calceolariae***

6. Mature female pale to greenish yellow, covered by white wax, except on the intersegmental lines; with 17 lateral pairs of waxy filaments that posteriorly become progressively longer, the anterior anal pair about one quarter width of body, whereas the caudal pair as long as body......... ..***Pseudococcus cryptus***

-. Mature female pink to light purple, covered by white wax through which the body colour may be seen; with 17 lateral pairs of wax filaments, 15 small, the posterior pairs progressively longer, one about a quarter and the other about half body length..................................... .....***Pseudococcus viburni***

3. MICROSCOPIC KEY TO CITRUS PSEUDOCOCCIDAE AND MONOPHLEBIDAE

1. Abdominal spiracles present (**Monophlebidae**)...***Icerya purchasi***

-. Abdominal spiracles absent (**Pseudococcidae**).......... 2

2. Anal bar present...…***Planococcus citri***

-. Anal bar absent................ 3

3. Dorsal setae conical.. .... ***Nipaecoccus viridis***

-. Dorsal setae filamentous.. 4

4. Some of the dorsal setae are long, longer than ceranial setae .... ***Pseudococcus cryptus***

-. All dorsal setae are short... 5

5. More than five oral collars on head between the antennae .... ***Pseudococcus viburni***

-. Less than five oral collars on head between the antennae..... 6

6. Ventral multiocular pores only on segments 7 and 8.................................... ***Pseudococcus longispinus***

-. Ventral multiocular pores on segments 6, 7, 8 and anteriorly ....***Pseudococcus calceolariae***

4. CITRUS PSEUDOCOCCIDAE SPECIES

4.1. *Planococcus citri* (Risso)

Diagnostic characters: This species is known as the citrus mealybug. Adult female about 3 mm long, 1.5 mm wide. The pink oval body is covered by white wax with visible body segmentation, with 17 pairs of small, subequal waxy filaments protruding from the periphery, and another, longer, in the anal region (Fig. **1**) The mature females produce a dense, soft, white ovisac, which is longer than the body and holds the eggs. The small red male is winged, about 1 mm in length, with two caudal filaments, 4-5 mm long.

Life history: This species is reported to be occasionally parthenogenetic [2], but a recent study demonstrated that it is obligate amphimictic [3]. A female lays 300-600 yellow, oblong eggs inside an ovisac. The first instar crawlers are small with yellow, oval body, red eyes and rather distinct antennae.

The subsequent nymphs are very similar to the females. Development time, from egg hatching to adult, depends on the temperature; at 25°C, on citrus, it is about 23 and 24.5 days for females and males, respectively [4]. The upper developmental threshold is around 30°C. To complete all nymphal stages 401 DD (day-degrees) are required at 7.7°C, which represent the lower developmental temperature threshold for nymphs [4]. Crawlers disperse on new plant parts and seek suitable feeding sites; once they settle, they insert their stylets into the host tissue to feed and then generally stay anchored, although they remain mobile throughout their lives. The citrus mealybug colonizes all plant parts; on roots they occur mainly in cold or warmer weather and in drier areas. On citrus the pest prefers the fruits (Fig. **2**), and usually remains there, clustered around the peduncle or where adjacent fruits touch, in the shady, inner, more humid tree parts. The pest overwinters as nymphs or adult females in protected areas, such as fruits, crevices in branches or roots. The pest may annually complete 2-3 generations in colder regions, up to 7-8 in warmer areas or in greenhouses.

Fig. (1). Female of *Planococcus citri* (Risso).

Economic importance: The citrus mealybug attacks several host plants, such as grapevine, persimmon, fig and many ornamentals in greenhouses or outdoors; it is however mainly known as a pest of citrus. Various citrus species and varieties are differently susceptible to the pest. In the Mediterranean Region it prefers varieties of orange [*Citrus* *sinensis* (Linnaeus) Osbeck], whereas in the Near East it attacks mainly grapefruit (*C. paradisi* Macf.), pummelo [*C. grandis* (Linnaeus) Osbeck], sweetie (*C. paradisi* x *C. grandis*), lemon (*C. limon* Burn), and the easy-peeling orange varieties (*C. sinensis*).

This extended list of preferred hosts is probably a consequence of the fruits' prolonged growth period [5]. The area of origin of this pest has not definitely been determined, as some authors believe it to be a native of the Far East [6], others of South America [7]. The citrus mealybug is nowadays of cosmopolitan distribution due to the international trade of agricultural products. Its damage consists of sucking plant sap and concurrently introducing a toxic saliva, which distorts tree growth, reduces development, discolors tissues and causes premature drop of leaves, flowers and early fruits. In addition the pest causes aesthetic damage, due to excreting honeydew and leaving cottony wax on the plant surface, particularly on the fruits, thus reducing their market value. Moreover, the honeydew attracts sugar-loving insects that increase the damage. Females of the lepidopterous *Cryptoblabes gnidiella* Millière, attracted to the honeydew, lay egg within mealybug colonies on the fruits. The first instar larvae initially feed on the honeydew and then pierce the fruit peel, inducing aesthetic damage, premature yellowing and drop [8]. Certain ants have a symbiotic relationship with the pest: in order to preserve the production of honeydew, on which they feed, and incidentally protecting it from predators and parasitoids.

Fig. (2). Colony of *Planococcus citri* (Risso) on orange fruit.

Management: Control of the citrus mealybug primarily relies on cultural practices that aim at preventing and/or reducing infestations. These techniques include correct irrigation and nitrogen fertilization; pruning and/or closely-spaced planting to allow light penetration and air circulation through the tree canopy; control of ant populations with chemicals spread on the ground or sprayed on the trunks, or by placing adhesive bands around the tree trunk. In the absence of ants, natural enemies effectively reduce the populations of the pest [9].

Fig. (3). Larva of *Cryptolaemus montrouzieri* Mulsant.

The most efficient biological control agents are *Cryptolaemus montrouzieri* Mulsant (Fig. **3**) (Coleoptera: Coccinellidae) and *Leptomastix dactylopii* Howard (Fig. **4**) (Hymenoptera: Encyrtidae).

Cryptolaemus montrouzieri is one of the most widely distributed natural enemies of mealybugs [10] and is being reared commercially for that purpose. This predator is sufficiently abundant in the Mediterranean Region, substantially contributing to the control of *P. citri*, as its larvae and adults are voracious feeders of the pest. However, in some regions it has been necessary to mass-release this predator, because its field populations cannot survive extremely low winter temperatures [11]. *Leptomastix dactylopii* is a solitary endoparasitic wasp that has been used in biological control programs around the world [12]. It is commercially reared for seasonal augmentative releases, because it usually fails to overwinter in the Mediterranean climate. The combined activity of beetle and wasp improves the biological control of the citrus mealybug [13]. The beetle feeds on the eggs or first instar of the pest, being most useful where mealybugs occur in large colonies. The wasp, on the other hand, prefers to oviposit into the bodies of third instar nymphs and young adults, being capable of locating small pest colonies due to its excellent searching abilities. Other control agents, *e.g.* the parasitoids *Anagyrus pseudococci* (Girault), *Leptomastoidea abnormis* (Girault), *Coccidoxenoides peregrinus* (Timberlake) and some native predators, including lady beetles and lacewings, aid in mealybug control. However, the ability of *P. citri* to encapsulate parasitoid eggs reduces the efficacy of these enemies [14].

Fig. (4). *Leptomastix dactylopii* Howard.

Chemical treatments with light mineral oils and organophosphates may be applied against the crawlers, the most susceptible stage. However, treatments should be carefully applied, because non-selective insecticides can cause citrus mealybug outbreaks due to disrupting the activities of natural enemies [5]. Economic thresholds for insecticide applications have been developed in the Mediterranean area; in general, treatments may be applied when visual inspections of citrus fruit (10 fruits >2cm diameter per plant, from 10 plants/ha) indicate that the threshold of 10-15% infested fruits has been reached. The sex pheromone of *P. citri*, identified about 30 years ago [15], is being commercially available but not widely used. Traps baited with this pheromone can improve monitoring techniques and be used for mass trapping. The pheromone delta trap can be exploited to monitor the flights of the males, and thus the levels of pest populations and crawler emergence, in order to plan chemical control treatments. One to three traps per hectare are usually placed in the orchard in March, before the assumed 1st male flights occur, and then monitored weekly. Recent studies have opened up new opportunities to apply pheromone traps for mass trapping [5].

4.2. *Pseudococcus longispinus* (Targioni Tozzetti)

Diagnostic characters: This species is known as the long-tailed mealybug. The adult female is yellow-grey, covered by white wax through which its color can be seen. Body up to 3 mm in length, oval-elongated with three dorsal longitudinal lines and 17 pair of waxy filaments that protrude from the periphery; 15 are long, up to half as long as body width, and two in the anal region, one as long as half body length, the other longer than body. Mature females secrete a shallow white waxy felt that sticks to the substrate, but no ovisac. The adult male is similar to that of *P. citri*, with minor morphological differences in the genitalia.

Life history: This species is ovoviviparous or viviparous; females either lay eggs about to hatch or give birth to live crawlers; a total of about 200 progeny are produced. On citrus the this mealybug prefers the abaxial leaf surfaces, the fruits around the peduncle or settling where adjacent fruits touch, in the inner, more humid parts of trees. The pest overwinters mainly as second instar nymphs, but all instars can be found in protected areas during winter. The pest has up to 4-5 annual generations.

Economic importance: *Pseudococcus longispinus*, native of Australia, is nowadays a polyphagous cosmopolitan species, attacking mainly ornamentals in greenhouses and outdoors. In the Mediterranean Region the pest usually infests citrus trees, and its damage is very similar to that of the citrus mealybug. It feeds by sucking out plant sap, thus reducing development, discoloring tissues, causing premature leaf and fruit drop. In addition, the pest secretes much honeydew that is colonized by dark sooty mould fungi which contaminate the plant surface and cause aesthetic blemishes, thus reducing the market value of the fruits.

Management: The use of suitable cultural techniques, such as proper irrigation, the moderate use of nitrogen fertilizers, correct pruning and the exclusion of ants are recommended in order to prevent and/or reduce pest infestations (for more details see above, *P. citri*).The most efficient natural enemies are the hymenopterous parasitoids *Anagyrus fusciventris* (Girault), *Leptomastoidea abnormis*, *Tetracnemoidea brevicornis* (Girault), *T. sydneyensis* (Timberlake) and certain predators, such as *C. montrouzieri*. Chemical treatments are not often needed against this pest in citrus orchards, although localized outbreaks, following the frequent use of broad-spectrum insecticides to control other pests have been recorded [16]. When chemical treatments become necessary, they should be applied against the crawlers, spraying with emulsions of light mineral oil and organophosphates. The pheromone of the long-tailed mealybug has been synthesized [17], but pheromone traps are not yet available in commerce. This is due to the fact that early in the season pest population levels can be detected only by the visual inspection of citrus fruits and leaves, utilizing the economic threshold used for *P. citri*.

4.3. *Pseudococcus calceolariae* (Maskell)

Diagnostic characters: The species is known as the citrophilus mealybug. The body of adult females is red, oval in shape, up to 3-3.5 mm long, unequally covered by white wax, dorsum with distinctive longitudinal lines. It bears 17 pairs of lateral waxy filaments, 15 relatively short, and two in the anal region, one slightly longer and the caudal filament as long as half the length of the body, and thick. The adult male is larger than the male of *L. citri*, with minor morphological differences in the genitalia.

Life history: This species is oviparous; females lay some hundred yellow eggs into a ventral ovisac, placed amongst its woolly wax filaments. Annually the pest completes up to 3 overlapping generations.

Economic importance: *Pseudococcus calceolariae*, of Australian origin, is currently of world-wide distribution. It attacks a wide range of cultivated and woody plants, including citrus, avocado, berries, sugar cane, cocoa, grape and apple. On citrus the pest colonizes leaves and fruits around the peduncle or at the contact point between adjacent fruits in the shady, humid parts of the trees. Its damage is similar to that of other mealybugs: sucking the plant sap and thus bringing about reduced tree development, discoloration of tissues and premature drop of leaves and fruits. In addition, the pest secretes much honeydew that is colonized by dark sooty mould fungi that contaminate plant surfaces, reduce photosynthesis and cause aesthetic blemishes, which reduce the market value of the fruits.

Management: The use of suitable cultural techniques, including proper irrigation, the moderate use of nitrogen fertilizers, correct pruning and the exclusion of ants are recommended in order to prevent and/or reduce infestations of the citrophilus mealybugs (for more details see *P. citri*). This pest is attacked by a suite of natural enemies, including parasitoid wasps, *e.g. Tetracnemoidea brevicornis*, *Coccophagus gurneyi* Compere, *Tetracnemoidea peregrine* (Compere), and *Anagyrus fusciventris*, as well as predators, like the coccinellids *C. montrouzieri* and *Scymnus loewi* Mulsant, some lacewings and predatory midges. This complex of natural enemies generally controls the level of pest populations, as long as broad spectrum pesticides are not used. Heavy infestations may be controlled by treating with light mineral oil emulsions and organophosphates, to be applied when the crawlers appear. Detection of *P. calceolariae* is possible by examining citrus fruits and leaves (for economic threshold see *P. citri*). The sex pheromone of *P. calceolariae* was recently synthesized [18], but is not yet available in commerce.

4.4. *Pseudococcus cryptus* Hempel

Diagnostic characters: The species is known as the citriculus mealybug or the cryptic mealybug. The female is up to 3-3.5 mm long, its oval body is pale yellow or green-yellow, slightly rounded in lateral view, covered by white wax, except at the intersegmental lines, and without dorsal longitudinal lines. It bears 17 pairs of lateral, straight or slightly curved waxy filaments that become progressively longer posteriorly. The anterior anal pair is about one quarter the width of the body, whereas the posterior pair is as long as the body. Mature females produce a ventral ovisac. The adult male is similar to that of *P. citri,* with minor morphological differences in the genitalia

Life history: A female lays about 100-200 pale yellow eggs. The emergence period of the crawlers may be calculated by using seasonal degree-day accumulations of 302 DD above a base of 10±C°, starting from the first male flight [19]. Developmental from egg hatching to adult at 25°C requires about 25 and 30 days for females and males, respectively [4]. The peak occurrence of *P. cryptus* occurs in the spring, concurrent with the period of major foliage growth [5]. The pest overwinters mainly as second instar nymphs, annually completing three generations [4].

Economic importance: *Pseudococcus cryptus* (= *P. citriculus* Green), which has originated in Southeast Asia, is currently of cosmopolitan distribution, having been recorded from more than 20 host families. It is a major citrus pest in the Near East, settling mainly on the foliage and stems, especially damaging to red grapefruit, pummelo, and easy-peeling varieties [20]. The direct and indirect damages it causes are similar to those due to the citrus mealybug, *i.e.* sucking out plant sap thus retarding development, tissue discoloration and premature leaf and fruit drop. The pest secretes honeydew that is colonized by dark sooty mould fungi, which contaminate plant surfaces, reduce photosynthesis and cause aesthetic blemishes, reducing the market value of the fruits.

Management: The use of appropriate cultural techniques, proper irrigation, the moderate use of nitrogen fertilizers, correct pruning and the exclusion of ants are recommended in order to prevent and/or reduce infestations of *P. cryptus* (for more details see *P. citri*). The pest is attacked by several natural enemies, including parasitoid wasps, *e.g. Anagyrus subalbipes* Ishii, *A. sawadai* Ishii, *Allotropa* sp. and *Clausenia purpurea* Ishii, as well as by predators, such as *C. montrouzieri* and other coccinellids. Heavy infestations should be treated with light mineral oil emulsions and organophosphates, to be applied when the crawlers appear. The sex pheromone of the *citriculus* mealybug has only recently been isolated and identified [21].

4.5. *Pseudococcus viburni* (Signoret)

Diagnostic characters: The species is known as the obscure mealybug. The adult female is oval, about 2.5 mm long, pink or light purple, covered by thin white wax through which the body can be seen. The body, which is devoid of dorsal longitudinal lines, bears 17 pairs of lateral smooth wax filaments; 15 are small, straight and thin, the two posterior pairs becoming progressively longer, one about one quarter and the other about one half the length of the body. Mature females produce an ovisac. The adult male is similar to that of *P. citri*, with minor morphological differences in the genitalia.

Life history: This oviparous species lays several hundred orange eggs inside its ovisac, from which orange-to-yellow crawlers emerge after 5–10 days. Under cold conditions the crawlers remain within the ovisac until temperatures increase, when they disperse on the host plants, colonizing all available surfaces. This mealybug completes 3-4 overlapping annual generations. The pest may over winter in all developmental stages, but mainly as eggs inside their ovisac and as first instar nymphs under the bark; however, these nymphs suffer much mortality during winter.

Economic importance: The obscure mealybug is thought to be native to Australia or to South America, but its exact area of origin is not clear because of several synonymous, *e.g. P. maritimus* (Ehrhorn), *P. longispinus*, *P. obscurus* Essig, *P. capensis* Brian, *P. malacearum* Ferris, and *P. affinis* (Maskell).

The pest infests plants in nearly 90 families, including grapevine, persimmon, citrus and many ornamentals; the Cactaceae are its more frequent hosts. The pest sucks out the phloem of plants, inducing direct damage such as reduced development, discoloration of tissues and premature leaf and fruit drop. Indirect damage is due to the excreted honeydew, which is colonized by dark sooty mould fungi that reduce photosynthesis and downgrade the market quality of the fruit.

Management: The use of appropriate cultural techniques, including proper irrigation, the moderate use of nitrogen fertilizers, correct pruning and the exclusion of ants are recommended in order to prevent and/or reduce infestations of the obscure mealybug (for more details see *P. citri*).

Efficient biocontrol agents of this pest are the encyrtid parasitoid *Pseudaphycus maculipennis* Merc., *P. flavidulus* (Brethes) and *Leptomastix epona* (Walker). Other parasitoids also attack *P. viburni*, but their potential of controlling the pest is questionable because the obscure mealybug is capable of encapsulating their eggs. Examples include *L. dactylopii* and *L. epona*, whose efficacy is thus reduced [14].

Heavy pest infestations may be treated with light mineral oil emulsions and organophosphates, to be applied when the crawlers appear. The sex pheromone of the obscure mealybug has recently been identified and synthesized [22], but is not yet available in commerce.

4.6. *Nipaecoccus viridis* (Newstead)

Diagnostic characters: The species is known as the spherical mealybug or Lebbeck mealybug. The adult female is about 3 mm long, 2.5 mm wide, with a round or broadly oval, purple or blue-green body that is slightly rounded in lateral view, and covered by white or pale yellow wax, without any bare dorsal areas. It bears 5-6 pairs of lateral wax filaments. Mature females produce an ovisac that covers the dorsum. The adult male is similar to that of *P. citri*, with minor morphological differences in the genitalia.

Life history: The spherical mealybug has been reported to reproduce parthenogenetically as well as sexually. Each female lays about 600 purple eggs at 25°C, after a preoviposition period of 3-14 days. The light to dark brown-red crawlers emerge after ca 10 days, and their development is completed in about 20 days at 25°C [23]. The spherical mealybug develops and reproduces continuously throughout the year, through multiple and overlapping generations, although slower in winter.

Economic importance: The spherical mealybug is probably a native of Asia, but the origin has not been precisely determined due to confusion in its taxonomic identification. The pest is currently widespread throughout tropical and subtropical regions [23], and known to attack hosts in more than 35 plant families, including tropical fruits, citrus, avocadoes, soybeans, cotton, and ornamentals [1]. Although the spherical mealybug prefers host leaves and fruits, the pest may colonize branches, shoots and flower buds. It sucks out the plant sap, causing direct damage such as reduced development, discoloration of tissues and premature leaf and fruit drop. Indirect damage consists of excreting honeydew that is colonized by dark sooty mould fungi which reduce photosynthesis and downgrade the market quality of the fruit [23].

Management: The use of appropriate cultural techniques, proper irrigation, the moderate use of nitrogen fertilizers, correct pruning and the exclusion of ants are recommended in order to prevent and/or reduce infestations of the spherical mealybug. Various parasitoids and predators have been identified as biological control agents [23], including the predator beetle, *C. montrouzeri* and the parasitic wasps *Anagyrus indicus* Subba Rao, *A. aegyptiacus* Moursi, *Leptomastix phenacocci* Compere. Heavy pest infestations may be treated with light mineral oil emulsions and organophosphates, to be applied when the crawlers appear (for more details see *P. citri*).

5. CITRUS MONOPHLEBIDAE SPECIES

5.1. *Icerya purchasi* Maskell

Diagnostic characters: This species is known as the cottony cushion scale. The bodies of nymphs and adults are red, with black antennae and legs, and are covered with a layer of white to cream-colored wax. The adult female carries a fluted ovisac that is produced behind and below the female and may be two to three times as long as the body (Fig. **5**).

Fig. (5). A colony of *Icerya purchasi* Maskell on an orange twig.

Life history: Nowadays this Australasian species has a cosmopolitan distribution. It is highly polyphagous, having been reported from plants in more than 40 families, in all zoogeographical regions. This species is a hermaphrodite and its males are rare. The females place 600-800 eggs in their ovisacs and the crawlers disperse immediately after hatching. The first and second instar nymphs feed on twigs and leaves, usually along the veins, whereas third instar nymphs and adults feed on branches and tree trunks. Cottony cushion scale rarely infests fruits. In the Mediterranean region it overwinters as adults, nymphs and eggs, and completes three annual generations [24].

Economic importance: Cottony cushion scale attacks reduce tree vigor (Fig. **6**). Heavy infestations result in much leaf and fruit drop, along with twig dieback. Furthermore, the excreted honeydew can reduce fruit quality and its market value.

Fig. (6). Citrus twig infested by *Icerya purchasi* Maskell.

Management: Cottony cushion scale was the first example of a successful classical biological control project. When this species invaded California in the late 19th Century, local authorities from the University of California facilitated the introduction of the coccinelid Vedalia beetle, *Rodolia cardinalis* Mulsant (Fig. **7**), from Australia.

Fig. (7). Adult of *Rodolia cardinalis* Mulsant associated with young instars of *Aleurothrixus floccosus* Maskell.

This predatory beetle brought about the complete control of the pest within a few months. The Vedalia beetle was then introduced into other parts of the world, including several countries in Mediterranean Region that were invaded by this pest [11] with similar beneficial results. Despite this success, localized, heavy outbreaks of the cottony cushion scale have been reported, especially in northern Mediterranean citrus orchards. In such cases organophosphate treatments are recommended at all periods, except during flowering.

REFERENCES

[1] Ben-Dov Y, Miller DR, Gibson GAP. ScaleNet: a Database of the scale insects of the world. 2009 http: //www.sel.barc.usda.gov/scalenet/scalenet.htm.

[2] Panis A. Observations faunistiques et biologiques sur quelques Pseudococcidae (Homoptera, Coccoidea) vivants dans le midi de la France. Ann Zool Ecol anim. 1969; 1(3): 211-44.

[3] Silva EB, Mendel Z, Franco JC. Can facultative parthenogenesis occur in biparental mealybug species? Phytoparasitica 2010; 38: 19-21.

[4] Arai T. Temperature-dependent development rate of three mealybug species, Pseudococcus citriculus Green, *Planococcus citri* (Risso), and *Planococcus kraunhiae* (Kuwana) (Homoptera: Pseudococcidae) on Citrus. Jpn J Appl Entomol Zool 1996; 40: 1, 25-34.

[5] Franco JC, Suma P, Borges da Silva E, Blumberg D, Mendel Z. Management strategies of mealybug pests of citrus in mediterranean countries. Phytoparasitica 2004; 32(5): 507-22.

[6] Flanders SE Citrus mealybug, four new parasites studied in biological control experiments. Calif Agric 1951; 5: 11.

[7] Bartlett BR. In: Clausen CP, Ed. Introduced Parasites and Predators of Arthropod Pests and Weeds: A World Review. US Dep Agric Handb 1978.; 480: pp. 137-70.

[8] Silva EB, Mexia A. The pest complex *Cryptoblabes gnidiella* (Milliere) (Lepidoptera: Pyralidae) and *Planococcus citri* (Risso) (Homoptera: Pseudococcidae) on sweet orange groves (*Citrus sinensis* (L.) Osbeck) in Portugal: interspecic association. Bol San Veg Plagas 1999; 25: 89-98.

[9] Moreno DS, Haney PB, Luck RF. Chlorpyrifos and diazinon as barriers to Argentine ant (Hymenoptera: Formicidae) foraging on citrus trees. J Econ Entomol 1987; 80: 208-14.

[10] Heidari M, Copland MJW. Host finding by Cryptolaemus montrouzieri (Col., Coccinellidae) a predator of mealybugs (Hom., Pseudococcidae). Entomophaga 1992; 37: 621-5.

[11] Katsoyannos P. Integrated insect pest management for citrus in Northern Mediterranean countries. Athens: Benaki Phytopathologique Institute 1996.

[12] Noyes JS, Hayat M. Oriental Mealybug Parasitoids of the Anagyrini (Hymenoptera: Encyrtidae). Wallingford: CAB International 1994.

[13] Copland MJ, Tingle CD, Saynor M, Panis A. In: Hussey NW, Scopes NEA, Eds. Biological Pest Control: The Glasshouse Experience. Poole: Blanford Press 1985; pp. 82-6.

[14] Blumberg D, Van Driesche RG. Encapsulation rates of three encyrtid parasitoids by three mealybug species (Homoptera: Pseudococcidae) found commonly as pests in commercial greenhouses. Biol Control 2001; 22: 191-9.

[15] Bierl-Leonhardt BA, Moreno DS, Schwarz M, Fargerlund J, Plimmer JR. Isolation, identification and synthesis of the sex pheromone of the citrus mealybug, *Planococcus citri* (Risso). Tetrahedron Lett 1981; 22(5): 389-92.

[16] Rosen D. Current status of integrated control of citrus pests in Israel. Bull OEPP 1974; 4: 363-8.

[17] Zou Y, Miller JG. Synthesis of the Pheromone of the Longtailed Mealybug, a Sterically Congested, Irregular Monoterpenoid J Org Chem 2009; 74(18): 7207-9.

[18] El-Sayed AM, Unelius CR, Twidle A *et al.* Chrysanthemyl 2-acetoxy-3-methylbutanoate: the. sex pheromone of the citrophilous mealybug, *Pseudococcus calceolariae*. Tetrahedron Letters. Vol. 2010; 51 (7): 1075-8.

[19] Arai T. Attractiveness of sex pheromone of *Pseudococcus cryptus* Hempel (Homoptera: Pseudococcidae) to adult males in a citrus orchard Appl Entomol Zool 2002; 37 (1): 69-72.

[20] Blumberg D, Ben-Dov Y, Mendel Z. The citriculus mealybug, *Pseudococcus cryptus* Hempel, and its natural enemies in Israel: history and present situation. Entomologica 1999; 33: 141-52.

[21] Arai T, Sugie H, Hiradate S, Kuwahara S, Itagaki N, Nakahata T. Identification of a sex pheromone component of *Pseudococcus cryptus* J Chem Ecol 2003; 29: 2213-23.

[22] Millar JG, Midlandb SL. Synthesis of the sex pheromone of the obscure mealybug, first example of a new class of monoterpenoids. Tetrahedron Letters 2007; 48: 6377-9.

[23] Shaaraf NS, Meyerdrik DE. A review on the biology, ecology and control of *Nipaecoccus viridis* (Homoptera: Pseudococcidae). Misc Publ Entomol Soc Am 1987; 66: 1-18

[24] Avidov Z, Harpaz I. Plant Pests of Israel. Jerusalem: Israel Univ. Press 1969.

CHAPTER 13

Coccidae

Apostolos Kapranas*

Department of Crop Production, Technological Educational Institute, 41110, Larissa, Greece

Abstract: Soft scale insects (Hemiptera: Coccidae) are important citrus pests. They are sessile insects that damage citrus by sucking plant sap, thus reducing plant vigor. They also excrete copious amounts of honeydew that promotes the development of sooty mold fungi, which hinder the photosynthetic ability of the plant and directly reduce the market value of the fruit. In the Mediterranean Region six species are of economic importance for citrus production: *Coccus hesperidum* (Linnaeus), *Coccus pseudomagnoliarum* (Kuwana), *Saissetia oleae* (Olivier), *Ceroplastes rusci* (Linnaeus), *Ceroplastes floridensis* Comstock, and *Ceroplastes sinensis* Del Guercio. Field and microscopic identification keys are provided and notes on their biology and current practices of pest management are also reviewed.

Keywords: Citriculture, Mediterranean Region, soft scale insects, bio-ecology, damage, control.

1. INTRODUCTION

The Coccidae (Hemiptera: Coccidae), or soft scale insects, include about 1,090 described species placed in 162 genera [1]. They are sexually dimorphic, the wingless adult females retaining an external morphology similar to that of the immatures (neoteny). The head, thorax and abdomen are fused, the insect having a flattened or globular/semi-globular sac-like appearance. The adult males are more insect-like and bear one pair of wings, their abdomen has a sclerotized genital organ. The first instar nymphs (crawlers) disperse either actively by walking or passively, by being wind-borne. The female life cycle consists of two or three nymphal instars, whereas males develop through two nymphal, followed by a prepupal and a pupal instar, before emerging as winged adults. All instars are covered with a soft waxy cover (the test). In the subfamily Coccinae the test is usually thin. In the subfamily Ceroplastinae, especially in the late nymphal and adult stages, the bodies are covered with a voluminous waxy test that sometimes attains characteristic shapes, such as a turtle shell or H-shaped shield. Most species possess functional legs in all instars and may move between various parts of the host-plants during development, despite their usual, typical sessile nature. They feed on the phloem of mainly perennial and occasionally annual plants; this feeding on citrus debilitates the plant directly by nutrient depletion and sometimes causes whole terminal die-back. Heavy infestations may result in fruit size reduction. Soft scales excrete copious amounts of honeydew, which is colonized by dark sooty mold fungi, significantly reducing the market value of the fruits.

2. FIELD KEY TO CITRUS COCCIDS

1. Mature females with an oval-shaped body, flat to slightly convex in appearance, shield smooth; nymphs flat, yellowish, sometimes transparent...... ..2

-. Mature females with an oval or rectangular shaped body, more convex in appearance, shield reticulate or covered with wax, sometimes plate-like; nymphs flat, at times bearing tubercles and appearing star-like.. 3

Mature females with a dorsal median ridge; individuals occur in colonies of more or less the same developmental stage and/or size...***Coccus pseudomagnoliarum***

-. Mature females without a dorsal median ridge; individuals occur in colonies of variable size and / oz developmental stage ....***Coccus hesperidum***

***Address correspondence to Apostolos Kapranas:** Department of Crop Production, Technological Educational Institute, *Larissa,* Greece; Tel: (30) 2410-684280; E-mail: akapranas@gmail.com

Vincenzo Vacante and Uri Gerson (Eds)

Mature females brownish, with H-shaped dorsal reticulations; shield of ovipositing scales hard and dark, with a rubber-like appearance; nymphs flat with a smooth yellowish shield .... ***Saissetia oleae***

-. Mature females covered with thick wax that sometimes has a plate-like appearance; nymphal wax secretions star-like..4

4. Shield of body divided into plates, or plate-like, with distinct nuclei..5

-. Shield of body not divided into plates, no nuclei... ***Ceroplastes floridensis***

Shield of body divided into 9 plates, each with distinct nucleus; color of body whitish-pink.................. .... ***Ceroplastes rusci***

-. Shield of body divided into 7 plates with distinct nuclei; in older specimens the plates might be absent; color of body reddish-brown.. .***Ceroplastes sinensis***

3. MICROSCOPIC KEY TO CITRUS COCCIDS

1. Submarginal tubercles present...2

-. Submarginal tubercles absent...3

2. Ventral tubular ducts distributed medially, near the middle and hind legs .... ***Coccus hesperidum***

-. Ventral tubular ducts forming a submarginal band around the body .... ***Saissetia oleae***

3. Antennae with 7 segments.. ***Coccus pseudomagnoliarum***

-. Antennae with 6-7 segments...4

4. Tibio-tarsal sclerosis absent..***Ceroplastes floridensis***

-. Tibio-tarsal sclerosis present...5

5. With one subapical seta.. ***Ceroplastes rusci***

-. With two subapical setae..***Ceroplastes sinensis***

4. CITRUS COCCIDS SPECIES

4.1. *Coccus hesperidum* (Linnaeus)

Diagnostic characters: The body is oval-shaped, sometimes asymmetrical, of variable size but not longer than 4-5mm. Yellow-brown in color, with some brown spots that increase in density and intensity as the insect matures (Fig. **1**). Immatures generally flat, becoming more convex with age. Different-sized individually are found together due to the concurrent occurrence of overlapping generations.

Life history: This species, usually known as brown soft scale, has a cosmopolitan distribution and occurs throughout the Mediterranean Region.

Its area of origin is hypothesized to be either the Afro-Ethiopian, Oriental, and/or Austro-Oriental regions [2]. It is very polyphagous, infesting about 510 hosts in more than 120 plant families. Determining the number of female stages or their differences is problematic due to the difficulty of observing the molting of the different stages and to the overall similarity between stages. It is believed [3] that the insect passes through two molts before becoming adult, but this might be erroneous, as other con-generic species develop through 3 nymphal instars. An intermediate nymphal stage was reported [4] between the crawler and the first instar nymph, in which conspicuous morphological changes were observed. In the Mediterranean area *C. hesperidum* is parthenogenetic; males have however been reported in northern Europe [4, 5]. This species is ovoviviparous; immediately after oviposition the ovisac breaks and the crawlers disperse to settle nearby. The phototaxic and negatively geotaxic crawlers settle on the upper surfaces of citrus leaves and young shoots [6]. Crawlers also disperse passively, by being wind-borne. In most parts of the Mediterranean Region *C. hesperidum* has 3-4 overlapping generations.

Fig. (1). Female of *Coccus hesperidum* (Linnaeus).

Economic importance: Heavy infestations of the brown soft scale reduce tree vigor and can kill the terminals of citrus shoots. The copious amounts of the excreted honeydew are colonized by dark sooty mold fungi, which reduce the fruit grade. Furthermore, the honeydew attracts ants that may interfere with the biological control of other pests, especially of hemipterous insects, like other soft scales, armored scales and mealybugs.

Management: Only seldom is it necessary to apply chemicals against this pest, because it is usually controlled by its various natural enemies. Oil sprays are recommended in early spring against crawlers, whereas organophosphates and carbamates are ineffective as well as being detrimental to natural enemies. The timing of the chemical treatment can be determined by trapping crawlers in transparent sticky band traps. Later on and through the season, visual inspections of green twigs and upper surfaces of leaves, near the mid-rib, should be conducted.

The most efficient parasitoids are *Coccophagus lycimnia* Walker (Hymenoptera: Aphelinidae) and other *Coccophagus* spp. [7]. A single attempt at classical biological control of *C. hesperidum* was undertaken in France, when the parasitoid *Aneristus ceroplastae* Howard (Hymenoptera: Aphelinidae) was imported and temporarily became established [8]. Ant exclusion is also necessary because it enhances parasitism and predation of the pest. The population status of coccinelid predators can be assessed by beating branches with a stick over a 1 m^2 cloth.

Fig. (2). *Metaphychus flavus* (Howard) parasitizing nymph of *Coccus hesperidum* (Linnaeus).

Preventative releases of insectary-reared *C. lycimnia* can be used against light infestations, and for heavier infestations curative releases of *Rhyzobius forestieri* Mulsant (Coleoptera: Coccinellidae) should be used within an augmentative biological control program.

Fig. (3). Nymphs of *Coccus pseudomagnoliarum* (Kuwana).

Pruning is recommended within IPM programs because it increases soft scale mortality due to exposure to sunlight.

4.2. *Coccus pseudomagnoliarum* (Kuwana)

Diagnostic characters: The species is usually known as the citricola scale. The adult female is oval-shaped and more convex than *C. hesperidum*, its length ranging between 2–7 mm. The shield of the adult female bears a longitudinal ridge across its body. The young nymphs are more transparent and flat than those of the brown soft scale, older nymphs becoming dark brown (Fig. **3**). Colonies of the citricola scale consist of individuals of similar size, due to the synchronization of developmental stages that is caused by its univoltine life cycle.

Life history: The origin of this scale is debatable. As it was initially thought to be native to the arid regions of Asia [9, 10], most of the classical biological control efforts against this pest in California were concentrated in Japan and neighboring regions [9, 11, 12]. Later attempts were focused on the Mediterranean basin [12-14], where the pest was first found on citrus in Turkey, the Greek islands and in Sicily during the 1970s [15-17]. More recently it was discovered in the interior, colder citrus areas of eastern Spain [18]. The citricola scale is parthenogenetic, beginning to oviposit during May–June, and producing about 2,000 eggs during one-two months. The crawlers hatch either immediately or within 3 days [7] and initially settle on the under surfaces of leaves, infesting braches only rarely, during severe infestations. They develop slowly in the summer and autumn, molting only once during that period. Around November their color changes to darker brown and they begin to migrate onto twigs, where they overwinter as nymphs. In early spring they develop faster, turn gray and become adults. The pest has only one annual generation and its host range is narrower than that of the brown soft scale. The preferred host is trifoliate orange [*Poncirus trifoliata* (Linnaeus)], although it has also been reported infesting some Ulmaceae [19].

Economic importance: Severe infestations of the citricola scale can reduce tree vigor, induce terminal die-back, and reduce flowering, fruit size and yield. Furthermore, the sooty mold developing on the excreted honeydew can interfere with photosynthesis and downgrade the quality of the harvested fruit.

Management: Chemicals should be applied locally after the appearance of crawlers, using sprays of a 1% mineral oil emulsion with an organophosphate (0.0250%). Pruning coupled with the moderate use of nitrogen fertilizers and irrigation is also recommended. The pest is usually under effective control by natural enemies, such as *Metaphycus flavus* Howard (Fig. **2**) (Hymenoptera: Encyrtidae), other *Metaphycus* spp. and *C. lycimnia,* in Greece and in Turkey [15, 16]. In Spain *Metaphycus helvolus* Compere attacks citricola scale early in the season, when the insects are small, causing up to 50% parasitism [20]. Native coccinelid (Coleoptera) predators, such as *Exochomus quadripustulatus* (Linnaeus) and *Chilocorus bipustulatus* (Linnaeus), also play an important role in controlling pest populations [7].

Practitioners applying augmentative control of the citricola scale by insectary-released parasitoids should take into account that the timing of release is important, because most pest individuals are at the same stage of development, their size being thus highly synchronized. Early spring temperatures do not allow much parasitoid activity whereas later, during mid- to late-spring, the insects are unsuitable for parasitization. Persistent or light to medium infestations may be controlled by augmentative releases of *Metaphycus* spp. during autumn, when the pests are small. Releases of *C. lycimnia* are recommended in the Mediterranean Region [7]. More recent studies in California indicate that the use of this natural enemy might interfere with the activity of other parasitoids like *Metaphycus* spp., because of the hyperparasitic nature of the males of *C. lycimnia* [21]. The evidence for the use of this parasitoid to control citricola scale is thus in conflict [22]. Adequate control of heavier pest infestations was obtained by curative releases of *R. forestieri* (20-30 adults/tree) [7]. Small colonies of the brown soft scale may assist citricola scale control, because they serve as alternative hosts for the parasitoids during periods when *C. pseudomagnoliarum* is unsuitable for parasitism.

4.3. *Saissetia oleae* (Olivier)

Diagnostic characters: The mature female is convex and brownish in color (Fig. **4**). An H-shaped ridge appears on the dorsum of its shield during the late second instar. The body later becomes dark, very convex, and almost circular in appearance (2-5 x 1-4 x 1.2-2.5 mm) and is known as the rubber stage.

Fig. (4). Female of *Saissetia oleae* (Olivier) parasitized by *Metaphycus* sp.

Life history: This species, commonly known as black scale, a native of Africa, is now cosmopolitan. It is polyphagous, whose list of hosts includes 315 plants (*ScaleNet* – see Note below). It is a major pest of olive and citrus in Mediterranean Region, California, South Africa and other citrus growing areas. Usually the pest completes only a single annual generation in the Mediterranean Region [6, 23-25]. However, in some areas and under favorable conditions (*e.g.* closer to humid coasts or in areas with excess irrigation), a second autumn generation may develop [26, 27]. The black scale is parthenogenetic, adult males being rarely encountered. A female may lay a few hundred to 2,000 eggs that aggregate under its body. Much of the hibernating populations of *S. oleae* consist of the second and third instar nymphal stages. Young females appear in early spring and onwards, depending on the specific region. The eggs and the crawlers suffer much mortality due to high summer temperatures and low humidities, which together frequently reduce the pest's populations.

Economic importance: The black scale is the most important coccid citrus pest in the Mediterranean Region. Its infestations reduce tree vigor and may cause leaf and fruit drop and even much terminal die-back. The dark sooty mold fungi that colonize the excreted honeydew interfere with photosynthesis and downgrade fruit quality.

Management: Partial natural biological control of the black scale is provided by a complex of native parasitoids. The most important are *Metaphycus flavus*, *M. lounsburyi* Howard, *Microterys* nr. *lunatus* Dalman, *Coccophagus pulchellus* Westwood, *C. lycimnia*, *C. scutellaris* Dalman, *Scutellista cyanea* Motschulsky, *Moranila californica* Howard and *Eblema scitula* Rambour [8, 28-31]. However, their combined level of control is inadequate, requiring the introduction of several parasitoids into some Mediterranean countries [7]. Most became established but the level of control they exert on black scale populations is moderate. The most important are *M. helvolus*, *Metaphycus bartletti* Annecke and Mynhardt, *Metaphycus swirskii* Annecke and Mynhardt.

Cultural practices, such as pruning in spring, ant exclusion of ants by banding trees, the moderate use of nitrogen fertilizers and irrigation are recommended within IPM programs. Visual inspections of braches and leaves in order to detect the pests should be continuous throughout the season. Furthermore, levels of parasitism, obtained by examining samples and predator activity by beating branches of trees over a 1 m^2 cloth should be included in the monitoring program. During spring when there is a light to medium infestation (<5 adults/40cm of twig; <5 nymphs/leaf), preventative releases of *R. forestieri* (2-5 adults/tree) or of *M. lounsburyi* and *M. bartletti* should be undertaken in areas where the pest has a second generation; this should be the major treatment. Heavier infestations can be adequately controlled by releases of *R. forestieri* (20-30 adults/tree).

During summer, when natural enemies are absent and *R. forestieri* is unavailable, cover sprays of a 1% mineral oil emulsion, and organophosphate pesticides should be applied locally onto braches. Augmentative releases of *M. helvolus* should be complemented with releases of *M. lounburyi* during autumn in areas where the black scale develops a second generation. If the populations persist then sprays with light mineral oil and a synthetic insecticide, as above, should be used. Cover sprays with a copper fungicide against sooty mold fungi are recommended in cases of severe infestations.

4.4. *Ceroplastes rusci* (Linnaeus)

Diagnostic characters: The insect, known as the fig wax scale, is covered by nine pinkish–white wax plates (Fig. 5). The largest covers the dorsum; three are on either side and two are at the anterior and posterior ends. Each plate has a central nucleus. Some of the wax of the anterolateral and mediolateral plates indicates the stigmatic wax bands. The body of the mature, reddish insect is tortoise-like. Young nymphs are star-shaped.

Fig. (5). *Ceroplastes rusci* (Linnaeus).

Life history: *Ceroplastes rusci* probably originated in the Afrotropical Region [32] and is known from many Mediterranean countries (Algeria, Cyprus, Egypt, Greece, Israel, Italy, Lebanon, Morocco, Spain, Tunisia and Turkey [33]. Nowadays it occurs in other citrus growing parts of the world, including the Austro-Oriental region and Florida. The fig wax scale has been recorded from 108 host plants, being an economic pest of citrus and figs. Detailed biological data about *C. rusci* infesting figs were provided from Palestine [34]. The fig wax scale has two annual generations, overwintering as adult females on twigs. The first generation crawlers appear in early spring and settle on the leaves. After a month they molt to second instar nymphs and move onto leaf petioles and young shoots, whereon they mature. The second generation is produced in the summer and their nymphs mature in late autumn.

Economic importance: Heavy infestations and the resultant sooty mold fungi that grow on the excreted honeydew reduce tree vigor and cause leaf drop. The honeydew attracts ants that interfere with the biological control of this and other pests.

Management: Several naturally enemies keep the fig wax scale under control. The most important parasitoids are *Tetrastichus ceroplastae* Girault, *C. lycimnia*, and *S. cyanea*. The most important predators are *C. bipustulatus, E. scitula* and *E. quadripustulatus* [35-37]. Heavy infestations may be controlled with light mineral or petroleum oils alone, or in combination with an organophosphate, to be applied after the appearance of the first-generation crawlers. Augmentative releases of parasitoids, such as *T. ceroplastae* in autumn, are also being tried to control the pest within an IPM program [7].

4.5. *Ceroplastes floridensis* Comstock

Diagnostic characters: The grayish to pinkish-white body is oval or rectangular in dorsal view, hemispherical in lateral view. The waxy shield consists of a single uniform plate, with nuclei. Stigmatic filamentous wax bands are present near, and confined to, both spiracles; the anterior band is directed dorsally.

Life history: This species, known as the Florida wax scale, is probably of oriental origin and is now widely distributed in tropical areas. It is polyphagous, infesting about 250 host plant species. In Greece and in Israel the pest has two annual generations [38-40]. It is parthenogenetic and oviparous, each female producing 1,000-2,000 eggs. The young infest leaves and twigs, whereas older stages are frequently found on twigs and branches. The insects settling on the leaves prefer the midrib.

Economic importance: *Ceroplastes floridensis* is considered a major citrus pest in Israel [39]. It excretes large amounts of honeydew that is colonized by sooty mold fungi which lessen fruit value. Heavy scale infestations reduce tree vigor.

Management: A complex of natural enemies usually reduces the pest populations to below economic damage levels. The most important are *T. ceroplastae*, *C. lycimnia*, *S. cyanea*, and *E. quadripustulatus*. In summer heavy infestations may be controlled by chemical cover sprays that should be applied locally. It has been suggested [41] that the pest could be controlled by thinning its populations with a nutritional spray of 4% potassium nitrate with 2% oil.

4.6. *Ceroplastes sinensis* Del Guercio

Diagnostic characters: The reddish brown body is elliptical in dorsal view, convex in lateral view, covered by thick wax. This waxy shield is divided into a dorsal and six lateral plates on young females, all with nuclei; these plates disappear on older insects. Stigmatic filamentous wax bands are present near, and confined to, both stigmatic areas. The star-shaped nymphs are pinkish-red.

Life history: This species has originated from Central–South America and now infests citrus in temperate climates in Australia, southern Europe, northern Africa, North America and Mexico [42]. It is polyphagus, infesting about 175 host plants. In Italy and in Greece it has a single annual generation [43, 44]. Oviposition in Greece starts in early summer and the crawlers appear in July. The females are highly fecund, each producing about 3000 eggs. Some individuals attain maturity during October-November, whereas the rest attain maturity in spring. Small numbers of male nymphs were reported during autumn [44].

Economic importance: *Ceroplastes sinensis* excretes large amounts of honeydew that are colonized by dark sooty mold fungi, which interfere with photosynthesis. The host plants are stressed, fruit yield and quality are reduced, and heavy infestations may lead to the death of branches.

Management: The Chinese wax scale is a sporadic pest of citrus in Italy and Spain. Its most important natural enemies in the Mediterranean region are *S. cyanea* and *E. scitula* [35, 45, 46]. Heavy infestations may be controlled by chemicals, applied in summer after the appearance of crawlers.

NOTES

Data on host plant species are derived from Scale Net: A Database for the Scale Insects of the World. http://www.sel.barc.usda.gov/SCALENET/SCALENET.HTM.

REFERENCES

[1] Ben-Dov Y. A Systematic Catalogue of the Soft Scale Insects (Homoptera: Coccoidea: Coccidae) of the World with Data on Geographical Distribution, Host Plants, Biology and Economic Importance. Gainesville: Sandhill Crane Press Inc. 1993.

[2] Kozár F, Ben-Dov Y. In: Ben-Dov Y and Hodgson CJ, Eds. Soft Scale Insects, Their Biology, Natural Enemies and Control. Elsevier, Amsterdam, 1997; pp. 213-28.

[3] Annecke DP. Biological studies on the immature stages of soft brown scale, *Coccus hesperidum* (Homoptera: Coccidae). South Afr J Agric Sci 1966; 9: 205-28.

[4] Saakyan-Baranova AA. On the biology of the soft scale *Coccus hesperidum* L. (Homoptera, Coccoidea). Entomol Rev 1964; 43: 268-96.

[5] Newstead R. Observations on scale-insects (Coccidae) - III. Bull Entomol Res 1917; 7**:** 343-80.

[6] Bodenheimer FS. Citrus Entomology in the Middle East. W. Junk, The Hague 1951.

[7] Katsoyannos P. Integrated Insect Pest Management for Citrus in Northern Mediterranean Countries. Benaki Phytopathological Institute, Athens, Greece 1996.

[8] Panis A. Note sur quelques insectes auxiliares regulateurs des populations de Pseudococcidae et de Coccidae (Homoptera, Coccoidea) des agrumes en Provence orientale. Fruits 1981; 36: 49-52.

[9] Bartlett BR. In: Clausen CP Ed. Introduced Parasites and Predators of Arthropod Pests and Weeds: A World Review. U.S.D.A. Agriculture Handbook 480 1978; pp. 57-74.

[10] Gill RJ. The Scale Insects of California. Tech Series Agric Biosyst Plant Path 1. California Department of Food and Agriculture, Sacramento, CA. 1988.

[11] Kennett CE. Results of exploration for parasitoids of citricola scale, *Coccus pseudomagnoliarum* (Homoptera: Coccidae), in Japan and their introduction to California. Kontyu 1988; 56: 445-57.

[12] Kennett CE, Hagen KS, Daane KM. In: Nechols *et al.*, Eds. Biological Control in the Western United States: Accomplishments and Benefits of Regional Research Project W-84, 1964-1989. Oakland, CA: University of California Division of Agriculture and Natural Resources 1995; pp. 148-49.

[13] Bernal JS, Luck RF, Morse JG. Host influences on sex ratio, longevity, and egg load in two *Metaphycus* species parasitic on soft scales: Implications for insectary rearing. Entomol Exp Appl. 1999a; 92: 191-204.

[14] Bernal JS, Luck RF, Morse JG. Augmentative release trials with *Metaphycus* species (Hymenoptera: Encyrtidae) against citricola scale (Homoptera: Coccidae) in California's San Joaquin Valley. J Econ Entomol 1999b; 92: 1099-1107.

[15] Öncüer C, Tunçyürek M. Observations sur la biologie et les ennemis naturels de *Coccus pseudomagnoliarum* Kuw. dans les vergers d'agrumes de la région égéenne. Fruits 1975: 30: 255-7.

[16] Argyriou LC, Ioannides AG. *Coccus aegaeus* (Homoptera, coccidea, coccidae) De Lotto: nouvelle espece de lecanine des citrus en Grece. Fruits 1973; 30: 161-62.

[17] Barbagallo S. Notizie sulla presenza in Sicilia di una nuova cocciniglia degli agrumi. Entomologica 1974; 10: 121-39.

[18] Guillten M, Breva M, Castaro A *et al. Coccus pseudomagnoliarum* (Hemiptera: Coccidae), nueva plaga de citricos en Espana. Bol Sanid Veg Plaga*s* (In press).

[19] Dreistadt SH. Citricola scale (Homoptera: Coccidae) abundance on Chinese hackberry and scale control with spray oil or acephate trunk implants. J Econ Entomol 1996; 89: 481-7.

[20] Tena A, Garcia-Mari F. Suitability of citricola scale *Coccus pseudomagnoliarum* (Hemiptera: Coccidae) as host of *Metaphycus helvolus* (Hymenoptera: Encyrtidae): Influence of host size and encapsulation. Biol Control 2008; 46: 341-7.

[21] Bernal JS, Luck RF, Morse JG, Drury MS. Seasonal and scale size relationships between citricola scale (Homoptera: Coccidae) and its parasitoid complex (Hymenoptera: Chalcidoidea) on San Joaquin Valley citrus. Biol Control 2001; 20: 210-21.

[22] Schweizer H, Morse JG, Luck RF, Forster LD. Augmentative releases of a parasitoid (*Metaphycus* sp. nr. *flavus*) against citricola scale (*Coccus pseudomagnoliarum*) on oranges in the San Joaquin Valley of California. Biol Control 2002; 24: 153-66.

[23] Argyriou LC. Studies on the morphology and biology of the black scale [*Saissetia oleae* (Bernard)] in Greece. Ann Inst Phytopathol Benaki 1963; 5: 353-77.

[24] Neuenschwander P, Paraskakis M. Studies on distribution and population dynamics of *Saissetia oleae* (Oliv.) (Hom., Coccidae) within the canopy of the olive tree. Z Angew Entomol 1980; 90: 366-78.

[25] Podoler H, Bar-Zacay I, Rosen D. Population dynamics of the Mediterranean black scale, *Saissetia oleae* (Olivier), on citrus in Israel. 1. A partial life-table. J Entomol Soc South Afr 1979; 42: 257-66.

[26] Llorens Climent JM. Las cochinillas de los agrios. Servicio de Protección de los Vegetales, Valencia, Spain 1984.

[27] Tena A, Soto A, Vercher R, Garcia-Mari F. Density and structure of *Saissetia oleae* (Hemiptera: Coccidae) populations on citrus and olives: relative importance of the two annual generations. Environ Entomol 2007; 36: 700-06.

[28] Argyriou LC, Stavraki HG, Mourikis PA. A list of recorded entomophagous insects of Greece. (In Greek; summary in English). Benaki Phytopathological Institute, Athens, Greece 1976.

[29] Carrero JM, Limon F, Panis A. Note biologique sur quelques insectes entomophages vivant sur olivier et sur agrumes en Espagne. Fruits 1977; 32: 548-51.

[30] Tuncyürek M. Les cochenilles nuisibles aux *Citrus* en Turquie. Al Awamia 1970; 37: 67-80.

[31] Viggiani G. Acclimato in Italia *Metaphycus helvolus* (Compere), parasite di *Saissetia oleae* (Oliv.) e di alter dannose cocciniglie. Bol Lab Entomol Agr "Fillippo Silvestri" Portici 1978; 32: 52-9.

[32] Qin TK, Gullan PJ, Beattie GAC. Biogeography of the wax scales (Insecta: Hemiptera: Coccidae: Ceroplastinae). J Biogeogr 1998; 25: 37-45.

[33] Talhouk AMS. Citrus Pests Throughout the World. Basel, Switzerland: Ciba-Geigy Agrochemicals, Technical Monograph No. 4, 1975.

[34] Bodkin GE. The fig wax scale (*Ceroplastes rusci* L.) in Palestine. Bull Entomol Res 1927; 17: 259-63.

[35] Panis A. Action predatrice D'*Eublema scitula* (Lepidóptera Noctuidae, Erastriinae) dans le Sud de la France. Entomophaga 1974;19: 493-500.

[36] Argyriou LC, Santorini AP. On the phenology of *Ceroplastes rusci* L. (Hom. Coccidae) on fig-trees in Greece. Med Rijksfacul Land Gent 1980; 45: 593-01.

[37] Inserra S. Il *Ceroplastes rusci* L. negli agrumeti della provincia di Catania. Bol Lab Entomol Agr "Fillippo Silvestri" Portici 1971 28: 77-97.

[38] Argyriou LC, Kourmadas AL. *Ceroplastes floridensis* Comstock an important pest of citrus trees in Aegean islands. Fruits 1980; 35: 705-8.

[39] Ben-Dov Y. Phenology of the Florida wax scale, *Ceroplastes floridensis* Comstock (Homoptera: Coccidae) on citrus in Israel. Phytoparasitica 1976; 4: 3-7.

[40] Podoler H, Dreishpoun Y, Rosen D. Population dynamics of the Florida wax scale, *Ceroplastes floridensis* (Homoptera: Coccidae) on citrus in Israel. 1. A partial life-table. Acta Oecolog 1981; 2: 81-91.

[41] Yardeni A, Shapira E. Thinning populations of the Florida wax scale, *Ceroplastes floridensis* Comstock (Coccidae), by use of potassium nitrate and spray oil, as an option in IPM of citrus groves in Israel. Israel J Entomol 1995; 29: 271-76.

[42] Qin TK, Gullan PJ, Beattie GAC *et al.* The current distribution and geographical origin of the scale insect pest *Ceroplastes sinensis* (Hemiptera: Coccidae). Bull. Entomol. Res. 1994; 84: 541-9.

[43] Monastero S, Zaami V. Le cocciniglie degli agrumi in Sicilia *Ceroplastes sinensis* D.G., *Pseudococcus citri* R., *Icerya purchasi* M. Boll Ist Entomol Agr Univ Palermo 1959; 3: 1-82.

[44] Stathas GJ, Kavallieratos NG. Eliopoulos A Biological and ecological aspects of Chinese wax scale, *Ceroplastes sinensis* Del Guercio (Hemiptera: Coccidae): a two-year study from central Greece. Austr J Entomol 2003; 42: 271-75.

[45] Limon F Melia A Blasco J Moner P [Study of the distribution, level of attack, parasites and predators of the lecanine scale insects (*Saissetia oleae* Bern. and *Ceroplastes sinensis* Del Guercio) on citrus in the Province of Castellon.] (In Spanish; summary in English). Bol Serv Def Cont Plag Insp Fitopatolog 1976; 2: 263-76.

[46] Monaco R, Sabatino A. Epidemiologia e piante ospiti del *Ceroplastes sinensis* in Puglia. Inf Fitopatol 1980; 30: 3-6.

CHAPTER 14

Diaspididae

Uri Gerson*

Department of Entomology, Faculty of Agricultural, Food and Environmental Quality Sciences, The Hebrew University of Jerusalem, P.O. Box 12, Rehovot, 76-100, Israel

Abstract: The Diaspididae, like all Coccoidea (Hemiptera), feed on plants and may at times become pests. Nine species are important pests of citrus in the Mediterranean Region. They consist of *Aonidiella aurantii* (Maskell), *Aspidiotus nerii* Bouché, *Chrysomphalus aonidum* (Linnaeus), *C. dictyospermi* (Morgan), *Lepidosaphes beckii* (Newman), *L. gloverii* (Packard), *Parlatoria cinerea* Hadden, *P. pergandii* Comstock, and *P. ziziphi* (Lucas). Field and microscopic keys for these pests are provided, along with diagnostic characters, life history details, economic importance data and management methods for each species.

Keywords: Citriculture, Mediterranean Region, armored scale insects, bio-ecology, damage, control.

1. INTRODUCTION

The Diaspididae (Hemimetabola: Hemiptera: Coccoidea), or armored scale insects, is a large family of nearly 2,500 described species placed in about 400 genera. The adult females are devoid of antennae, legs and wings, and the body is covered by a hard, separable (except during molt) shield. The male nymphs are covered by a smaller, elongated shield, and the short-lived adults have legs, antennae and wings. The legged first-instar nymphs ("crawlers") reach new sites on host plants by walking, spreading to other hosts by winds. Like all Coccoidea, they feed on plants and may at times become pests [1]. Nine species are known to be important pests of citrus in the Mediterranean Region.

2. FIELD KEY TO THE CITRUS DIASPIDIDS

1. Body and shield rounded or oval; body lilac or yellow 2

-. Body and shield elongate; body white 9

2. Body of adult female lilac 3

-. Body of adult female yellow 4

3. Shield black, with two longitudinal ridges and a caudal fringe of whitish wax .... ***Parlatoria ziziphi***

-. Shield gray, without longitudinal ridges or a posterior fringe of whitish wax.......... ***Parlatoria cinerea*** and ***Parlatoria pergandii****

4. Body of adult female separable from shield 5

-. Body of adult female inseparable from shield .... ***Aonidiella aurantii***

5. Shield of female yellow, grey or brown 6

-. Shield of female dark-red to black.......... ***Chrysomphalus aonidum***

***Address correspondence to Uri Gerson:** Department of Entomology, Faculty of Agricultural, Food and Environmental Quality Sciences, The Hebrew University of Jerusalem, P.O. Box 12, Rehovot, 76-100, Israel; E-mail: Gerson@agri.huji.ac.il

6. Shield brown-yellow, male body with reddish spots .... ***Aspidiotus nerii***

-. Shield grey-white, male body without reddish spots .... ***Chrysomphalus dictyospermi***

7. Shield narrow, almost parallel sided; anterior part of adult body sclerotic .... ***Lepidosaphes gloveri***

-. Shield broadening towards posterior end, anterior part of adult body not sclerotic .... ***Lepidosaphes beckii***

**Parlatoria cinerea* and *Parlatoria pergandii* can only be separated by microscopic characters.

3. MICROSCOPIC KEY TO THE CITRUS DIASPIDIDS

1. Dorsal macroducts two-barred 2

-. Dorsal macroducts one-barred 6

2. Pygidial openings of dorsal macroducts with sclerotic basal rings 3

-. Pygidial openings of dorsal macroducts without sclerotic basal rings 5

3. Without ear-like lobes laterad of anterior spiracles 4

-. With ear-like lobes laterad of anterior spiracles .... ***Parlatoria ziziphi***

4. With 15-30 pores on either side of vulva; 3-4 macroducts around anus .... ***Parlatoria cinerea***

-. With 10-16 pores on either side of vulva; no macroducts around anus .... ***Parlatoria pergandii***

5. Abdominal segments II to IV with strong lateral lobes .... ***Lepidosaphes gloverii***

-. Abdominal segments II to IV without strong lateral lobes .... ***Lepidosaphes beckii***

6. Body of adult female not kidney-like, with perivulvar pores 7

-. Body of adult female kidney-like, without perivulvar pores .... ***Aonidiella aurantii***

7. Dorsal macroducts long, 6-10 times their diameters 8

-. Dorsal macroducts short, 4-5 times their diameters .... ***Aspidiotus nerii***

8. A cluster of short submarginal ducts located on dorsum of 2nd abdominal segment. Often with 5 groups of perivulvar pores .... ***Chrysomphalus aonidum***

-. Without a cluster of short submarginal ducts on dorsum of 2nd abdominal segment. With 4 groups of perivulvar pores .... ***Chrysomphalus dictyospermi***

4. CITRUS DIASPIDID SPECIES

4.1. *Aonidiella aurantii* (Maskell)

Diagnostic characters: Shield round, reddish; exuvium subcentral (Fig. **1**); body yellow-red, the anterior pre-pygidial parts becoming kidney-like and sclerotic at maturity and inseparable from the shield. With one-barred dorsal macroducts; without perivulvar pores. Male shield pale-yellow, with yellow exuvium located off-center. Body color orange.

Life history: This species, known as the California red scale, is viviparous, giving birth to crawlers, whose emergence is affected by temperature and light. They emerge mostly in the morning, at temperatures above 12°C, but not in the dark at low temperatures. Under warmer conditions they emerge also in the dark, mostly during morning [2]. First-generation crawlers tend to settle on exposed fruit surfaces, where they are joined by those of the second generation, the progeny of females already on the fruits, [3]. The California red scale raises three or four annual generations, the peaks in crawler production occurring in February, June, September and October. In warmer Regions it may complete another 1-2 generations. Pest populations usually undergo two large cycles, peaking in the autumn, ebbing during winter and spring, then increasing in mid-summer to decline again in late summer. The settlement sites of the crawlers are affected by the prevailing light and humidity. Under humid conditions they prefer to settle on young citrus trees or on the outer rows of mature orchards, being rare within dense plantations. But in warmer and drier areas the pest is more abundant within trees, where humidity is somewhat higher.

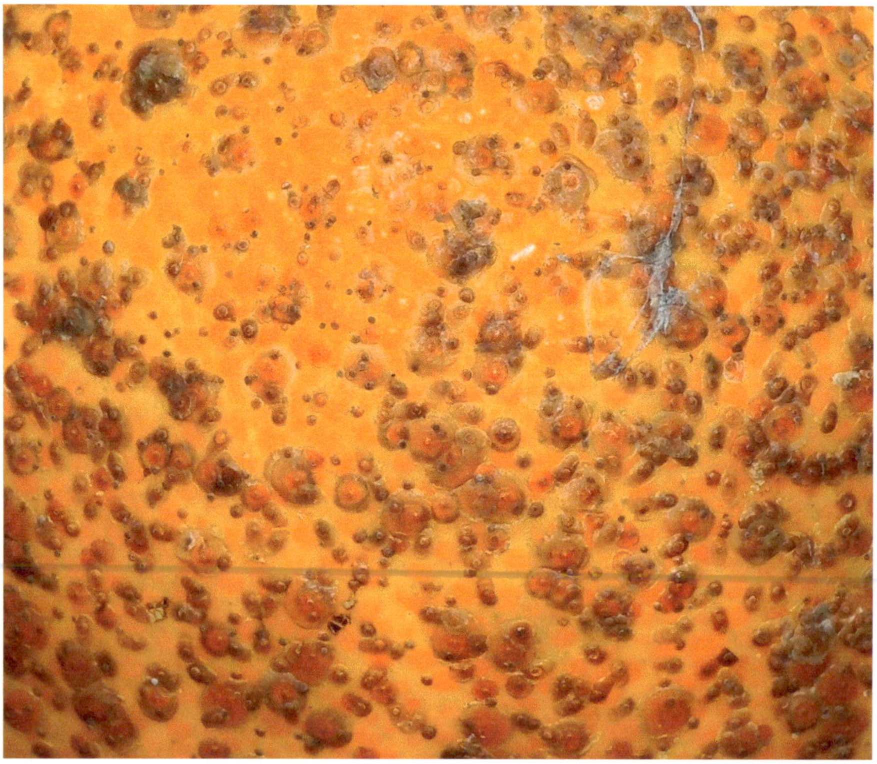

Fig. (1). A heavy infestation of *Aonidiella aurantii* (Maskell).

Crawlers prefer to settle on citrus species and varieties that have relatively few oil glands (sour orange, *Citrus aurantium* Linnaeus) as compared to those with abundant oil glands (mandarin, *Citrus reticulata* Blanco), a factor that affects the damage caused to these fruits. The threshold of development was calculated to be 11°C and 825 days-degrees were required for one generation. The optimal conditions for development seem to be 24°C and air humidity of above 60%. The fecundity of the pest is affected by the season and by the parts of the host plant. The wind-borne crawlers, which appear to be relatively tolerant to desiccation, then establish new colonies. California red scale may be mass-reared on green lemons, banana squash fruit or on potato tubers, whereas individuals can be reared on disks of lemon leaves floating on water.

Economic importance: This species is a major pest of citrus throughout the world, attacking all above-ground parts of the trees. On trunks of young trees the pest may form thick incrustations, and it also infests the leaves (Fig. **2**) and fruits (Fig. **3**). At feeding the pest injects toxins into the leaves, which causes them to yellow and drop. Heavy defoliation retards and stunts tree growth, affecting the yield; young trees are especially at risk. Although individuals that settle on young fruit in early summer often die, the pit that remains around them detracts from the market value of such fruit. Large pest numbers on the face of fruits are difficult to remove even by descaling machines, often further reducing the fruits' value.

Fig. (2). Orange leaves infested by *Aonidiella aurantii* (Maskell).

An average of 220 offspring was produced on lemon fruit in the spring, 108 in the summer and only 38 on the same host in autumn. On leaves fecundity came to 19 progeny in spring, 80 in summer and 44 in autumn. The crawlers (and males) may be dispersed by winds to a distance of over 300 m, their numbers decreasing with increasing distances from the source.

Management: Natural enemies often suffice to control the California red scale. Except in very dry and hot areas, and unless disrupted by pesticides, the natural enemies consist mainly of a suite of hymenopterous parasitoids of the family Aphelinidae, especially ectoparasitoids of the genus *Aphytis* [4].

The endoparasitoid *Comperiella bifasciata* Howard (Encyrtidae) has been introduced into and became established in Cyprus [5]. The coccinellid beetle *Chilocorus bipustulatus* (Linnaeus) is a major predator.

Other options include descaling machines, which can remove the pests (unless infesting fruits in very large numbers) with high-pressure water sprays [6]. The isolation of the female sex pheromone of the California red scale led to efforts to disrupt its mating by attracting the short-lived males, efforts that were later abandoned. The pheromone is still used to monitor the periods of mass male flights, hence of subsequent crawler emergence, in order to facilitate their chemical control. Many pesticides were used to control the pest. These included oils and organophosphates; the use of the latter has been discontinued due to environmental concerns and the widespread development of resistance. When biological control is disrupted insect growth regulators may be applied in the spring against the crawlers, the most susceptible stage [6]. If windbreak trees (*e.g.* cypress) are also infested, they too should be treated.

Website: http://www.inra.fr/Internet/Produits/HYPPZ/RAVAGEUR/6aonaur.htm

4.2. *Aspidiotus nerii* Bouché

Diagnostic characters: Female shield round, yellow-brown, exuvium central (Fig. **4**); body yellow; dorsal macroducts one-barred, short, about 4-5 times their diameters. Pygidium with four groups of perivulvar pores. Shield of male whitish, with yellow exuvium off center; body yellow with reddish spots.

Life history: This wide-spread, very polyphagous species is also known as ivy scale or oleander scale. When grown on potato tubers, development (egg to egg) required about six weeks at 24 and 28°C, almost 13 weeks at 19°C, and the insect produced ca 100 progeny at 24°C, during seven weeks [7].

Fig. (3). Severe attack of *Aonidiella aurantii* (Maskell) on lemon fruit.

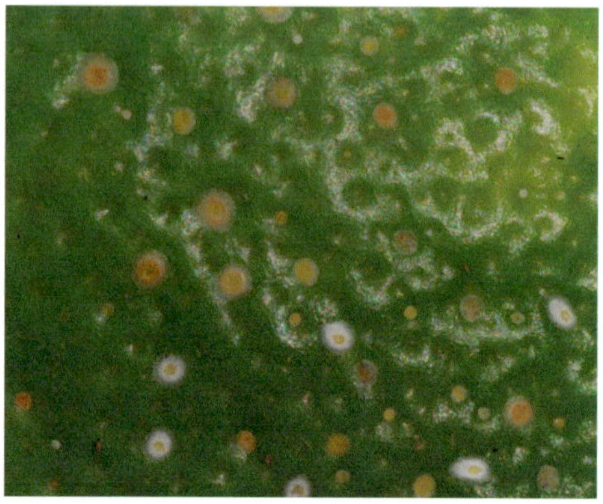

Fig. (4). *Aspidiotus nerii* Bouché.

The scale completes 3-5 annual generations [8], usually overwintering as young females [9]. The male crawlers often prefer lighted plant Regions, clearly apart from the settlement sites of the females. Improved plant conditions appear to predispose citrus to ivy scale attack.

The oleander scale occurs in two forms: one, the more common, has both males and females, whereas the other form is uniparental, with females only [7].

Economic importance: The ivy scale is a minor pest of citrus, mostly in the western Mediterranean Region, whereas in its eastern countries it seldom occurs on citrus, mainly on lemon. Damage is due to sucking out plant sap, which causes leaf drop as well as reducing plant growth. Pest presence on the fruits disfigures them, reducing their market value [10] (Fig. **5**).

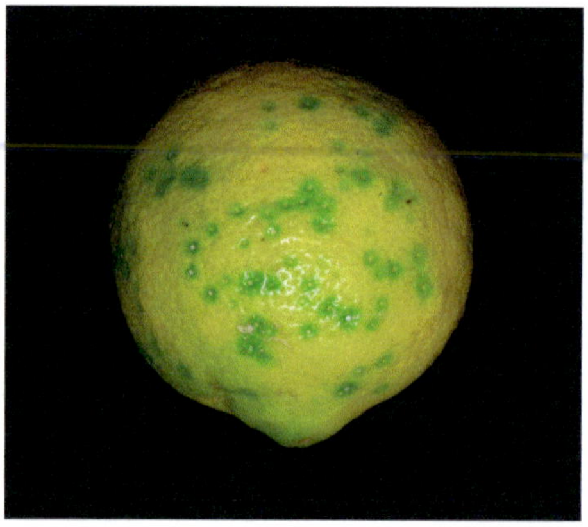

Fig. (5). Lemon fruit infested by *Aspidiotus nerii* Bouché.

Management: Natural enemies often suffice to control the ivy scale, unless disrupted by pesticide applications. Pruning of heavily-infested branches may reduce the populations. Summer mineral oils and various pesticides control the pest, especially when aimed at the crawlers. The ivy scale is attacked and often controlled by a group of hymenopterous parasitoids of the family Aphelinidae, especially ectoparasitoids of the genus *Aphytis*. The pest is also attacked by several predatory beetles, especially of the family Coccinellidae (lady beetles) like *C. bipustulatus* [11].

Website: http://www.inra.fr/Internet/Produits/HYPPZ/RAVAGEUR/6aspner.htm

4.3. *Chrysomphalus aonidum* (Linnaeus)

Diagnostic characters: Female shield round, dark-red to black, exuvium central (Fig. **6**). Body yellow; dorsal macroducts one-barred, long, about 6-10 times their diameters. A cluster of submarginal dorsal ducts is located on the 2nd abdominal segment. Pygidium often with five groups of perivulvar pores, 7-12 on each side and another 2-3 in anterior position. Male shield dark, oval in shape, exuvium submarginal; body orange.

Fig. (6). *Chrysomphalus aonidum* (Linnaeus).

Life cycle: This pest is also known as the Florida red scale, circular black scale and circular purple scale. In the Mediterranean Region it usually raises 4-5 annual generations, in February, during mid-summer, September and October. Pest populations attain their largest numbers in the autumn, drop during winter, increase in mid-summer, decline again in late summer and then peak. The crawlers are very sensitive to the dry and hot spring (*e.g.* khamsin) winds, preferring to settle in the inner tree parts, often not far from their mothers. The optimal conditions for development seem to be around 24°C and air humidity of above 60%; this pest does not tolerate arid conditions. Its fecundity is affected by the season and by the host plant. About 220 offspring were produced on lemons in spring, 110 in summer but only 40 on the same host in autumn. On citrus leaves the pest had 20 progeny in spring, 80 in summer and 44 in autumn [9]. The pest may be mass-reared on banana squash fruit.

Economic importance: The pest occurs in most tropical and subtropical parts of the world, and although polyphagous, it injures mainly citrus, preferring grapefruits and lemons over oranges (Figs. **7** and **8**).

It occurs on the leaves and fruit, rarely on branches. Although attacking young as well as established citrus trees, it prefers the latter. During feeding the pest secretes toxins into the affected leaves, which then turn yellow and fall. Heavy leaf drop can retard and stunt tree growth, reducing yields. Young infested fruits shrink, fail to develop and may die; if attacked at a later stage, the black, conspicuous, hard-to-remove shields cover the fruit and greatly reduce its market value.

Fig. (7). Lemon fruit infested by *Chrysomphalus aonidum* (Linnaeus).

Management: Florida red scale is usually completely controlled by the hymenopterous aphelinids *Aphytis holoxanthus* DeBach (an ectoparasitoid) and *Pteroptrix smithi* (Compere) (an endoparasitoid), introduced into Israel during 1956/57 [12]. Together they provide complete control of the pest, unless disrupted by pesticides and/or by hot and dry weather [13]. Other options include descaling machines, which can remove the pests (unless infesting fruits in very high numbers) with high-pressure water sprays [6], and oil sprays. These should best be applied when pest populations ebb, in late summer, or (on Valencia trees) in the autumn. If windbreak trees (*e.g.* cypress) are also infested, they should likewise be sprayed.

Fig. (8). Orange fruit infested by *Chrysomphalus aonidum* (Linnaeus).

Website: http://primera.tamu.edu/kcchome/images/floridaredscale.htm#Linnaeus

4.4. *Chrysomphalus dictyospermi* (Morgan)

Diagnostic characters: Female shield round, grey-white, exuvium central. Body yellow, dorsal macroducts one-barred, long, about 6-10 times their diameters. Without a cluster of submarginal dorsal ducts located on the 2nd abdominal segment; pygidium with four groups of perivulvar pores, about 5-10 on either side. Male shield grey, exuvium submarginal; body pale-orange.

Life history: This species, also known as the Spanish Red or dictyospermum scale, is polypahgous, has 3-4 annual generations, from March-April on. Each female lays 100-150 eggs, the emerging crawlers settling first on elder leaves, later also on the young foliage, shoots and fruits [8].

Economic importance: Yellow spots appear around the pest's feeding sites, causing leaf drop. Heavy infestations result in the wilting of branches and the distortion of fruits.

Management: The pest is usually controlled by the hymenopterous encyrtid *Comperiella lemniscata* Compere and Annecke and by introduced ectoparasites of the genus *Aphytis* [14, 15].

Website: http://www.inra.fr/Internet/Produits/HYPPZ/RAVAGEUR/6chrdic.htm

4.5. *Lepidosaphes beckii* (Newman)

Diagnostic characters: Female shield elongated, brown-red, broadening towards the posterior end, exuvium at narrow end (Fig. **9**). Anterior part of whitish body remains membraneous at maturity. Dorsal macroducts two-barred; perivulvar pores in five groups, about 20-30 on each side and another 4-8 in an anterior position; abdominal segments II-IV without lateral lobes. Body of male whitish, shield brown-red, elongated, exuvium terminal.

Life cycle: This pest, known as the purple scale or the citrus mussel scale, infests mostly citrus leaves and fruits, being less common on green twigs (which it may kill), seldom occurring on wooden branches. It prefers the inner parts of trees and usually begins to be noticed in citrus orchards as they reach their 10-12th year. The first-generation crawlers settling on young fruit in the spring prefer the underside of the calyx, thus initially remaining unnoticed. The second generation crawlers mostly originate from other tree substrates or from females that had settled under the fruit calyx.

Thus they colonize the upper parts of the fruits and then its entire face [3]. Females living on fruit produce about 250-300 eggs, those on leaves only about 170. The threshold of development is at 8°C and about 1,100 day-degrees are required for one generation. Annually the pest usually completes four generations [9].

Economic importance: In many parts of the world the purple scale is a major pest of citrus, damaging mostly the leaves (Fig. **10**) and fruit (Fig. **11**). Heavily infested leaves turn yellow (probably from a combination of toxic secretions and losing assimilates) and may drop.

Fig. (9). *Lepidosaphes beckii* (Newman).

As a result tree growth is stunted. The sites where the fruit-infesting scales had fed remain green at color change, detracting from their market value. Purple scale can be mass-reared on green lemons, potatoes and banana squash.

Management: The pest may be controlled with white oils or with organophosphate insecticides, but usually it is completely controlled by the introduced hymenopterous aphelinid ectoparasitoid *Aphytis lepidosaphes* Compere [16], which is attracted to the pest by a kairomone [17].

Fig. (10). *Lepidosaphes beckii* (Newman) on citrus leaf.

Fig. (11). *Lepidosaphes beckii* (Newman) on citrus fruit.
Website: http://www.inra.fr/Internet/Produits/HYPPZ/RAVAGEUR/6mytbec.htm

4.6. *Lepidosaphes gloverii* (Packard)

Diagnostic characters: Female shield elongated, parallel-sided, exuvium terminal (Fig. **12**). Anterior part of body white, becoming sclerotic at maturity. Dorsal macroducts two-barred; perivulvar pores in five groups, 15-20 on each side, another 2-4 in an anterior position; abdominal segments II-IV with lateral lobes. Male shield brown-red, elongated, exuvium terminal; body whitish.

Life cycle: This pest, also known as the Glover scale, infests mostly citrus twigs and small branches, less commonly leaves and fruit.

Fig. (12). *Lepidosaphes gloverii* (Packard).

It prefers the inner, more humid parts of trees as it is unable to survive hot, dry summers [18]. The pest usually begins to be noticed in citrus orchards as they mature. When reared on lemons at 30°C and 75 ± 50% RH, the shortest mean generation time (crawler to crawler) was 67 days; each female produced about 200 eggs under these conditions [19]. The pest usually has five annual generations [8].

Economic importance: Heavy infestations of *L. gloverii* can damage fruit in maturing citrus orchards because the area around the pest does not change color, leaving green spots that reduce fruit market value [19]. Although usually not regarded as a major pest of citrus, its co-occurrence with purple scale (*L. beckii*) can cause compounded damage. Another direct result of heavy infestations is widespread leaf drop and twig dieback [19] that weaken trees, reducing their productivity. The Glover scale seems to be more common in the western Mediterranean Region.

Management: The Glover scale is usually controlled by several species of the hymenopterous aphelinid genera *Aphytis* and *Encarsia*.

Website:
http://scholar.google.co.il/scholar?hl=en&q=website+Lepidosaphes+gloveri&btnG=Search&as_sdt=2000&as_ylo=&as_vis=0

4.7. *Parlatoria cinerea* Hadden

Diagnostic characters: Female shield oval, grey, exuvium terminal, without dorsal ridges or a posterior fringe of white wax. Body of young female white, later turning lilac, dorsal macroducts two-barred, the pygidial apertures with sclerotic rings. Three-four macroducts are located around anus, and with 15-30 pores on either side of vulva. Male shield whitish, elongate, with the dark exuvium at one end; body lilac.

Life history: The biology of this pest, known as the tropical grey chaff scale, has not been studied, but it is probably quite similar to that of the chaff scale, *Parlatoria pergandii* Comstock (below), except that *P. cinerea* seems to prefer warmer and more humid regions. In orchards where it coexists with the chaff scale, the latter dominates the leaves during summer, the tropical grey chaff scale in winter.

As both pests are attacked by the same suite of natural enemies, these shifts probably reflect the pests' sensitivity to prevalent humidities, with *P. pergandii* being the less susceptible [20]. Both species undergo 3-4 annual cycles.

Economic importance: The pest infests all above-ground parts of citrus trees aged 10-12 years or more. In South America it was also found on citrus roots.

Most damage occurs on the fruit, which is infested shortly after setting. Puncture wounds inflicted by the pest remain undetected until the fruit begins to change color. The sites around these wounds then remain green, resulting in many greenish, disfiguring blemishes, which reduce the fruits' market value. Late-ripening citrus varieties, like Valencia, suffer heavier damage because the pest raises larger populations thereon. Damage is estimated to be below 10% of overall costs. Very heavy feeding on leaves may cause foliage twisting. The pest also settles on the woody parts of citrus trees (especially on the trunk), where they form thick encrustations. The insects living there are hidden under the shields of the dead pests, which do not drop off. However, there is no indication that these encrustations cause any damage; tropical grey chaff scale does not secrete toxins into the host plants.

Management: High-pressure water sprays (of 10-20 l per tree at 300 psi) may reduce 60-90% of the pest populations on the trunk and main branches in about one year [21]. Another option is control with mineral oils; a single application of the insect growth regulator Tiger (pyriproxyfen) may suffice to eliminate the pest for one year [22].

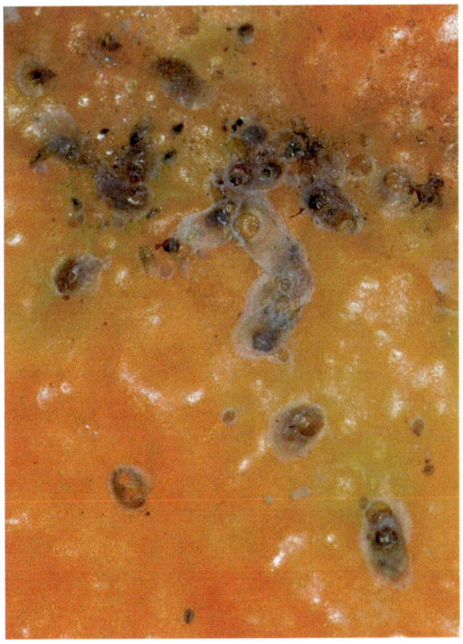

Fig. (13). *Parlatoria pergandii* Comstock.

The tropical grey chaff scale is attacked by several predatory insects and mites and by parasitic Hymenoptera. The important predators are the beetle *C. bipustulatus* and the mite *Hemisarcoptes coccophagus* Meyer (Hemisarcoptidae). The Hymenoptera are the ectoparasitoid *Aphytis hispanicus* (Mercet) and the endoparasitoid *Encarsia inquirenda* (Silvestri), both Aphelinidae. The predators are most active on the wooden parts of the trees during summer, whereas the parasitoids attack the pest mostly in the spring, especially on the leaves and fruits. Neither group of natural enemies appears to reduce the level of pest damage. However, the predators, acting during the pest's population build-up in summer, appear to be more important in reducing its numbers [23].

Website:
http://scholar.google.co.il/scholar?hl=en&q=website+Parlatoria+cinerea&btnG=Search&as_sdt=2000&as_ylo=&as_vis=0

4.8. *Parlatoria pergandii* Comstock

Diagnostic characters: Female shield oval, grey, exuvium terminal; without dorsal ridges or a posterior fringe of white wax (Fig. **13**). Body of young females white, later becoming lilac; dorsal macroducts two-barred, their pygidial apertures with sclerotic rings. No macroducts located around anus, and with 10-15 pores on each side of vulva. Male shield whitish, elongate, with dark exuvium at one end; body lilac.

Life cycle: This pest, known as the chaff scale, completes four overlapping generations in the coastal plain of Israel. Its populations on fruits and leaves peak in spring and then decline rapidly, possibly due to the hot spring winds and to natural enemies, some of which are most active at that time. Pest numbers later increase, building up during summer and winter. The size of these populations appears to be stable from year to year [24].

In orchards where the pest coexists with *P. cinerea*, the latter dominates on leaves during winter, the chaff scale in summer.

Fig. (14). *Parlatoria pergandii* Comstock on mature citrus fruit.

As both are attacked by the same suite of natural enemies, these shifts probably reflect the insects' sensitivity to prevalent humidities, with *P. cinerea* being the more susceptible [20]. Chaff scale can be mass-reared on Butternut squash fruit, *Cucurbita moschata* Duchesne.

Economic importance: The pest infests all above-ground parts of citrus trees aged 10-12 years or more; heavy attacks on such trees may retard their development. Most damage occurs on the fruits, where feeding sites remain undetected until the fruit begins to mature, due to the fact that these sites remaining green. As a result many greenish, disfiguring blemishes appear on the yellow or orange-colored fruit, reducing their market value (Figs **14** and **15**). Late-ripening citrus varieties, like Valencia, suffer heavier damage because the pest remains on the fruits for longer periods, raising larger populations thereon. Damage is estimated to be below 10% of overall costs. Rare, very heavy infestations on leaves cause loss of color and twisting [24].

Management: High-pressure water sprays (of 10-20 liter per tree at 300 psi) may reduce pest populations on the trunk and main branches by over 60% within three months, by about 90% after one year [21]. The pest can be controlled by mineral oils; a single application of the insect growth regulator Tiger (pyriproxyfen) sufficed to control the pest for one year [22].

Fig. (15). *Parlatoria pergandii* Comstock on mature orange fruit.

The chaff scale is attacked by several predatory insects and mites and by parasitic Hymenoptera. The important predators are the beetle *C. bipustulatus* and the mite *H. coccophagus*. The Hymenoptera are the aphelinids *A. hispanicus* and *E. inquirenda.* The predators are most active on the wooden parts of the trees during summer, whereas the parasitoids attack *P. pergandii* mostly in the spring, especially on the leaves and fruits. Neither group of natural enemies substantially diminishes the level of fruit damage, but the predators, acting during summer when pest populations build up, seem to be more important in reducing its numbers.

Website: http://www.seea.es/divulgac/galima/Plagas/Parlatoria_pergandii/Parlatoria_pergandii.htm

4.9. *Parlatoria ziziphi* (Lucas)

Diagnostic characters: Female shield oval, black, covering most of body, with two dorsal longitudinal ridges and a posterior fringe of white wax (Fig. **16**). Body of young female initially white, later becoming lilac; dorsal macroducts two-barred, their pygidial apertures with sclerotic rings. No macroducts are located around anus, and with 10-20 pores on either side of vulva. Ear-like lobes are located laterad of anterior spiracles. Male shield whitish, elongate, with the dark exuvium at one end and a longitudinal ridge; body lilac.

Life history: The pest, also known as the black parlatoria scale, completes 3-5 annual overlapping generations in the Mediterranean Region.

Economic importance: The feeding of this pest may cause leaf drop, dieback of twigs and branches, fruit deformation and drop. In addition, the presence of the conspicuous shields on the fruit (from which they are hard to remove) greatly detracts from their market value. In Upper Egypt it is the most important pest of citrus [28].

Fig. (16). *Parlatoria ziziphi* (Lucas).

Although leaves are the preferred settling sites, this insect also feeds on fruit and branches [26] (Fig. **17**). Most of the pests settle in the lower, shady parts of grapefruit trees, showing a clear preference for its western aspects [27].

Management: In some Mediterranean countries the pest is often controlled by organophosphates, to which it is developing resistance and which may interfere with the activity of natural enemies. In Upper Egypt pest populations were much (up to 99%) reduced by white emulsion oil sprays [28]. Two hymenopterous endoparasitoids, the aphelinid *Encarsia citrinae* (Craw) and the encyrtid *Habrolepis aspidioti* Compere and Annecke, together attacked about 20% of the pest populations.

Fig. (17). *Parlatoria ziziphi* (Lucas) on leaf, twig and citrus fruit.

Website:
http://scholar.google.co.il/scholar?hl=en&q=Parlatoria+ziziphi&btnG=Search&as_sdt=2000&as_ylo=&as_vis=0

REFERENCES

[1] Danzig EM, Pellizzari G. In: Kozár F, Ed. Catalogue of Palaearctic Coccoidea. Hungarian Academy of Sciences. Budapest: Akaprint Nyomdaipari Kft. 1998; pp. 172-370.

[2] Willard JR. The rhythm of emergence of crawlers of California red scale, *Aonidiella aurantii* (Mask.), (Homoptera: Diaspididae). Austral J Zool 1972; 20**:** 49-65.

[3] Rodrigo ME, Garcia-Mari F, Rodriguez-Reina JM, Olmeda T. Colonization of growing fruit by the armored scales *Lepidosaphes beckii, Parlatoria pergandii* and *Aonidiella aurantii* (Hom. Diaspididae). J Appl Entomol 2004; 128: 569-75.

[4] Uygun N, Karaca I, Şekeroğlu E. Population dynamics of *Aonidiella aurantii* (Maskell) (Homoptera: Diaspididae) and its natural enemies on citrus in the Mediterranean Region of Turkey from 1976 to 1993. Israel J Entom 1995; 29**:** 239-46.

[5] Orphanides GM. Establishment of *Comperiella bifasciata* (Hym: Encyrtidae) on *Aonidiella aurantii* (Hom: Diaspididae) in Cyprus. Entomophaga 1996; 41**:** 53-7.

[6] Peleg BA, Bar-Zakay I. The pest status of citrus scale insects in Israel (1984-1994). Israel J Entom 1995; 29**:** 261-4.

[7] Gerson U, Hazan A. A biosystematic study of *Aspidiotus nerii* Bouché (Homoptera: Diaspididae), with the description of one new species. J Nat Hist 1979; 13: 275-84.

[8] Monastero S. Le Cocciniglie degli Agrumi in Sicilia. Ist Entomol Agrar Boll 1962; 4: 65-151.

[9] Avidov Z, Harpaz I. Plant Pests of Israel. Jerusalem: Israel Universities Press 1969.

[10] Tumminelli R, Perrotta G, Raciti E, Colazza S. Cellophane tape with adhesive on both sides to monitor the emergence of armored scale insect crawlers in Eastern Sicily lemon orchards. Bull IOBC/WPRS 2006; 29(3): 251-4.

[11] Alexandrakis V, Neuenschwander P. Le rôle *d'Aphytis chilensis* (Hym.: Aphelinidae), parasite d'*Aspidiotus nerii* (Hom.: Diaspididae) sur olivier en Créte. Entomophaga 1980; 25: 61-71.

[12] Rosen D. The hymenopterous parasites of citrus armored scales in Israel (Hymenoptera: Chalcidoidea). Ann Entomol Soc America 1965; 58: 388-96.

[13] Steinberg S, Podoler H, Rosen D. Biological control of the Florida red scale, *Chrysomphalus aonidum*, in Israel by two parasite species: current status in the coastal plain. Phytoparasitica 1986; 14: 199-204.

[14] Pina T, Martinez B, Verdu MJ. Presencia en la Peninsula Iberica de *Comperiella lemniscata* (Hym.: Encyrtidae) sobre el piojo rojo de los citricos, *Chrysomphalus dictyospermi* (Hemiptera: Diaspididae). [Presence of *Comperiella lemniscata* (Hym.: Encyrtidae) in the Iberian Peninsula on red scale *Chrysomphalus dictyospermi* (Hemiptera: Diaspididae)]. Bol San Veg Plag 2001; 27: 29-34.

[15] Pina T, Verdu MJ. 2007. Establecimiento y dispersion de *Aphytis melinus* y *A. lingnanensis* (Hym.: Aphelinidae), dos parasitoides introducidos para el control de *Chrysomphalus dictyospermi* Morgan y *Aonidiella aurantii* (Diaspididae) en citricos de la Comunidad Valenciana. [Establishment and dispersal of *Aphytis melinus* and *A. lingnanensis* (Hym: Aphelinidae), two parasitoids introduced to control *Chrysomphalus dictyospermi* Morgan and *Aonidiella aurantii* (Maskell) (Hem.: Diaspididae) in citrus of the Valencian Region (Spain)]. Bol San Veg Plag 2007; 33: 311-20.

[16] Abdel-Fattah MI, El-Saadany G. The role of parasitoids in the control of the purple scale, *Lepidosaphes beckii* (New.) in Egypt. Zeit Angew Entomol 1979; 87: 154-9.

[17] El-Kareim AIA, Shanab LM. Host-seeking stimulants (kairomones) for parasitoids of purple scale insects. Acta Phytop Entomol Hungarica 1992; 27: 29-34.

[18] Gill RJ. The scale insects of California. Part 3. The armored scales (Homoptera: Coccoidea: Coccidae). Technical Series in Agricultural Biosystematics and Plant Pathology No. 3. Sacramento: California Department of Food and Agriculture 1997.

[19] Bruwer IJ. In: Bedford ECG, Van den Berg MA, De Villiers EA, Eds. Citrus pests in the Republic of South Africa. Nelspruit: Dynamic Ad1998: pp. 153-7.

[20] Gerson U. Interrelationships of two scale insects on citrus. Ecology 1967a; 48: 872-3.

[21] Cesnik R, Medina CL. High pressure water sprays to control chaff scale (Homoptera: Diaspididae) in citrus. J Entomol Sci 1995; 30: 93-4.

[22] Bar-Zakay I. Using IGRs in citrus: achievements and problems. Phytoparasitica 1992; 20: 80.

[23] Gerson U. The natural enemies of the chaff scale, *Parlatoria pergandii* Comstock, in Israel. Entomophaga 1967b; 12: 97-109.

[24] Gerson U. Studies of the chaff scale on citrus in Israel. J Econ Entomol 1967c; 60: 1145-51.

[25] Gerson U, Izraylevich S. A review of host utilization by *Hemisarcoptes* (Acari: Hemisarcoptidae) parasitic on scale insects. Syst Appl Acarol 1997; 2: 33-42.

[26] Blackburn VL, Miller DR. Pests not know to occur in the United States or of limited distribution. No. 44: Black Parlatoria scale. USDA, APHIS 1984; 84-10.

[27] Salama HS, Abdel-Salam AL, Donia A, Megahed MI. Studies on the population and distribution pattern of *Parlatoria zizyphus* (Lucas) in citrus orchards in Egypt. Insect Sci Appl 1985; 6: 43-7.

[28] Coll M, Abd-Rabou S. Effect of oil emulsion sprays on parasitoids of the black parlatoria, *Parlatoria ziziphi* in grapefruit. BioControl 1998; 43: 29-37.

CHAPTER 15

Tephritidae

Gavino Delrio* and Arturo Cocco

Department of Plant Protection, University of Sassari, via De Nicola 1, 07100 Sassari, Italy

Abstract: Two tephritid species are important pests of citrus in the Mediterranean Region. The Mediterranean fruit fly, *Ceratitis capitata* (Wiedemann), is widely distributed, whereas the peach fruit fly, *Bactrocera zonata* (Saunders), has only recently been recorded in Egypt. Both are multivoltine and polyphagous that can cause direct losses in fruit production and impede international trade. Descriptions, host plants, life history, effects of hosts and climatic factors, biotic mortality factors and population dynamics are discussed, along with the most important control methods of both species.

Keywords: Citriculture, Mediterranean Region, fruit flies, bio-ecology, damage, control.

1. INTRODUCTION

Tephritidae, or fruit flies, is a large family of nearly 4,500 described species arranged in about 500 genera [1]. Fruit flies are characterized by the long extendible ovipositor of the female that is used to lay eggs under the skin of fruits and vegetables, which then provide food for the developing larvae. A number of fruit fly species attack citrus fruit in the most important growing areas in the world. These pests are tropical and subtropical multivoltine species with a broad host range of pulpy fruits, characterized by a great dispersal ability and high reproductive potential.

Among the fruit flies that infest citrus, only a few are considered important, either because of their wide distribution area or their constant attacks on citrus fruits. Some are still confined almost entirely to their areas of origin: the Mexican fruit fly *Anastrepha ludens* (Loew), the South American fruit fly *A. fraterculus* (Wiedemann), the Oriental fruit fly *Bactrocera dorsalis* (Hendel), the Queensland fruit fly *B. tryoni* (Froggatt), the Chinese and Japanese orange fruit fly *B. tsuneonis* (Miyake) and the Natal fruit fly *Ceratitis rosa* Karsch. Two fruit flies are present in the Mediterranean Region: the Mediterranean fruit fly (medfly) *Ceratitis capitata* (Wiedemann), the most serious and widespread species attacking citrus, and the peach fruit fly *Bactrocera zonata* (Saunders). *Ceratitis capitata* is also an important pest of citrus in South Africa, Australia, Argentina, Brazil and Central America. On the other hand, the geographical range of *B. zonata* extends from Iran to South-East Asia, through its native areas (India and Pakistan). This pest is expanding its geographical distribution and has recently been detected in the South-East Mediterranean Region.

In addition to large direct losses in fruit production, fruit flies seriously impede international trade due to severe quarantine regulations designed by importing countries to avoid cross-border introductions.

2. *CERATITIS (CERATITIS) CAPITATA* (WIEDEMANN)

The genus *Ceratitis* comprises more than 90 species, all native to the Afrotropical Region. They are small flies, carry blotches or stripes on their wings and have a rounded scutellum with yellow and black patches. The genus name *Ceratitis* [Greek: horns, antlers] and the specific name *capitata* [Latin: enlarged at the tip] refer to a character of the male, the rhomboid expansion at the apex of the anterior pair of the orbital setae.

Description

Adult. A small fly (about two-thirds the size of the housefly), well characterized by the peculiar colours of body and wings, and by the spatulate diamond-shaped orbital setae of the male (Fig. **1**).

***Address correspondence to Gavino Delrio:** Department of Plant Protection, University of Sassari, via De Nicola 1, 07100 Sassari, Italy; Email: gdelrio@uniss.it

Fig. (1). Male of *Ceratitis capitata* (Wiedemann).

Head yellow. Antennae yellow with the two basal segments darker; arista blackish with short hairs. Eyes reddish-purple, with a blue iridescence. Frontal and ocellar bristles black; lower orbital modified in male, stem pale with apical end dark and diamond-shaped. Thorax shiny black on the dorsal surface, with a characteristic pattern of silvery markings, ventral surface yellow; scutellum inflated, black with a pale yellow band at the base. Legs yellow; fore femur of males posterodorsally with a bush of long orange-coloured hairs along entire length. Wings, usually held in a drooping position, broad, transparent, with black, brown and yellow markings. Abdomen orange yellow, with two reddish-brown bands on the second and fourth segments. Male abdomen wider than long. Female abdomen longer than wide (Fig. **2**); oviscape shorter than abdominal terga, ovipositor (aculeus) short and flattened (Fig. **3**).

Fig. (2). Female of *Ceratitis capitata* (Wiedemann).

Medflies are sexually dimorphic, with males characterized by more brightly coloured eyes, front legs with more and longer hairs, brighter white labrum and a pair of supra-fronto-orbital bristles spatula-shaped at the tip.

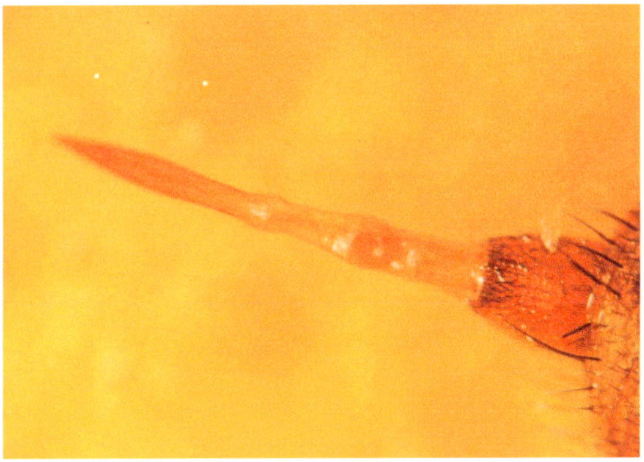

Fig. (3). Ovipositor of female of *Ceratitis capitata* (Wiedemann).

Body length: 3.45-4.60 mm; wing length: 3.65-4.55 mm; ovipositor length: 0.9-1.3 mm.

Egg. Smooth and shiny white, oblong in shape with a chorionic protrusion at the anterior pole containing the micropylar hole. Length 0.9-1.1 mm, width 0.2 mm (Fig. **4**).

Larva. Cream coloured, typically elongated, cylindrical maggot-shaped (Fig. **5**). Anterior end narrowed and somewhat curved ventrally; caudal end flattened with distinctive tubercles in the intermediate area. Head with 9-13 rows of short rounded teeth at oral ridges; mouthhooks black, heavily sclerotized, without preapical teeth. Thoracic and abdominal segments with large creeping welts of 9-13 rows of small spinules. Anterior spiracles with 8-10 tubules; posterior spiracles each with 13 slits. Body length of third instar: 6.5-9 mm; width 1.2-1.5 mm.

Fig. (4). Eggs of *Ceratitis capitata* (Wiedemann).

Rapid identification of citrus-infesting larvae of the genera *Anastrepha, Bactrocera* and *Ceratitis,* is possible using optical microscopy. The larvae of these genera can be discriminated by a combination of features, principally: shape of the antennomaxillary complex, shape of the cephalopharyngeal skeleton, anal lobes and the hairs of the posterior spiracles [2].

Puparium. Straw to dark brown in color, rounded at the anterior end, lateral, dorsal and ventral surfaces slightly convex, with a distinct segmentation, the posterior end rounded (Fig. **6**). Length 4.0-4.5 mm, width 2.0-2.5 mm.

Host Plants

The medfly is a highly polyphagous pest with a host range of over 400 fruit species, its distribution ranging from sub-Saharan Africa (where it originated) to numerous adventive areas.

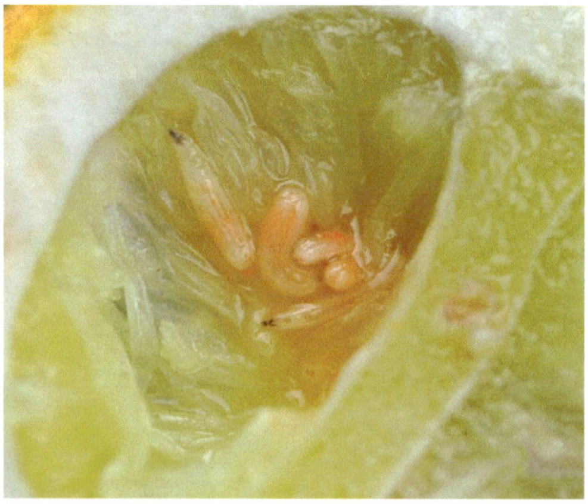

Fig. (5). Larvae of *Ceratitis capitata* (Wiedemann).

The suitability of many fruits as larval hosts has mostly been determined in the laboratory by exposing females to fruits. However, reports of field infestation are known for less than 50% of the known medfly host species [3].

Fig. (6). Puparia of *Ceratitis capitata* (Wiedemann).

In the Mediterranean Region, the most common sequence of hosts, from spring to winter is: sour orange, grapefruit, late ripening varieties of sweet orange, loquat, apricot, peach, fig, pear, apple, persimmon, kiwi, prickly pear, table grape, clementine, tangerine and early ripening varieties of sweet orange. A number of alternative hosts are known, such as guava, mango, plum, date, pomegranate, pepper, *Lycium subglobosum* Dunal, *Aberia caffra* Hook. f. and Harv., *Argania spinosa* (Linnaeus) Skeels and others, which are of great local importance. Stone fruits, figs and ripe citrus fruits are the most susceptible hosts, but in periods of host shortage, immature fruits and fruits with a high mechanical resistance to oviposition penetration (*e.g.* apple) can also be attacked. The nature of the hosts directly influences egg and larval mortality, the rate of development and the quality of adults. Compounds present in some fruits, *e.g.* apple (tannins) and persimmon (citrus essential oils), are highly toxic to eggs and newly-hatched larvae.

Life History

Adult medfly females lay eggs beneath the skin of suitable hosts, mostly ripening or ripe fruits. The larvae tunnel directly into the pulp of the fruit and moult twice as they feed and grow. Mature third-instar larvae leave the fruit and usually burrow into the soil, where they form puparia and pupate. The emerging adults open the puparia; crawl upward out of the soil, mate, and a new cycle begins.

The adults are long-lived and can, in favourable conditions, survive for several months. Attainment of sexual maturity depends on climatic conditions and ingestion of foods containing essential nutrients during the post-emergence period. They acquire proteins from bird faeces and decomposing fruits, whereas carbohydrates are obtained from honeydew and ripe fruit juices. Sugar is essential for the production of viable eggs, and oviposition increases with protein ingestion.

Tephritids ingest a variety of micro- organisms, mostly Enterobacteriaceae, while feeding on natural food sources with high nitrogen content (*i.e.* bird faeces and insect frass on plant surfaces). *Ceratitis capitata* has a relatively conserved bacterial community in its guts. Its digestive system microbiota include various species of *Klebsiella, Enterobacter*, *Pectobacterium* and *Citrobacter* (Enterobacteriaceae). Most fix nitrogen in the host gut, thus providing this mineral to the host. The microbial community comprises also pectinolytic bacteria in all fly developmental stages. They are more abundant in the larval guts, suggesting the importance of pectynolysis in the earlier stages of medfly development. During oviposition, the females transmit the enterobacterial community to their offspring and to the fruits, where the bacteria proliferate, causing fruit decay [4].

The medfly has a lek-mating system in which males aggregate beneath neighbouring leaves on a fruit tree and release a pheromone to attract females [5]. Natural leks occur on a small subset of available plants and citrus appears to be a preferred host for lek formation. The males aggregate in leks in the morning and early afternoon, depending on climatic conditions, and defend the abaxial surface of a leaf as their mating territories. They call females by everting their rectal epithelium and releasing an attractant pheromone. Attracted females land within a male's territory and are courted with distinctive vigorous wing fanning, wing buzzing and head rocking, after which copulation occurs [6] (Fig. **7**). Copulation generally lasts 2-3 hours. The female may reject the male at any mating stage, by simply dropping from the leaf. The functions of the complex mixture of volatile compounds produced from the male epithelium to attract females are not fully understood. Over 60 components have been detected within the male pheromone, five of which constitute the major quantitative emissions. Live calling males were more attractive to virgin females than the blend of the five major components in flight tunnels [7]. In field tests, traps baited with the synthetic blend captured fewer flies than protein-baited traps.

Fig. (7). Mating adults of *Ceratitis capitata* (Wiedemann).

Male medflies may mate many times in their lifetime, whereas females are rather oligogamous, with an estimated female re-mating frequency of 4 to 21% in wild populations [8]. The duration of copula, sperm replenishment and male accessory gland fluids appear to influence female sexual receptivity [9]. The male accessory gland secretions contain biologically active compounds that switch the behaviour of the wild female, from attraction to pheromones to attraction to fruit odors, implying a shift from mating to oviposition.

To locate potential hosts, medfly females are attracted to the following stimuli: fruit volatiles; plant color, form and size; form, size and color of individual fruits, which together elicit attraction after fly arrival on the host plants. They are also attracted to contact fruit stimuli (surface structure and chemical factors), which elicit boring attempts and oviposition/ and pheromones deposited on the fruit surface after egg-laying, which deters repeated oviposition. Females land on fruits, locate a suitable ovipositing site and lay a clutch of 1-10 eggs in a perforated cavity underneath the skin of fruits (Figs. **8** and **9**).

Fig. (8). Female of *Ceratitis capitata* (Wiedemann) ovipositing on peach fruit.

Throughout their life the females can produce 300-1000 eggs, frequently laid in fruit cracks, wounds or pre-existing oviposition punctures. On citrus fruits, the black oviposition site, 1 mm in diameter, is easily detectable, as the skin surrounding the wound forms a pale chlorotic blotch, about 1 cm in diameter. Oviposition in injured sites on the fruit surface increases the survival of female offspring in citrus fruits, because the larvae are less exposed to the toxic essential oils of the flavedo.

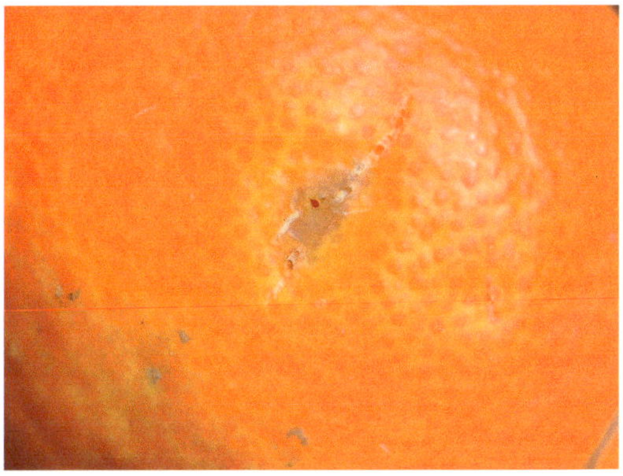

Fig. (9). Oviposition puncture of *Ceratitis capitata* (Wiedemann) on citrus fruit.

After egg laying, ovipositing females mark the fruit with a pheromone by dragging their ovipositor over the fruit surface. This pheromone effectively deters oviposition attempts for several days [10]. The presence of young larvae also has oviposition-deterring effect. However, the pheromone has a time-limited effect, as several females were observed to oviposit in the same fruit and up to 80 larvae have been found in a peach fruit. First-instar larvae burrow in the peel and actively migrate into the pulp of the fruit, feeding and macerating the tissues. Mature, last-instar larvae then move to the fruit skin and jump out to pupate in the soil, usually at dawn. Pupating larvae and pupae in the soil are exposed to several natural enemies, particularly ants.

The physical structure and moisture of the soil significantly affect pupation depth and survival. Mature larvae pupate at a depth of 0.5-2 cm in wet soils, whereas in dry soils pupation occurs in up to 5 cm depths. During the pupal stage, the main biotic and abiotic mortality factors are predators and extremely high or low temperature, respectively [11]. Peak adult emergence takes place in the early morning and teneral adults are subjected to predation.

Overwintering

The medfly is a non-diapausing species whose development slows down at low temperatures. Overwintering may occur at any stage, but adults can only be found in areas with warm winter climates. When moving toward the northern distribution boundary, the medfly overwinters mostly as larvae or pupae. In the warmer Mediterranean Regions, overwintering adults tend to aggregate on evergreen plants (*e.g.* citrus), whereas larvae develop during the winter mostly on prickly pears, sour and sweet oranges, as well as on other minor secondary hosts [12].

Dispersive Movements

Medfly adults possess high dispersal capabilities and can fly for several kilometers, even among non-host plants. However, concentrations of released flies are relatively high around the release point, because they mostly fly only for a few hundred metres. Mark-recapture studies of releasing sterile males indicated a short median distance flown, with 90% moving only 400-700 m, although the trend of declining catch rates was consistent up to 9.5 km [13]. Short distance movements are strongly affected by the search for food and host plants, as well as by sexual activity. Adults tend to remain where hosts are available, but disperse rapidly when such hosts are absent or become unavailable where the flies emerge. The attractive influence of host plants is especially important because adults can appear suddenly in a grove coinciding with the fruit ripening period. Migratory movements have been reported from the coastal to the interior zones in Spain, on the basis of trap catches of the first adults, and in Israel from the coastal plain to the central mountains, where the medfly does not overwinter [14, 15].

Effects of Climatic Factors and Hosts

Climate plays a fundamental role in determining medfly abundance, influencing development, survival, and the behaviour of individuals, thus limiting the pest's geographical distribution.

Mean adult longevity depends on the temperature and may vary from 10 days at 38°C to 100 days at 16-19.5°C [16]. A wild Hawaiian population, kept at alternating temperatures of 24:13 and 35:24°C (maximum: minimum), had an average life span of 116 and 40 days for females and 118 and 40 days for males, respectively [17]. Mediterranean populations of *C. capitata* are in general longer-lived than tropical ones, probably because the seasonality of fruit hosts in temperate ecosystems determined the evolution of a longer lifespan in order to survive periods of host absence [18].

The lower temperature threshold is about 16°C for ovarian maturation and oviposition, and about 16-18°C for mating. The pre-oviposition period varies greatly among individuals and is extended by the lowering of temperatures. Laboratory-reared populations have a mean pre-oviposition period that ranges from 4.7 days at 29°C constant temperatures to 35.1 days at 16°C. At alternating temperatures the mean pre-oviposition period for wild females was of 37.4 days at a 24:13°C regime and of 9.5 days at 29:18°C [17]. The optimal temperature for egg production ranges between 25 and 30°C and females can lay up to one thousand eggs. However, fecundity varies greatly among the geographically isolated medfly populations. Daily fecundity ranges from 20 to 30 eggs per female under optimum conditions. The oviposition rate is influenced by the same factors as those for the pre-oviposition period.

Temperature and host fruit affect the developmental rate and determine the number of annual generations completed by the medfly in each area. The relationship between constant temperatures and developmental rate has been clearly defined and the lower developmental threshold was found to be around 10-12°C. Larval development is usually faster in maturing fruits, when they are rich in sugars (*e.g.* figs, peaches) than in unripe fruits with compact pulp (apples) or rich in acids (citrus fruits) [19]. Duration of the larval stage of wild populations at 25°C, in 15 different host fruits, ranged between 1 to 3 weeks [20]. Adult body size, longevity and fecundity are affected by the nature of the larval host fruits as well as by the presence of larval conspecifics within the same fruit. For example, *C. capitata* females reared on plum have higher fecundity than those that had developed within citrus or pear [21].

The availability of temperature thresholds and thermal constants, appropriate for each life stage, facilitates the development of models for estimating the medfly life cycle under various climatic conditions [22, 23]. The duration of a generation in the Mediterranean Region varies from about 20 days in summer to 2-3 months in winter. The medfly completes 6-7 annual generations in areas with typical Mediterranean climate (*e.g.*, Algeria, Southern Italy), whereas in more northern regions there are only 2-3 cycles [12].

Resistance of Citrus Fruit

Although the medfly severely damages citrus crops, it does not seem to be highly adapted to develop in citrus fruits. Chemical and physical characteristics of the citrus peel are strong barriers against medfly attacks. The secretion of gum, chemical compounds of the flavedo and the formation of calluses on the egg cavity cause high egg and larval mortality. Moreover, the mechanical resistance of the albedo prevents larvae from reaching the fruit pulp. The short and sharp ovipositor is the main constraint limiting medfly females in their efforts to avoid the toxic properties of the citrus flavedo layer. However, ovipositing females are successful in placing eggs into the albedo of certain citrus fruits (*e.g.* clementine). Furthermore, the duration of larval development in citrus fruits is longer, and the quality of developed adults is lower, than in many other attacked fruits [19, 22]. The ability of the pest to attack citrus fruits successfully is dependent on fruit species and cultivar, with bitter orange, grapefruit and clementine being the most susceptible and lemon and sour lime almost completely resistant. Orange cultivars show a variable susceptibility to medfly attacks. Rind thickness and texture, density and volume of oil glands and a high concentration of limonene in the peel oil are important factors contributing to the relative resistance of citrus fruit.

Citrus resistance to the development of the of the medfly larvae is related to the degree of fruit ripening. Unripe fruits are more resistant because the flavedo layer has a high concentration of essential oil glands per surface unit and the albedo is thicker and more compact than in ripe or over-ripe fruits, up to the point that medfly development is possible even in over-ripe lemons.

Although citrus fruits are temporary resistant to medfly attacks, severe damage may occur. The punctures in the fruit can cause premature colouring in unripe fruits and early drop. The bacteria that are introduced during egg-laying promote fruit rot, and punctures and larval galleries open the way for moulds (particularly *Penicillium* spp) and other organisms that increase decay. Furthermore, punctures cause skin blemishes, reducing the fruit marketability.

Evaluation of economic losses caused by the medfly on citrus is very difficult to determine due to many factors, such as population fluctuations, abundance, severity of damage and different cultural situations. Economic losses to citrus fruits due to *C. capitata* attacks, when no control measures had been taken, were estimated in Spain as 1-2.5%, in Turkey about 3.5-6.2%, in Sardinia as 0.7% for oranges and up to 46% for clementines [12].

Abiotic Mortality Factors

The tolerance of *C. capitata* to thermal extremes is very high during its different stages. Temperatures of 40°C for a few hours represent the upper limits for eggs and larvae [12]. Third-instar larvae are more heat-resistant than eggs, and 100% mortality is achieved by exposure to temperatures of 46, 48, 50 and 52°C for 60, 15, 4, and 1 minute, respectively [24]. Ground temperatures greatly affect the mortality of burrowing larvae, and none can survive when temperatures exceed 48°C [25]. Pupae are more heat-resistant, tolerating daily exposures of 3 hours of 42°C, of one hour at 46°C and 15 minutes at 50°C [16]. Adults die within a few minutes of exposure to 40°C [12]; the critical thermal maximum for adults is around 43°C. Adults may however escape high temperatures by taking refuge in cooler microclimates, *e.g.* under shaded leaves; their survival is dependent on age and feeding status.

Temperatures slightly above 0°C, lasting several days, kill eggs and larvae in the fruit. No medfly eggs or larvae survive exposure to 4-7°C for 7 weeks, 0.5-4°C for 3 weeks, 0°C for 2 weeks or -2.5°C for 1 week [26]. Resistance to low temperatures is particularly high in pupae and adults, which may withstand temperatures below 0°C and survive for a few hours even at -4°C [12].

The geographical distribution of the medfly is limited by winter climatic conditions: average monthly temperatures lower than 2°C for 1-3 consecutive months or below 10°C for 4 months prevent the establishment of the pest [22]. Summer infestations of the medfly in many temperate areas should therefore be attributed to recurrent invasions, because winter conditions do not permit the insect's overwintering.

Moisture and temperature influence pupal survival, adult longevity and fecundity. Soil saturation during seasons of heavy rainfall can in particular cause high pupal mortality. Relative air humidity and high temperatures are key factors for population abundance in dry climate regions, *e.g.* Israel, where dry winds kill the adults. Models based on average monthly temperatures and humidities (ecoclimograms) are used to describe medfly population abundance, showing adult optimum longevity and fecundity between 60% and 75% RH [12, 22].

Biotic Mortality Factors

In its native sub-Saharan Africa the medfly is attacked by several species of opiine Braconidae (Hymenoptera) and other parasitoids. However, despite the long history of the pest's establishment in the Mediterranean Region, there are few records of field parasitism by either native or introduced parasitoids. Only the braconid *Psyttalia concolor* (Szepligeti) and the hymenopterous eucoilid *Aganaspis daci* (Weld), which are larval-pupal parasitoids, occur in the Mediterranean Region, attacking the pest almost exclusively on figs.

Ceratitis capitata spends part of its life cycle in the soil, where late third-instar larvae, pupae and newly-emerged adults can be found. These stages are attacked by a large number of ground dwelling predators, such as ants, carabid and staphylinid beetles, earwigs and spiders. In Spanish citrus groves, pupal predation by ants is an important regulatory factor of medfly populations, particularly in winter months [27].

Population Dynamics

Population densities of the medfly in different fruit-growing areas are affected by host species and variety, crop sanitation practices, climate factors and type of cultivation (specialized or mixed). Pest densities are usually highest in subtropical climates and in areas of mixed fruit cultivation, where adults occur throughout the year. On the other hand, populations are reduced in temperate climates with low winter and spring temperatures and in wide monoculture areas, where the females cannot find ovipositing substrates in certain periods of the year.

Medfly populations are relatively unstable even in the most favourable areas. Adults show a clear preference for certain hosts that offer better ecological conditions for their life and fruits more suitable for oviposition. The medfly is much attracted to ripening fruits, whereon adults quickly increase their populations until the harvest, then dispersing to other groves with suitable fruits. However, low adult populations are also found in hosts without fruits and non-host plants.

The seasonal fluctuations of the medfly populations in the Mediterranean Region show a characteristic pattern, determined by the climate and host sequence, availability and suitability. The lowest population density is in winter because of unfavourable climatic conditions. In spring, adults develop and mate, but the growth rate is low because of limited susceptible hosts and sub-optimal temperatures. However, survival is high and adults show good dispersal ability, overcoming the scattered distribution of host plants. In summer, climate and larval foods are optimal and the populations reach their highest growth rate. Larval mortality is low because they develop within very suitable fruits, resulting in high quality adults. Populations reach their highest densities at the end of summer, but by that time the decrease of susceptible hosts increases larval competition and species dispersal. In autumn, medflies aggregate on citrus and can severely infest susceptible varieties, especially in warm-winter areas.

Medfly population densities fluctuate dramatically during the year. Densities estimated by the mark-recapture method resulted in the following values: 10-100 flies/ha during the unfavourable season and 5,000 in summer in Tunisia; 8 flies/ha in spring and 15,000 in summer in the island of Procida (Italy); 4 flies/ha in August and above 6,000 in November in a Sardinian citrus area [12].

Medfly populations show strong annual variations in accordance with climatic conditions. Dry summers and cold/wet springs strongly reduce populations in hot climates (*e.g.* Israel) and temperate areas, respectively [12, 22].

Economic Importance

The medfly is not a key pest of citrus as it is of other fruits (*e.g.* peach, fig, persimmon), because in the areas susceptible to its attacks most citrus fruits mature in winter when climatic conditions are unfavourable for the pest. In temperate areas of the Mediterranean Region, the pest infests early-ripening citrus species (Satsuma, tangerine, clementine, some orange varieties) from October to December (Fig. **10**). Cold temperatures then lower the population density under the damage threshold until the following May-June, when most citrus fruits have already been picked. However, late-ripening sweet orange varieties, such as 'Valencia late', might be infested before harvest, in late spring.

Fig. (10). Navel orange with oviposition punctures of *Ceratitis capitata* (Wiedemann).

Citrus cultivated in warmer areas of the Mediterranean Region (Morocco, Egypt, Israel) are more exposed to medfly attacks because the pest develops there throughout the year.

Management

Attractants. Medflies are attracted to colours and shapes that mimic host fruits and foliage. The preferred colour is yellow, which reflects light between 500 and 580 nm, like bright leaves. The drawback of the yellow colour is that it attracts many different insect species, including natural enemies. Yellow sticky traps have a limited range of attraction and are less affected by climatic factors than olfactory traps. For these characteristics, yellow traps are useful in ecological investigations involving the sampling of a population within a limited spatial range.

Glass fly traps baited with molasses or with fermenting protein were used in the 20th century to control the medfly. Kerosene was first used as an attractant for males, being later replaced by the seed oil of the garden angelica, *Angelica archangelica* Linnaeus, and by the synthetic compound trimedlure. The active component of angelica oil was identified as α-copaene and α-ylangene, two structurally-related sesquiterpenes. Alpha-copaene is also a minor component of the essential oils of many medfly host plants, including citrus, guava and mango. This compound might serve as a chemical signal to locate the rendezvous site for courtship and mating. Although α-copaene is more attractive than trimedlure, it is not used in medfly detection programmes because its concentration in plants is very low and the chemical synthesis is difficult and expensive.

The most-often used compound in medfly monitoring and detection programmes is the synthetic attractant trimedlure [tert-butyl 4 (and 5)-chloro-trans-2-methylcyclohexane-1-carboxylate] [28]. This is a parapheromone, as it is not produced by the insect itself, but the male response is similar to that of a natural pheromone. Although it is not as attractive as other tephritid parapheromones, trimedlure and its analogues are very effective for medfly detection and population monitoring. Trimedlure and α-copaene share the 1-methyl-cyclohexene moiety, which is a common feature of plant kairomones that attract male medflies.

Food odour attractants have been used for a long time for medfly population monitoring and bait-spray control applications. Traps baited with protein hydrolysates are effective in attracting adults and are still in common use. Ammonia is the essential volatile attractant of hydrolyzed proteins, but other volatile compounds also affect the response of medflies. In the last decade, more practical, longer-lasting dispensers of food odour volatiles have been developed to replace the water solution of protein hydrolysates. A dry, synthetic dispenser based on three combined odour lures [ammonium acetate, 1,4 diaminobutane (putrescine) and trimethylamine] has been developed, which proved to be more powerful and relatively more female-sex-specific than protein hydrolysates [29].

Medfly control programmes largely employ this female-biased trapping system, because it enables the early detection of medfly populations, thus improving the effectiveness of suppression measures. However, the female-biased attractant is still more expensive than the conventional male-specific trapping system based on trimedlure.

Monitoring. Reliable methods for detecting medfly populations are essential for effective decision-making in control programmes. Traditionally the medfly was monitored by sampling fruits and by deploying traps that used a combination of olfactory and visual cues. Trapping is applied to evaluate the effectiveness of measures such as bait sprays, sterile insect technique (SIT) and biological control, used to reduce the pest population below the economic threshold.

Most commercially produced traps contain trimedlure, which is male-specific and captures an extremely low percentage of females. This parapheromone may be mixed with a sticky material and applied to plastic panels or formulated in longer-lasting laminate and polymeric plugs in combination with various types of traps. The most widely used is the Jackson trap, which is a hollow, delta-shaped white-waxed cardboard, with a sticky insert on the inner bottom side.

McPhail traps were used for almost a century to monitor male and female medflies in the Mediterranean Region. This trap is a bell-shaped glass container with an invaginated opening at the bottom and a water reservoir, baited with various liquid proteins. The trap type dramatically affects the attraction of medflies to food-based compounds. In recent years, plastic dry traps with a toxicant as killing agent became preferred over water-based traps, because they can be serviced with ease. The traps are baited with synthetic female-biased attractants that are more specific and have a better controlled release rate than the liquid attractants. These food-type lures have an estimated duration of attraction of 2-3 months, far longer than the shelf life of protein-based compounds.

Biological control. Parasitoids, employed either by classical introductions or augmentative releases in conjunction with SIT, could be used for medfly control. Several biological control attempts were made during the last 50 years by importing parasitoids into the Mediterranean Region. Among these, the larval-pupal parasitoid *Diachasmimorpha longicaudata* (Ashmead) and the egg-pupal parasitoid *Fopius arisanus* (Sonan) (Hymenoptera: Braconidae), which are important control agents of the medfly in Hawaii, were released in citrus groves but were not subsequently recovered. Another four braconid species were recently imported and released in Israel and two, *Fopius ceratitivorus* Wharton and *Diachasmimorpha krausii* (Fullaway) seem to be established [30]. However, the main problem of biological control in commercial fruit groves is that parasitoids are unable to control medfly infestations below an extremely low economic threshold. The control of *C. capitata* is usually with protein bait sprays incorporating synthetic insecticides. More environmentally friendly products, *e.g.* plant- and microorganism-derived insecticides have been developed for pest control. Spinosad, for instance, is a compound extracted from the soil bacterium *Saccharopolyspora spinosa* Mertz and Yao. It is less hazardous to the environment than chemicals in current use because of its moderate toxicity to mammals, birds and beneficial insects. In addition, the entomopathogenic fungi *Metarhizium anisopliae* (Metschnikoff), *Entomophthora muscae* (Cohn) Fresenius, and *Beauveria bassiana* (Balsamo) Vuillemin were effective against adults and pupae of *C. capitata* and other fruit flies under field and laboratory conditions [31].

Mass-trapping and the 'lure and kill' technique. The mass-trapping technique exploits the attraction of medfly adults to baited traps where the flies are killed by an insecticide. Alternatively, fruit flies can be captured by attractive sticky traps. Several attempts were made to develop mass-trapping systems of the medfly. Their effectiveness depends on a specific and powerful attractant that can be deployed in localized traps.

Many combinations of traps and attractants have been developed and tested during the years. Trimedlure and its analogue ceralure, which is more attractive and longer-lasting, were assayed in male annihilation programmes, but were not effective as a stand-alone strategy for medfly control.

Mass-trapping has recently been more effectively applied because of the development of a synthetic, highly selective and effective combination of three chemical compounds developed for attracting medfly females [29]. The reduction of female density, and therefore of the number of eggs laid, is a critical factor in medfly management. Plastic McPhail traps with a piece of dichlorvos or deltamethrin strip as toxicants are commonly used. Traps should be hung individually on citrus trees 1-2 months before the beginning of fruit ripening, until the end of the harvest. To reduce the cost of mass-trapping, it is essential to have dispensers with a lifespan of more than 3 months, because it allows the deployment of a single trap and dispenser per year. The density of traps may vary, depending on population abundance and type of trapping device. In field trials satisfactory mass-trapping results were obtained by using 50 traps/ha; however, this technique is not effective enough to control high-density populations. Mass-trapping is currently being applied in the Mediterranean Region; in Spain more than 30,000 ha of citrus groves are being treated with this technique [32].

Adult fruit flies can be attracted to bait stations and subjected to a contained toxic compound ('lure and kill' technique). With respect to mass-trapping, in 'lure and kill' fruit flies are not trapped, but instead poisoned by exposure to insecticides. There are two main types of bait stations: 1) stations where insecticide and bait are mixed in a single formulation; adults are lured to the mixture and ingest a lethal dose of the insecticide, and 2) devices with bait separated from the insecticide; flies are attracted to the bait and exposed to contact insecticides (*e.g.* pyrethroids). Several types of bait stations have been developed locally, but their effectiveness is not often well documented. However, commercial bait stations, with an average lifetime of 4 months, are currently available.

Chemosterilisation has also been proposed to sterilize medflies directly in the field. This technique relies on the attraction of adults to bait devices where they are exposed to the Insect Growth Regulator (IGR) lufenuron. This compound is effective in reducing the viability of eggs laid by both chemosterilised females and unexposed females that had mated with lufenuron-treated males. The device contains both female and male attractants (ammonia-based and trimedlure dispensers, respectively) and a proteinaceous phagostimulant gel mixed with lufenuron. Flies landing on the trap come in contact with the gel, which induces a feeding response and the uptake of lufenuron. Bait stations deployed at a density of 24 units/ha. In citrus groves reduced medfly populations [33].

Sterile insect technique. The Sterile Insect Technique (SIT) shows significant potential when employed in the frame of area-wide integrated pest management programmes and has been successfully used against the medfly. This technique should be applied to entire medfly populations at a regional scale within a delimited geographical area. Its application therefore requires the coordinated collaboration of many agencies and subjects (farmers' associations, local and Regional governments, agricultural and development agencies). SIT can play a considerable role in all major pest management strategies by: (1) suppression of local or total populations, (2) eradication of well-established populations, and (3) containment and prevention of invasions. SIT requires the production of large numbers of the target insects in specialized mass-rearing facilities, sterilization of one or both sexes, and their release in numbers large enough to effectively compete with the wild pest population. The target pest population could theoretically be eradicated within a few generations by continuous and systematic releases of sterile male insects.

Wheat-base diets fortified with yeasts and sugar are used worldwide for rearing *C. capitata* larvae. Local availability influences selection of bulking agents (*e.g.* sugarbeet and sugarcane bagasse, corn cobs, *etc*) that serve as larval feeding substrates. Pupae are sterilized by γ-radiation, treated with a powdered fluorescent dye to mark the adults during their emergence, and transferred to emergence towers. A mixture of sugar and agar-thickened water is provided as a food source for the newly-emerged adults. Sterile flies can be released in the field by either aerial or ground release methods. Ground distribution may be accomplished either at fixed points or by roving releases from moving vehicles. Aerial fly releases are used in large scale SIT programmes. The 3-4 days-old adults are cooled to about 4°C and loaded into chilled adult release machines installed in aircrafts. They are then released and revive before reaching the ground.

The basis of the SIT approach is the sterile/wild fly over-flooding ratio. Based on mathematical models, some programmes have set a target ratio of 80-100 sterile males to one wild fly, of either sex. The release density of sterile flies depends on the size of the wild populations, which is reliably assessed by trimedlure or food-based traps. Such data provide information on population amplitude and the sex ratio of both sterile and wild populations.

Each of the steps in a SIT programme, from mass-rearing to the release of sterile males in the field, is critical in order to obtain successful medfly control.

Medfly rearing procedures have been improved and automated over the years, reducing the overall cost of the SIT process and making it competitive with other control strategies. Significant economies of scale derive from mass-rearing in factories with high production capability. For example, a mass-production facility in Guatemala produces over 2.5 billion sterile male flies each week and several thousand sterile flies are produced for USD 1.

The quality of sterile insects is strongly affected by many factors in the mass-rearing facilities. As a result of artificial mass-rearing, laboratory populations can develop genetic traits that differ from wild populations. Quality and physical characteristics of artificial larval diets can influence traits and behaviour of adults, such as body size, survival, longevity, flight and mating ability. To address these quality problems of the mass-reared flies, a filter rearing system (FPR) has been developed. Its principle is maintaining mother colonies with low adult and larval densities under semi-natural environmental conditions, thus with reduced selection pressures. These stock colonies are checked every generation and undesirable individuals/populations are eliminated. These colonies are used to produce adults which are to be sterilized following 3-4 generations of sequential amplification [34].

The reproductive biology of the medfly has a female-choice mating system, indicating that only males should be released for the success of the SIT. Rearing facilities currently use genetic sexing strains, based on a temperature-sensitive lethal mutation and a male linked translocation, in order to mass-produce sterile males [35]. The females of these strains are killed with a high-temperature treatment during embryo development, so that only males are produced. This technique doubles the production capabilities of mass-rearing facilities and avoids releasing sterile females, which could damage fruits with oviposition punctures.

The sterilization of the medfly is induced by irradiation that causes breaks in the chromosomes leading to a dominant lethal mutation. However, the use of radiation to sterilize insects does have some adverse effects on their competitiveness, which in turn reduces the efficacy of the technique. Several strategies are employed to minimize somatic damage, thus preserving insect quality standards: irradiation of insects during late-stage pupae and within a reduced-oxygen atmosphere, *etc.* Recently, the RIDL system (Release of Insects carrying a Dominant Lethal), based on transgenic embryonic sterility, was successfully developed in *C. capitata* and might provide a valuable alternative to medfly sterilization by irradiation. The transgenic strain induces embryonic lethality in eggs fertilized by released transgenic males, which carry a dominant, repressible, lethal gene [36]. Mass-rearing of this strain requires an antibiotic in the larval diet that suppresses the lethality; however, adults are long-lived and competitive in the absence of the repressor. Transgenic males mate with wild females in the field and induce the embryonic mortality of the female progeny in the absence of the antibiotic. That prevents egg hatching and thus the introgression of transgenes into wild populations. The release of sterile transgenic insects is probably one of the lowest risk strategies for transgenic organisms.

The lek-mating system of the medfly is characterized by a high level of female discrimination among potential males. The mating of females with sterile males can be enhanced by exposing sterile males to ginger root oil or essential oils derived from fruits of orange and guava, all containing the sesquiterpene α-copaene. Several sterile-male release programmes now include aromatherapy, exposing sterile males to ginger root oil prior to release, in order to increase the effectiveness of medfly control.

A number of SIT-based suppression programmes against the medfly are currently being applied in the Mediterranean Region, in Israel and Jordan, Tunisia and Spain. Almost all SIT programmes have been instituted by public-funded organizations, but private companies have recently started to commercially produce and market sterile medflies.

Chemical control. Eggs and larvae of *C. capitata* develop within fruits, and thus are protected from non-systemic insecticides. Products currently registered for medfly control in citrus crops in Europe include malathion, chlorpyrifos-methyl, phosmet, lambda-cyhalothrin, deltamethrin, lufenuron and spinosad, which are all non-systemic insecticides. Due to the fact that adults are the only exposed life stage, the aim of chemical control treatments is to prevent egg-laying in the fruit.

Insecticidal control of medfly adults is possible by using cover or bait sprays. Organophosphates and pyrethroids were the products commonly applied to whole trees to kill adults. However, these insecticides are highly toxic to beneficial arthropods and disrupt biological control agents in citrus groves.

Baits based on the organophosphate compound malathion, mixed with hydrolyzed proteins acting as attractants and phagostimulants were the most widely used and effective control strategy against the medfly. Protein sources strongly attract both male and female medflies, so that bait sprays can be applied in spots and there is only minimal impact on non-target insects. Ground and aerial applications of malathion-bait sprays were used in some countries in area-wide control projects. Medfly populations were thus effectively suppressed, favouring fruit exports. However, the extensive use of malathion bait sprays adversely affected the activity and survival of natural enemies in the treated orchards. In consequence bait sprays with the newly-developed insecticide spinosad were developed. This insecticide is combined with a protein-based attractant in a bait spray that is also registered in organic agriculture. Although this insecticide seems to be the less hazardous of the medfly bait sprays currently used and is moderately toxic to parasitoids, the development of resistance to spinosad was observed in olive fruit fly populations [37, 38].

Another compound that is accepted in organic farming is kaolin clay, which on trees forms a particle film that deters medfly oviposition. Kaolin cover sprays successfully protected citrus fruits from medfly attacks under field conditions [39]. However, treatments with this product might have negative effects on natural enemies.

Quarantine. Exporting citrus from production areas where the medfly is present into regions free of this pest necessitates effective quarantine treatments, which should combine complete insect kill, minimal damage to plant tissues and reasonable cost. The use of methyl bromide for phytosanitary purposes has become problematic and its use as a quarantine treatment may soon be prohibited. Currently, mostly high- or low-temperature treatments are used for quarantine fruit fly disinfestations of fruit crops. Ripe Valencia oranges can be disinfested by exposure to forced air so that the interior of the fruit reaches 47°C for about 0.5 hours. However, the primary obstacle to the widespread use of heat treatments is the sensitivity of many citrus fruits to the high temperatures required for disinfestations.

The pre-imaginal stages of *C. capitata* are killed by temperatures below 10°C; however, only temperatures near 0°C are of practical use because of the long time required for total medfly kill at higher temperatures. Cold treatments are applied to transported citrus fruits whenever transit times are 10 days or longer. Citrus sent to the USA are subjected to an expensive treatment during shipment, in cold chambers or in refrigerated containers. Fruits are exposed to 0, 0.55, 1.11, 1.66, or 2.22°C for 10, 11, 12, 14, or 16 days, respectively, as specified by the United States Department of Agriculture (USDA/APHIS/PPQ) regulations. The treatment may be phytotoxic depending on the fruit species and varieties. Low temperatures should therefore be kept accurately constant and recorded by special devices throughout the treatment period.

3. *BACTROCERA* (*BACTROCERA*) *ZONATA* (SAUNDERS)

The genus *Bactrocera* comprises about 500 species, mostly native to the Asian, Pacific and Australasian Regions. The peach fruit fly *Bactrocera zonata* was previously known as *Dacus zonatus*, but the genus *Dacus* was restricted to those species of the tribe Dacini whose abdominal terga are fused together into a single sclerotised plate. *Bactrocera zonata* was accidentally introduced into Egypt from its native area, India and Pakistan, in 1998 and has since spread throughout the country [40]. It was later detected in Israel and at the port of Valencia, where it was successfully eradicated.

Description

Adult. About the size of a house fly, orange-brown in color. Head without ocellar and post-ocellar setae; face with a spot in each antennal furrow; first flagellomere at least 3 times longer than wide. Thorax pale orange-brown with the postpronotal lobes without setae, sometimes with small setulae; scutum with narrow lateral yellow-orange stripes (vitta), anterior supra-alar setae and prescutellar acrostichal setae; scutellum not bilobed, pale colored with a narrow black line across the base, with only 2 marginal setae. Wings without a cross band and the costal band reduced to an apical spot. Abdomen pale orange-brown with all tergites separate; a pair of slightly depressed areas (ceromata) on tergite 5; male with a pecten on tergite 3. Body approximately 6 mm in length; wing length: 5.2-6.1 mm; ovipositor (aculeus) length: 1.0-1.2 mm [1].

Egg. White, elliptical with a micropyle at one end, tapering at both ends. Length 1.1 mm, width 0.2 mm.

Larva. Whitish to yellowish, slender, elongate. Head with 10-11 rows of teeth; mouthhooks moderately sclerotised, with a long thin apical tooth. Thoracic segments with dorsal spinules. Abdominal segments with rows of ventral spinules, forming creeping welts. Anterior spiracles with 13-15 short tubules; posterior spiracles with slits 3.0-3.5 times longer than wide. Body length of third instar: 9-12 mm.

Puparium. Cylindrical, dark-brown, length about 5 mm.

Host Plants

The peach fruit fly is polyphagous, having been recorded from over 50 species of cultivated and wild fruits and vegetables [1]. It is a key pest of peach, guava, mango and custard apple in the native areas and can also severely damage citrus, persimmon, pear, apricot, fig and cucurbits.

Because of its polyphagous habits, *B. zonata* attacks different hosts throughout the year. In India, females infest ber fruits (*Zizyphus* spp.) in spring, whereas in May and June the preferred oviposition substrates are peaches and loquats. In the rainy season, when the population density is highest, the pest damages mango, guava and citrus.

Life History

The peach fruit fly is a non-diapausing, polyphagous, multivoltine species. Adults are active throughout the year in their native area except for a short period in January and February. Overwintering occurs mostly in the larval or pupal stages.

Adults are strong fliers and can move many kilometres searching for hosts for oviposition. Passive dispersal is mainly by means of winds and movement of infested fruits.

Under laboratory conditions of 27°C, 60-70% RH and a 15:9 (L:D) photoperiod, adults reached sexual maturity 8-16 days after emergence, whereas maximum mating activity was achieved 10-15 days after adult eclosion [41]. Both males and females are polygamous; mating preferentially takes place at dusk and lasts, on average, 10 hours. Females start laying eggs after a pre-oviposition period of 10-23 days, ovipositing 20-564 eggs in batches of 2-9 under the skin of fruits. Eggs hatch within 1-3 days, although their developmental time may be longer under unfavourable conditions. Larvae feed on the fruit pulp, converting the flesh into an inedible mass. Larval development lasts for 4-30 days, depending on the temperatures [42]. Mature larvae leave the fruit and pupate in the soil, in depths up to 5 cm. Duration of the pupal stage varies from 4 to 16 days in summer to up 46 days in winter. Adults emerge from the soil and feed on plant nectar and other secretions, honeydew and rotting fruits. Adult longevity depends on temperature and quality and availability of hosts; under laboratory conditions, *B. zonata* adults lived for 79 days [43]. The peach fruit fly can complete 3-9 annual generations in different areas of its geographical range. In Egypt it completed 8.5 generations, with generation time ranging from 34 to 56 days [44].

Influence of Hosts and Climatic Factors

Climatic parameters strongly affect peach fruit fly development, thus influencing its abundance and geographical range. The developmental temperature thresholds of eggs, larvae and pupae range from 12.6 to 12.8°C, which are higher than those of *Ceratitis capitata* (10.2-11.6°C) [42]. Egg, larval and pupal survival is highest at 25°C, as are female fecundity and fertility [42, 43]. Developmental rates of all pre-imaginal stages were positively correlated with temperature over the range 15-30°C, in accordance with previous studies on other fruit fly species. Developmental periods were very similar at 30 and 35°C, whereas mortality increased with temperatures, suggesting that the upper temperature threshold might be close to 35°C [42]. Moisture is a key factor during pupation, which occurs in the soil. These pupae are more tolerant to immersion in water and to desiccation than those of *C. capitata* [45].

Host quality strongly affects the pest's pre-imaginal mortality, fecundity and other life parameters. Egg, larval and pupal mortality and developmental time were lower on peach and mango than on orange, plum and apple [43]. The reproductive parameters were also influenced by larval host fruits. Females reared on peach, mango and guava had the highest fecundity and oviposition rates.

The influence of host-range and climatic factors on fruit fly coexistence was studied in the island of Reunion, where several species of the genus *Ceratitis* coexist with *B. zonata* [45]. The peach fruit fly was detected in Reunion in 1991 and appears to be displacing *C. capitata* in warm and dry areas. Fruit flies exploit almost the same fruits, indicating that the host-range does not seem to affect niche differentiation. The peach fruit fly also displaced other fruit flies in some areas of its native range.

Management

In fruit orchards the pest is usually controlled by insecticidal cover sprays. However, more sustainable and environmentally safer techniques are gaining importance. The most effective include bait applications and the male annihilation techniques.

Protein-baited traps are widely used in management programmes against many tephritids. Fruit fly adults need a protein source in their diet to attain sexual maturity and are therefore attracted to traps that emanate ammonia. Protein hydrolysate, mixed with insecticides to form baits, applied as spot treatments, effectively reduce peach fruit fly infestations with minimal impact on beneficial insects.

In Egypt, promising results were obtained by replacing malathion with the bioinsecticide spinosad in bait applications against the peach fruit fly.

The male annihilation technique relies on the extermination of males so that females cannot reproduce. Males of many *Bactrocera* spp. are strongly attracted to parapheromones, such as methyl eugenol. This compound has both olfactory and phagostimulatory action to males and can attract males from distances of up to 1 km. Adults are killed by wooden blocks soaked with a parapheromone and an insecticide, or captured in plastic traps containing cotton wicks soaked with methyl eugenol. The male annihilation technique was used to control many *Bactrocera* species and has successfully been applied in several peach fruit fly eradication programmes (Israel, California). This technique is being applied in Egypt in all the country's fruit-producing areas. The population density of the pest was reduced about 4-fold from 2008 to 2009 by deploying bait stations separated by distances of 50 m [46].

Similarly to other tephritid species, like *C. capitata*, studies are being conducted to apply the Sterile Insect Technique (SIT) to *B. zonata* [41]. However, no eradication programmes have yet been developed for the peach fruit fly.

Biological control of fruit flies whose larvae feed in large fruits is difficult to achieve, because the larvae move inwards, thus being protected from parasites. Some parasitoids, such as the chalcidid *Dirhinus giffardii* Silvestri and the pteromalids *Spalangia cameroni* Perkins and *Pachycrepoideus vindemiae* Rondani have been recorded on pupae of the peach fruit fly in Egypt [47].

REFERENCES

[1] White IM, Elson-Harris MM. Fruit flies of economic significance: their identification and bionomics. Wallingford: CAB International 1992.

[2] Frias D, Selivon D, Hernandez-Ortiz V. Taxonomy of immature stages: new morphological characters for Tephritidae larvae identification. Proceedings of the 7th International Symposium on fruit flies of economic importance; 2006: Salvador, Brasil; pp. 29-44.

[3] Copeland RS, Wharton RA, Luke Q, De Meyer M. Indigenous hosts of *Ceratitis capitata* (Diptera: Tephritidae) in Kenya. Ann Entomol Soc America 2002; 95: 672-94.

[4] Yuval B, Ben-Ami E, Ben-Yosef M, Jurkevitch E. The Mediterranean fruit fly and its bacteria – potential for improving sterile insect technique operations. J Appl Entomol 2010; doi: 10.1111/j.1439-0418.2010.01555.x.

[5] Whittier TS, Kaneshiro KY, Prescott LD. Mating behaviour of Mediterranean fruit flies (Diptera: Tephritidae) in a natural environment. Ann Entomol Soc America 1992; 85: 214-8.

[6] Feron M. L'instinct de réproduction chez la mouche Mediterraneenne des fruits *Ceratitis capitata* Wied. (Dipt. Trypetidae). Comportement sexuel – comportement de ponte. Rev Path Veg Entomol Agr France 1962; 41: 1-129.

[7] Jang EI, Light DM, Binder RG, Flath RA, Carvalho LA. Attraction of female Mediterranean fruit flies to the five major components of male-produced pheromone in a laboratory flight tunnel. J Chem Ecol 1994; 20: 9-20.

[8] Bonizzoni M, Katsoyannos BI, Marguerie R, *et al.* Microsatellite analysis reveals remating by wild Mediterranean fruit fly females, *Ceratitis capitata*. Mol Ecol 2002; 11: 1915-21.

[9] Delrio G, Cavalloro R. Influenza dell'accoppiamento sulla recettività sessuale e sull'ovideposizione in femmine di *Ceratitis capitata* Wied. Entomologica 1979; 15: 127-43.

[10] Prokopy RJ, Ziegler JR, Wong TTY. Deterrence of repeated oviposition by fruit-marking pheromone in *Ceratitis capitata* (Diptera: Tephritidae). J Chem Ecol 1978; 4: 55-63.

[11] Cavalloro R, Delrio G. Fattori edafici influenzanti l'impupamento di *Ceratitis capitata* Wiedemann. Boll Lab Ent Agr Filippo Silvestri 1975; 32: 190-5.

[12] Delrio G. Tephritid pests in citriculture. In: Cavalloro R, Di Martino E, Eds. Integrated pest control in citrus-groves 1986; 135-49.

[13] Meats A, Smallridge CJ. Short- and long-range dispersal of medfly, *Ceratitis capitata* (Dipt., Tephritidae), and its invasive potential. J Appl Entomol 2007; 131: 518-23.

[14] Ros JP. Control genetico contra *C. capitata* Wied. por el metodo de insetos esteriles: trabajos realizados en Espana (1969-1973). Proceedings of the Symposium on the sterility principle for insect control; 1974: Innsbruck, Austria: International Atomic Energy Agency 1975; pp. 57-92.

[15] Israely N, Ritte U, Oman SD. Inability of *Ceratitis capitata* (Diptera: Tephritidae) to overwinter in the Judean Hills. J Econ Entomol 2004; 97: 33-42.

[16] Rivnay E. The Mediterranean fruit fly in Israel. B Entomol Res 1950; 41: 321-41.

[17] Vargas RI, Walsh WA, Kanehisa D, Stark JD, Nishida T. Comparative demography of three Hawaiian fruit flies (Diptera: Tephritidae) at alternating temperatures. Ann Entomol Soc Am 2000; 93: 75-81.

[18] Diamantidis AD, Papadopoulos NT, Nakas CT, Wu S, Muller H-G, Carey JR. Life history evolution in a globally invading tephritid: patterns of survival and reproduction in medflies from six world Regions. Biol J Linn Soc 2009; 97: 106-17.

[19] Mourikis P. Data concerning the development of the immature stages of the Mediterranean fruit fly (*Ceratitis capitata* (Wiedemann) (Diptera: Trypetidae) on different host-fruits and on artificial media under laboratory conditions. Ann Inst Phytopathol Benaki 1965; **7**:59-105.

[20] Carey JR. Host-specific demographic studies of the Mediterranean fruit fly Ceratitis capitata. Ecol Entomol 1984; 9: 261-70.

[21] Krainacker DA, Carey JR, Vargas RI. Effect of larval host on life history traits of the Mediterranean fruit fly, *Ceratitis capitata*. Oecologia 1987; 73: 583-90.

[22] Bodenheimer FS. Citrus entomology in the Middle East. The Hague: W Junk Publishers 1951.

[23] Shoukry A, Hafez M. Studies on the biology of the Mediterranean fruit fly *Ceratitis capitata* (Wied.). Entomol Exp Appl 1979; 26: 33-9.

[24] Gazit Y, Rossler Y, Wang S, Tang J, Lurie S. Thermal death kinetics of egg and third instar Mediterranean fruit fly (Diptera: Tephritidae). J Econ Entomol 2004; 97: 1540-6.

[25] El Keroumi A, Naamani K, Dahbi A *et al.* Effect of ant predation and abiotic factors on the mortality of medfly larvae, *Ceratitis capitata*, in the Argan forest of Western Morocco. Biocontrol Sci Techn 2010; 20: 751-62.

[26] Back EA, Pemberton CE. Effect of cold-storage temperatures upon the Mediterranean fruit fly. J Agric Res 1916; 5: 657-66.

[27] Urbaneja A, Garcìa Marì F, Tortosa D, *et al.* Influence of ground predators on the survival of the Mediterranean fruit fly pupae, *Ceratitis capitata*, in Spanish citrus orchards. BioControl 2006; 51: 611-26.

[28] Beroza M, Green N, Gertler SI, Steiner LF, Miyashita DH. New attractants for the Mediterranean fruit fly. J Agr Food Chem 1961; 9: 361-5.

[29] Epsky ND, Hendrichs J, Katsoyannos BI, *et al.* Field evaluation of female-targeted trapping systems for *Ceratitis capitata* (Diptera: Tephritidae) in seven countries. J Econ Entomol 1999; 92: 156-64.

[30] Argov Y, Gazit Y. Biological control of the Mediterranean fruit fly in Israel: introduction and establishment of natural enemies. Biol Control 2008; 46: 502-7.

[31] Ortu S, Cocco A, Dau R. Evaluation of the entomopathogenic fungus *Beauveria bassiana* strain ATCC 74040 for the management of *Ceratitis capitata*. B Insectology 2009; 62: 245-52.

[32] Navarro-Llopis V, Alfaro F, Dominguez J, Sanchis J, Primo J. Evaluation of traps and lures for mass trapping of Mediterranean fruit fly in citrus groves. J Econ Entomol 2008; 101:126-31.

[33] Alemany A, Gonzales A, Juan A, Tur C. Evaluation of a chemosterilization strategy against *Ceratitis capitata* (Diptera: Tephritidae) in Mallorca island (Spain). J Appl Entomol 2008; 132: 746-52.

[34] Caceres C. Mass rearing of temperature sensitive genetic sexing strains in the Mediterranean fruit fly (*Ceratitis capitata*). Genetica 2002; 116: 107-16.

[35] Vreysen MJB, Robinson AS. Ionising radiation and area-wide management of insect pests to promote sustainable agriculture. A review. Agron Sustain Dev 2010; doi: 10.1051/agro/2010009.

[36] Schetelig MF, Caceres C, Zacharopoulou A, Franz G, Wimmer EA. Conditional embryonic lethality to improve the sterile insect technique in *Ceratitis capitata* (Diptera: Tephritidae). BMC Biology 2009; **7**(4): 10.1186/1741-7007/7/4.

[37] Urbaneja A, Chueca P, Monton H, *et al.* Chemical alternative to malathion for controlling *Ceratitis capitata* (Diptera: Tephritidae), and their side effects on natural enemies in Spanish citrus orchards. J Econ Entomol 2009; 102: 144-51.

[38] Kakani EG, Zygouridis NE, Tsoumani KT, Seraphides N, Zalom FG, Mathiopoulos KD. Spinosad resistance development in wild olive fruit fly *Bactrocera oleae* (Diptera: Tephritidae) populations in California. Pest Manag Sci -2010; 66: 447-53.

[39] Braham M, Pasqualini E, Ncira N. Efficacy of kaolin, spinosad and malathion against *Ceratitis capitata* in citrus orchards. B Insectology 2007; 60: 39-47.

[40] Amro MA, Abdel-Galil FA. Infestation predisposition and relative susceptibility of certain edible fruit crops to the native and invading fruit flies (Diptera: Tephritidae) in the New Valley Oases, Egypt. Ass Univ Bull Environ Res 2008; 11: 89-97.

[41] Qureshi ZA, Ashraf M, Bughio AR, Hussain T. Rearing, reproductive behaviour and gamma sterilization of fruit fly, *Dacus zonatus* (Saund.) (Diptera- Tephritidae). Entomol Exp Appl, 1974; 17: 504-10.

[42] Duyck PF, Sterlin JF, Quilici S. Survival and development of different life stages of *Bactrocera zonata* (Diptera: Tephritidae) reared at five constant temperatures compared to other fruit fly species. B Entomol Res 2004; 94: 89-93.

[43] Hussain T. Demography and population genetics of *Dacus zonatus* (Saunders). PhD thesis, Faculty of Sciences, University of the Punjab 1995.

[44] Shehata NF, Younes MWF, Mahmoud YA. Biological studies on the peach fruit fly, *Bactrocera Zonata* (Saunders) in Egypt. J Appl Sci Res 2008; 4: 1103-06.

[45] Duyck PF, David P, Quilici S. Climatic niche partitioning following successive invasions by fruit flies in La Réunion. J Anim Ecol 2006; 75: 518-26.

[46] FAO/IAEA (Food and Agriculture Organization/International Atomic Energy Agency). Peach fruit fly (*Bactrocera zonata*) eradication programme in Egypt. Insect pest control newsletter 2010; 75: 32-3.

[47] Badr El-Sabah AF, Afia Y. New records for parasitoid species of cucurbit fly, *Dacus ciliatus* and peach fly, *Bactrocera zonata* in Egypt. Egypt J Biol Pest Co 2004; 14.

CHAPTER 16

Gracillariidae, Yponomeutidae and Pyralidae

Yael Argov[1] and Uri Gerson[2*]

[1]Israel Cohen Institute for Biological Control, Plant Production and Marketing Board, Citrus Division, POB 80 Bet Dagan, 50250, Israel and [2]Department of Entomology, Faculty of Agricultural, Food and Environmental Quality Sciences, The Hebrew University of Jerusalem, P.O.Box 12, Rehovot, 76-100, Israel

Abstract: Four species of moths are harmful to citrus in the Mediterranean Region. They are assigned to the families Gracillariidae (*Phyllocnistis citrella* Stainton), Yponomeutidae (*Prays citri* Millière) and Pyralidae [*Cryptoblabes gnidiella* Millière and *Ectomyelois ceratoniae* (Zeller)]. The first two are direct pests, whereas the others inhabit and damage citrus only when attracted there by mealybug honeydew. A key to the identification of the three families and to the larval stages of the 4 species are provided. Diagnostic characters, life history data, economic damage and management recommendations are provided for each species.

Keywords: Citriculture, Mediterranean Region, citrus leafminer, citrus flower moth, honeydew moth, catab moth damage, control.

1. INTRODUCTION

Four species of moths (Lepidoptera) are serious pests of citrus in the Mediterranean Region. Two, *Phyllocnistis citrella* Stainton and *Prays citri* Millière, are direct pests, whereas the other two, namely *Cryptoblabes gnidiella* Millière and *Ectomyelois ceratoniae* (Zeller), inhabit and damage citrus only when attracted there by mealybug honeydew. The four species are assigned to three families: *Ph. citrella* to the Gracillariidae, *P. citri* to the Yponomeutidae, and *C. gnidiella* and *E. ceratoniae* to the Pyralidae. The adults of these families are similar, and the following key may serve to distinguish between them.

2. KEY TO IDENTIFY THE FAMILIES

1. Antennae shorter than forewings, without a fringe of long hairs on hind wings.. 2

-. Antennae longer than forewings; with a fringe of long hairs on hind wings .... **Gracillariidae**

2. Forewings and hind wings subequal in size, or hind wing narrowing distally; ocelli absent (or much reduced).. **Yponomeutidae**

-. Forewings much larger than hind wings, which are rounded distally; ocelli present.. **Pyralidae**

The economic damage that these moths cause to citrus is due only to the larvae (commonly called caterpillars) of these four species, which may be separated by the following key:

3. KEY OF THE IDENTIFICATION OF THE LARVAE

1. Body red-brown to pink, final length 11-18 mm.. 3

-. Body light-green, final length 3-6 mm 2

***Address correspondence to Uri Gerson:** Department of Entomology, Faculty of Agricultural, Food and Environmental Quality Sciences, The Hebrew University of Jerusalem, P.O.Box 12, Rehovot, 76-100; E-mail: Gerson@agri.huji.ac.il

Vincenzo Vacante and Uri Gerson (Eds)

2. Head compressed dorso-ventrally; mining within leaves .... ***Phyllocnistis citrella***

-. Head not thus compressed; feeding on blossoms and flowers .... ***Prays citri***

3. Body pink, final length 18 mm .... ***Ectomyelois ceratoniae***

-. Body red-brown, final length 11-13 mm .... ***Cryptoblabes gnidiella***

4. GRACILLARIIDAE

This family, called leaf-blotch miners, has about 1,800 described species [1]. The moths are slender and small, with a wingspan of 5-20 mm. The antennae are as long as or longer than the forewings. The wings are narrow, with long hairs (a fringe) along the hind wing margins. When at rest, the moths can be distinguished by their habit of sitting up on their front legs, the body being steeply raised. The larvae, whose heads are compressed dorso-ventrally, form tunnels and cavities between the upper and lower surfaces of leaves, wherein they live and feed. Later the larvae of some species may roll the leaf to make a shelter and to pupate therein. This family is represented by one major citrus pest, *Phyllocnistis citrella* Stainton in the Mediterranean Region.

4.1. *Phyllocnistis citrella* Stainton

Diagnostic characters: The adult is about 2-3 mm long, silvery-white with black spots, and with four dark stripes on each forewing (Fig. **1**). Hind wings with a fringe of long hairs. The larvae are light-green in color, almost transparent, up to 3 mm in length (Fig. **2**).

Fig. (1). Adult of *Phyllocnistis citrella* Stainton.

Life history: Each female deposits about 50-130 eggs, placing them on the youngest leaves and twigs (Fig. **3**).

Emerging larvae burrow into leaves, their tunneling forming uneven mines; towards pupation the larvae cause the ends of the leaves to roll up (Figs. **4** and **5**).

Fig. (2). Larva of *Phyllocnistis citrella* Stainton.

The pest (known as the citrus leafminer) has several annual cycles. It completes a generation within a fortnight under long-day conditions and at 25°C; it has no diapause. Its phenology is determined by the host's growth and flashing cycles. The citrus spring flash is rarely affected due to low winter temperatures and/or because of natural enemy activity in the preceding year. The smaller, mid-summer and autumn flashes are more heavily attacked.

Fig. (3). Young leaf attacked by *Phyllocnistis citrella* Stainton.

Economic importance: This species is a major pest of citrus throughout its distribution. The main damage is to seedlings in nurseries, to young trees (1-3 years old) and to new growth after grafting. Larval mining within young citrus leaves distorts them, destroys the epidermis and reduces the photosynthetic capability of the foliage; damage to leaves may exceed 50%. Mining within the fruit's peel, although uncommon (except in pummeloes), causes direct damage. Injury to fruit-bearing trees is mostly cosmetic. This pest infests various species and varieties of citrus.

Fig. (4). Larval mine of *Phyllocnistis citrella* Stainton on citrus leaf.

Fig. (5). Damage of *Phyllocnistis citrella* Stainton.

Management: As far as known, no species or varieties of citrus are resistant to the pest. Chemical treatments are to be initiated as soon as the leafminer is noted on young (1-3 years old) trees, in the spring. The trees should then be treated with abamectin and by a concurrent soil treatment with imidacloprid. An alternative is to brush the systemic acetamiprid on the trunks; the precise method depends on tree age and trunk diameter. These pesticides can also be applied to control infestations that occur during later growth flashes. Fruit-bearing trees do not need to be treated.

General predators, like lacewings, spiders, ants and pirate bugs reduce pest numbers, but parasitic Hymenoptera are the main controlling factor (Fig. **6**). Some generalist indigenous parasitoids attack the pest, including *Pnigalio* sp. and *Cirrospilus brevis* Zhu, LaSalle and Huang in Spain, along with the introduced *Citrostichus phyllocnistoides* (Narayan) [2, 3]. In Western Algeria, Morocco and Sicily the most active are *Semilacher petiolatus* (Girault) and *C. phyllocnistoides* [4-6]. Three imported parasitoids usually control the citrus leafminer in Israel: *Cirrospilus ingenuus* Gahan, *C. phyllocnistoides* and *S. petiolatus*. The latter two appear to be the most important species that control the citrus leafminer in the Mediterranean area [7].

Fig. (6). Larva of *Phyllocnistis citrella* Stainton and associated young larva of ectoparasitoid (*Semielacher* sp. ?).

5. YPONOMEUTIDAE

The members of this family, known as the ermine moths, consist of several hundred described species. Their antennae are shorter than the forewings, which are as large as the hind wings. At rest the wings are held roof-like over the body. The larvae often form communal webs.

5.1. *Prays citri* Millière

Diagnostic characters: The adults of this pest, commonly called the citrus flower moth, are dull grey to brown in color, about 4-5 mm in length. Their wingspan is 10-14 mm and both pairs of wings are heavily fringed. The fore wings are a grainy grey-brown, with many spots, darker on the lower side and at the apex. The hind wings are very narrow, uniformly grey-brown and smoky at the tips. The antennae are fairly short. The larva is whitish, almost hairless, up to 6 mm in length (Fig. **7**).

Life history: Annually the pest completes 4-5 generations in the Mediterranean Region without undergoing a winter diapause. Complete development is achieved in less than 30 days in summer, within 60 days in winter. The emerging females begin to oviposit after a short time, laying about 60-165 eggs, mostly on the blossoms of lemon and citron trees [8]. The emerging larvae invade the blossoms, burrow and feed within, moving to new buds or flowers after devouring the early buds. They produce copious amounts of webbing that may join the infested plant parts together and within which they pupate. Alternately, they drop from the plants to pupate in the soil.

Fig. (7). Larva of *Prays citri* Millière on damaged lemon bottom.

The discovery that the citrus flower moth can be reared and mass-produced on lemon callus (which can be grown under sterile conditions on an artificial medium *in vitro*) considerably facilitates studies on the biology of this pest [9].

Economic importance: Lemon and citron buds suffer most from the citrus flower moth, but other citrus trees may also be attacked, although the resulting damage is negligible. Infested buds dry up and die; heavy attacks can result in loss of many flowers and even of small fruit (Figs. **8** and **9**). Due to the continuous blossoming of lemon trees, they may be attacked in autumn, although to a lesser degree than in spring. A different form of damage can be caused to grafts, as the larvae penetrate therein and feed on the cambium, thus sometimes killing the graft [10]. *Prays nephelomima* Meyrick, a closely related species, with a very similar damage pattern, has recently become a pest in New Zealand [11].

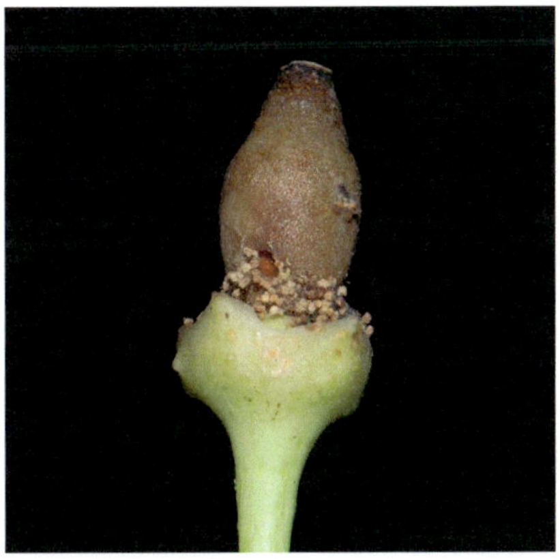

Fig. (8). A young lemon fruit damaged by a larva of Prays *citri* Millière.

Fig. (9). Young lemon fruit damaged by a larva of *Prays citri* Millière.

Management: Bacterial and fungal insecticides are being tried against *P. citri* in Egypt [12], but the main control strategy applied against the pest is mass trapping. This was facilitated by identifying the male sex pheromone and baiting traps with this pheromone (Z-7-tetradecenal). In Sicily pest numbers were reduced by mass trapping [13]. Flower infestation was significantly reduced by placing 120 traps per ha in a commercial lemon grove [14]. The rubber septa dispensers were loaded with 0.5–0.6 mg synthetic pheromone and refilled every four months. This method of control was significantly ($P<0.05$) cheaper and somewhat more efficient than spraying 3–6 times/year with insecticides (*e.g.* the organophosphate azinphos-methyl).

6. PYRALIDAE

A large family of about 30,000 described species, commonly called snout or pyralid moths. The adults have elongate forewings and wide, rounded and often pale hind wings. At rest their wings form a triangular shape, with the first pair of their long legs placed in front. The eggs are laid in suitable substrates, usually fruits or nuts. This family contains two major citrus pests, *Cryptoblabes gnidiella* and *Ectomyelois ceratoniae* in the Mediterranean Region. Both are sometimes assigned to the family Phyticinae, now considered a subfamily of Pyralidae [15].

6.1. *Cryptoblabes gnidiella* Millière

Diagnostic characters: The adult is about 7 mm in length, forewings grey-brown with red longitudinal strips, hind wings pale-grey. Larva red-brown, with two longitudinal darker bands, the dorsum bears darker spots that carry setae, length up to 14 mm (Fig. **10**).

Life history: The females are attracted to citrus fruits by the honeydew excreted by mealybugs (Pseudococcidae: Coccoidea), where they lay about 120-150 eggs [16]. The larvae initially feed on the honeydew and may then nibble on the fruit peel, and pupate at the feeding sites. Infested fruits may secrete a gum that kills the larvae. In Middle Eastern citrus groves the pest undergoes 5-6 annual generations [10]. Ascher *et al.* [17] studied the life history of the larvae when raised on an artificial diet, and the pest's reproductive behavior was described by Wysoki *et al.* [18].

Fig. (10). A larva of *Cryptoblabes gnidiella* Millière in a colony of the mealybug *Planococcus citri* (Risso) on an orange fruit.

Economic importance: *Cryptoblabes gnidiella* (commonly called the honeydew moth) is a minor pest of citrus throughout its distribution, the presence of the moth being often dependent on mealybug infestations. By itself the moth often causes only negligible fruit drop. In Portugal, however, it caused serious injury to sweet oranges, often compounding the mealybug damage, especially to the sweet orange CV Navel [16]. The pest may be monitored by pheromone traps [19]; in grape vineyards its mating can be disrupted by pheromones [20].

Management: Mealybug control in citrus orchards is usually followed by declines in honeydew moth numbers, and no control measures are needed. In avocado plantations the pest can be controlled by applications of bacterial pesticides (*e.g.* of *Bacillus thuringiensis* Berliner origin) [21].

6.2. *Ectomyelois ceratoniae* (Zeller)

Diagnostic characters: The adults of this pest, commonly called the carob moth, are 8-10 mm in length; the body is dorsally silvery dark. The forewings are silver-grey with two transverse, paler stripes. Hind wings grey, their venation brown. The larvae are pink, up to 18 mm in length, with a brown head and pronotum. Some authors place this species in the genus *Spectrobates* (*e.g.* Blumberg) [22].

Life history: During the year the carob moth develops in various fruits, like almonds, carobs, dates, figs and pomegranates, all of which can be damaged [23]. It infests citrus fruits (especially grapefruits in the eastern Mediterranean Region, other citrus fruits in the western area) in the autumn. The females, which are attracted to oviposition sites by the honeydew secreted there by mealybugs, deposit about 100-350 eggs. The young larvae initially feed on the honeydew and then nibble and damage the grapefruit peel, later burrowing deeper, beneath the calyx, next to the stalk. Usually only a single larva invades a fruit, often preferring the site where two fruits touch. However, these larvae do not survive on citrus and invariably die without maturing. A generation requires about several weeks, and 4-5 generations (feeding on other substrates) may annually be completed [10]. The life history and fertility of this pest when reared on carobs was studied by Gothilf [23]; its development is strongly affected by substrate moisture content, the optimum being around 20% fruit moisture [24]. Under conditions of low temperatures and short days the fully grown larvae of *E. ceratoniae* go into diapause [25]. Studies on the pest's life history are much facilitated by rearing the larvae on artificial diets (*e.g.* Al-Izzi *et al.*) [26].

Economic importance: The caterpillars tunnel into grapefruits and thereby cause artificial ripening. The fruits turn prematurely yellow and exude gum from the entry wounds, gum that often kills the larvae, but the fruits still drop prematurely. Losses may reach 10-20% per tree. Such damage to citrus occurs mostly in the fall, fruits being occasionally infested even without mealybug presence.

Management: Chemical treatments to control carob moths infesting date palms consist of organophosphates [22]. Efforts to control the pest on citrus with various hymenopteran parasitoids have not been successful, although on other fruits the caterpillars are attacked by several local and introduced natural enemies [27, 28]. Mating disruption with a sex pheromone is another management strategy that is being applied in date palm groves [22, 29]. Gothilf [23] suggested that by eliminating the various host-fruits (*e.g.* carobs, acacia pods and pomegranates) of the carob moth before autumn could reduce its damage to citrus fruits.

REFERENCES

[1] de Prins W, de Prins J. Gracillariidae (Lepidoptera). World Catalogue of Insects, Volume 6. Stenstrup: Apollo Book 2005.

[2] Garcia-Marí F, Granda C, Zaragoza s, Augustí M. Impact of *Phyllocnistis citrella* (Lepidoptera: Gracillariidae) on leaf area development and yield of mature citrus trees in the Mediterranean area. J Econ Entomol 2002; 95(5): 966-74.

[3] Karamaouna F, Pascual Ruiz S, Urbaneja A, Jacas J. Citrus leafminer *Phyllocnistis citrella* (Lepidoptera: Gracillariidae) and its parasitoids. Ten years after the implementation of classical biological control in Spain. Bull IOBC/WPRS 2008; 38: 177.

[4] Boualem M, Berkani A, Villemant C. Bio-ecological study of the parasitoid complex of *Phyllocnistis citrella* Stainton (Lepidoptera: Gracillariidae) in Western Algeria. Bull IOBC/*WPRS* 2008; 38: 183-8.

[5] Siscaro G, Caleva V, Reina P, Rizzo MC, Zappalà L *et al.* 2003. Current status of the biological control of the citrus leafminer in Sicily. Bull IOBC/WPRS 2003; 26(6): 29-36.

[6] Lo Genco A, Ciotta C, Lo Pinto M. Evolution of *Phyllocnistis citrella* Stainton (Lepidoptera: Gracillariidae) and its parasitoids in the last five years in citrus orchards of western Sicily (Italy). Bull IOBC/WPRS 2008; 38: 178-82.

[7] Argov Y, Rössler Y. Introduction, release and recovery of several exotic natural enemies for biological control of the citrus leafminer, *Phyllocnistis citrella*, in Israel. Phytoparasitica 1996; 24: 33-8.

[8] Liotta G, Mineo G. Osservazioni sulla biologia del Prays citri Mill. In Sicilia (II note). Boll Ist Ent Agr Univer Palermo 1963; 5(32): 1-32.

[9] Carimi F, Caleca V, Mineo G, Pasquale F, Crescimanno FG. Rearing of *Prays citri* on callus derived from lemon stigma and style culture. Entomol Exp Appl 2000; 95: 251-7.

[10] Avidov Z, Harpaz I. Plant Pests of Israel. Israel Jerusalem: Universities Press 1969.

[11] Gibb AR, Jamieson LE, Suckling DM, Stevens PS. Sex pheromone of the citrus flower moth *Prays nephelomima*: pheromone identification, field trapping trials, and phenology. J Chem Ecol 2005; 31: 1633-44.

[12] Shetata WA, Nasr FN. Laboratory evaluation and field application of bacterial and fungal insecticides on the Citrus Flower moth, *Prays citri* Miller (Lep, Hyponomeutidae) in lime orchards in Egypt. Anz Schadlingskd Pfl 1998; 71: 57-60.

[13] Nucifora A, Calabretta C. The technique of mass-trapping and its results against the citrus flower moth (*Prays citri* Mill.) - third contribution. In: Atti XIV Congr Naz Ital Entomol, Palermo 1985; pp. 861-5.

[14] Sternlicht M, Barzakay I, Tamim M. Management of *Prays citri* in lemon orchards by mass trapping of males. Entomol Exp Appl 1990; 55: 59-67.

[15] Gordh G, Hedrick DA. Dictionary of Entomology. Wallingford: CABI 2001.

[16] Silva EB, Mexia A. The pest complex *Cryptoblabes gnidiella* (Milliére) (Lepidoptera: Pyralidae) and *Planococcus citri (Risso) (Homoptera: Pseudococcidae)* on sweet orange groves *(Citrus sinensis* (L.) Osbeck) in Portugal: Interspecific association. Bol San Veg Plagas 1999; 25: 89-98.

[17] Asher KRS, Eliyahu M, Gurevitz E, Renneh S. Rearing the honeydew moth, *Cryptoblabes gnidiella*, and the effect of diflubenzuron on its eggs. Phytoparasitica 1983; 11: 195-8.

[18] Wysoki M, Ben Yehuda S, Rosen D. Reproductive behavior of the honeydew moth, *Cryptoblabes gnidiella*. Invertebrate Reproduction and Development 1993; 24: 217-24.

[19] Anshelevich L, Kehat M, Dunkelblum E, Greenberg S. Sex pheromone traps for monitoring the honeydew moth, *Cryptoblabes gnidiella*: effect of pheromone components, pheromone dose, field aging of dispenser, and type of trap on male captures. Phytoparasitica 1993; 21: 189-98.

[20] Gordon D, Anshelevich L, Zahavi T, Ovadia S, Dunkelblum E, Harari A. Integrating mating disruption techniques against the honeydew moth and the European grapevine moth in vineyards. Bull OILB/WPRS 2003; 26: 131-4.

[21] Wysoki M, Izhar Y, Gurevitz E, Swirski E, Greenberg S. Control of the honeydew moth *Cryptoblabes gnidiella* Mill. (Lepidoptera: Phyticidae) with *Bacillus thuringiensis* Berliner in avocado plantations. Phytoparasitica 1975; 3: 103-11.

[22] Blumberg D. Date palm arthropod pests and their management in Israel. Phytoparasitica 2008; 36: 411-48.

[23] Gothilf S. The biology of the carob moth *Ectomyelois ceratoniae* (Zell.) in Israel. III. Phenology on various hosts. Isr J Entomol 1970; 5: 161-75.

[24] Nay JE, Perring TM. Effect of fruit moisture content on mortality, development, and fitness of the carob moth (Lepidoptera: Pyralida). Environ Entomol 2006; 35: 237-44.

[25] Cox PD. The influence of photoperiod on the life cycle of *Ectomyelois ceratoniae* (Lepidoptera: Pyralidae). J Stored Prod Res 1979; 15: 111-16.

[26] Al-Izzi MAJ, Al-Maliky SK, Jabbo NF. Culturing the carob moth, *Ectomyelois ceratoniae* Zeller (Lepidoptera: Pyralidae), on an artificial diet. J Econ Entomol 1987; 80: 277-80.

[27] Biliotti E, Daumal J. Biology of *Phanerotoma flavitestacea* Fisher [*Hym.: Braconidae*]. Improvement of a method of permanent mass breeding toward biological control of *Ectomyelois ceratoniae* Zell. Annals Zoologie Ecologie Animal 1969; 1: 379-94.

[28] Gothilf S, Mazor M. Release and recovery of imported parasites of the carob moth, *Spectrobates ceratoniae* (Lepidoptera: Pyralidae), in Israel. Isr J Entomol 1987; 21: 10-23.

[29] Vetter RS, Millar JG, Neil NJ, Baker TC. Mating disruption of carob moth, *Ectomyelois ceratoniae*, with a sex pheromone analog. Southwest Entomol 2006; 31: 33.

CHAPTER 17

Formicidae

Andrea Lentini[1*] and Marcello Verdinelli[2]

[1]Department of Plant Protection, Faculty of Agricultural Sciences, University of Sassari, via E. De Nicola, 07100 Sassari, Italy and [2]Institute of Ecosystem Study, National Research Council, Traversa La Crucca 3, 07100 Sassari, Italy

Abstract: The economic importance of the Formicidae (ants) in agriculture is mainly due to their interactions with honeydew-producing Hemipterans, which reduce the abundance of beneficial arthropods in the canopies, thus hindering the biological control of pests. Argentine ant [*Linepithema humile* (Mayr, 1868)] and native ants (of the genera *Camponotus*, *Formica*, *Lasius*, *Pheidole*, *Plagiolepis*, *Tapinoma, Tetramorium*) are the most common ant species that aggressively protect and attend Hemipterans in the Mediterranean Region. Although many ants occur in citrus groves, very few species directly damage the plants. A brief review of the important species and guidelines for their management in citrus groves are presented.

Keywords: Citriculture, Mediterranean Region, citrus ants, bio-ecology, damage, control.

1. INTRODUCTION

The family Formicidae (Hymenoptera: Vespoidea) (ants) comprises over 12,000 described species placed in 287 genera. The number of extant subfamilies has risen from the five recognized in the earliest classifications to the 23 reported by Bolton *et al.* [1]. All ants are eusocial, forming perennial colonies with a wingless worker caste. The female castes (workers and queens) have antennae with 4-12 segments, whereas those of males are often elongated with 9-13 segments. Antennae are geniculate between the scape and the funicular segments. The second abdominal segment is reduced, forming a node or scale (petiole), which separates the alitrunk from the remaining posterior abdominal segments. The third abdominal segment is frequently reduced and isolated (postpetiole). Wings of alate queens are shed after mating. The metapleural glands are complex structures located at the extreme corner of the propodeum. Although it is generally accepted that these glands constitute a single diagnostic character separating the Formicidae from all other aculeate Hymenoptera, they have been secondarily lost in some genera, *e.g. Camponotus*, which are specialized occupiers of arboreal environments. In addition, these glands are reduced or absent in the males of many ant species as well as in the social parasites [2].

Many ants are considered agricultural pests, but very few cause any direct damage to plants by their phytophagous action. The economic relevance of ants is due mostly to the myrmecophily phenomenon and to the symbiosis that they establish with many Hemipterans (trophobionts), which may be key pests.

Many ants attend Hemipterans to some extent, because they produce honeydew, a sweet excretion derived from the digestive process. This kind of symbiosis is of ancient origin; the first evidence of such association between aphids and ants of the genus *Linepithema* dates to the early Oligocene. Honeydew consists mainly of carbohydrates and numerous ants, in the subfamilies Myrmicinae, Dolichoderinae and Formicinae, feed on sugar or other similar substances from which derive energy. Apart from aphids, honeydew is produced by several other Hemiptera, including soft scale insects (Coccidae), mealybugs (Pseudococcidae), jumping plant lice (Psyllidae), treehoppers (Membracidae), leafhoppers (Cicadellidae), froghoppers or spittle insect (Cercopidae), and members of the family Fulgoridae. Most myrmecophilous Hemipterans, especially aphids, have special structural and behavioral adaptations to live with ants [3]. A truly mutualistic relationship is thus established, the ants rendering a hygienic service to their partners by removing large quantities of the sugary and sticky material produced. Ants also provide protection by building refuges around Hemipteran colonies. In addition, their presence can help aphids to reach

***Address correspondence to Andrea Lentini:** Department of Plant Protection, Faculty of Agricultural Sciences, University of Sassari, via E. De Nicola, 07100 Sassari, Italy; E-mail: lentini@uniss.it

Vincenzo Vacante and Uri Gerson (Eds)

maturity more quickly. Of great economic significance is the fact that ants can disturb and/or kill the natural enemies of pest Hemiptera [4]. Moreover, ants can indirectly protect some non-producing honeydew pests [5, 6], interfere with the regulation of insect populations under biological control [7] and could be a key factor in biological control programmes based on the inoculative releases of natural enemies [8, 9].

2. CITRUS ANTS

About 300 species of ants have been observed in citrus orchards around the world and doubtless play an important role in the citrus ecosystem. Because ants have been one of the favorable subjects of study for entomologists since antiquity, the general literature on ants is vast. However, only few papers discuss applied research on the interactions between the ants, plants and other insects in citriculture. A tally of ants in citrus orchards in the Mediterranean Region is not available, and the paper about citrus ants throughout the world provides a good example of the complexity of the ant assemblages occurring in citrus groves [10]. The composition of ant communities in natural environments is affected by interspecific competition and by tolerance to physical factors. Highly competitive ants generally exclude potential competitors from their territory and reduce the foraging success of subordinate species. However, in open Mediterranean habitats, where thermal tolerance can reduce the importance of interspecific competition, the mutual exclusion usually found between dominant and subordinate species appears to be influenced by the physiological specialization of ants to different temperature ranges. The composition of ant communities in citrus groves is normally extremely poor because the tree canopy reduces the presence of heat-tolerant ants and increases that of a few heat-intolerant dominant species [11]. Moreover, the pesticide-driven simplification of the citrus ecosystem engenders a loss of biodiversity, reducing the richness of ant species.

Some ants found in citrus orchards are beneficial as general predators that kill a wide range of insects, including citrus pests, and exert an interspecific competitive pressure on potentially harmful ant species. In Florida, predation by ants is an important mortality factor affecting the young instars of the citrus leafminer *Phyllocnistis citrella* Stainton (Lepidoptera: Gracillariidae) [12], whereas in Spain such predation reduces the number of the pupae of the medfly, *Ceratitis capitata* (Wiedemann) (Diptera: Tephritidae), thus being an important factor in regulating the populations of this pest [13]. In the Mekong delta of Vietnam, citrus farmers consider the weaver ant, *Oecophylla smaragdina* Fabricius, to be a beneficial insect for its role not only in the protection of produce from pest damage, but also in the improvement of fruit quality, due to the excretory products deposited while ants patrol the fruit [14].

The underground nest activities of ants can benefit soil development, by increasing water infiltration and reducing runoff discharge. On the other hand, ant activities can be harmful, due to increased erosion losses in unconsolidated soils [15]. Some ant species (*e.g. Tapinoma nigerrimum* (Nylander) damage plants when feeding on the bark and the tender flush, often killing young citrus trees by girdling their trunks [16]. A particular damage to cultivated areas may be caused by ants that chew and enlarge orifices of polyethylene tubing used in drip irrigation systems [17].

The economic importance of ants in Mediterranean citrus groves is mainly due to their attending honeydew-producing Hemipterans. Such activities reduce the abundance of beneficial arthropods in the canopies, hindering the biological control of pests.

Much effort is nowadays being focused on understanding the role of ants and developing ant control programs in citrus orchards. Taking into account the most recent data on citrus ants found in the Mediterranean Region, mainly in Spain and Italy, we list about 60 species, either native (*e.g. Camponotus*, *Formica*, *Lasius*, *Pheidole*, *Plagiolepis Tapinoma*, *Tetramorium*) or exotic (*Linepithema*), of which only a few have been studied in some detail and are of economic relevance (Table **1**).

Table 1. Ant species found in citrus orchards in the Mediterranean region (species and location).

Aphaenogaster pallida (Nylander, 1849). Italy [18]	*Messor barbarus* (Linnaeus, 1767). Cyprus, Italy, Spain, Libya [10, 11, 19]
Aphaenogaster semipolita (Nylander, 1856). Italy [18]	*Messor capitatus* (Latreille, 1798). Italy [18]
Aphaenogaster senilis Mayr, 1853. Spain [11, 19]	*Messor semirufus* (André, 1883). Israel, Jordan [10]
Camponotus aethiops (Latreille, 1798). Italy [9, 18]	

Camponotus compressus (Fabricius, 1787). Israel, Jordan, Algeria, Libya, Morocco, Tunisia [10]
Camponotus cruentatus (Latreille, 1802). Spain [20]
Camponotus foreli Emery, 1881. Spain [11, 19, 20, 21]
Camponotus lateralis (Olivier, 1792). Italy [18]
Camponotus libanicus André, 1881. Turkey [10]
Camponotus nylanderi Emery, 1921. Italy [18]
Camponotus piceus (Leach, 1825). Italy [18]
Camponotus pilicornis (Roger, 1859). Spain [11, 19]
Camponotus sylvaticus (Olivier, 1792). Spain [20, 21]
Cardiocondyla mauritanica Forel, 1890. Spain [21]
Cardiocondyla obscurior Wheeler W.M., 1929. Israel [10]
Cataglyphus bicolor (Fabricius, 1793). Israel, Jordan, Syria, Morocco [10]
Cataglyphus viaticus (Fabricius, 1787). Egypt, Libya, North of Africa [10]
Crematogaster inermis Mayr, 1862. Israel, Jordan [10]
Crematogaster jehovae Forel, 1907. Israel, Jordan [10]
Crematogaster scutellaris (Olivier, 1792). Cyprus, Italy [10, 18]
Dolichoderus thoracicus (Smith F., 1860). Spain [10]
Formica cinerea Mayr, 1853. Italy, Mediterranea Basin [10]
Formica cunicularia Latreille, 1798. Spain, Italy [9, 11, 18-20]
Formica gerardi Bondroit, 1917. Spain [21]
Formica subrufa Roger, 1859. Spain [20]
Hypoponera eduardi (Forel, 1894). Spain, Italy [11, 18, 19]
Lasius alienus (Foerster, 1850). Italy, Turkey [10, 18]
Lasius grandis Forel, 1909. Spain [20]
Lasius niger (Linnaeus, 1758). Italy, Spain [9, 11, 19, 21]
Lepisiota bipartita (Smith F., 1861). Israel, Jordan [10]
Lepisiota frauenfeldi (Mayr, 1855). Algeria, Libya, Tunisia [10]
Lepisiota nigra (Dalla Torre, 1893). Cyprus, Italy [10]
Lepisiota nigrescens (Karavaiev, 1912). Algeria, Libya, Tunisia [10]
Linepithema humile (Mayr, 1868). Italy, Spain [10, 11, 18, 19, 21]
Messor structor (Latreille, 1798). Italy [18]
Monomorium destructor (Jerdon, 1851). Cyprus [10]
Monomorium subopacum (Smith F., 1858). Israel, Jordan, Syria, Libya, Cyprus, Italy, Spain, Tunisia [10]
Myrmica scabrinodis Nylander, 1846. Spain [11, 19]
Paratrechina jaegerskioeldi (Mayr, 1904). Cyprus, Israel, Jordan, Egypt, Libya [10]
Paratrechina longicornis (Latreille, 1802). Israel [10]
Pheidole pallidula (Nylander, 1849). Spain, Egypt, Libya, Morocco, Italy [10, 11, 18-21]
Plagiolepis pallescens Forel, 1889. Israel, Jordan [10]
Plagiolepis pygmaea (Latreille, 1798). Spain, Italy, Turkey, Mediterranean Basin [10, 18, 20, 21]
Plagiolepis schmitzii Forel, 1895. Algeria, Morocco, Libya, Tunisia, Spain, Italy [10, 11, 18, 19, 31]
Prenolepis nitens (Mayr, 1853). Italy, Mediterranea Basin [10]
Proceratium algiricum Forel, 1899. Italy [18]
Pyramica membranifera (Emery, 1869). Italy [10]
Solenopsis fugax (Latreille, 1798). Italy [18]
Solenopsis geminata (Fabricius, 1804). Italy [10]
Solenopsis robusta Bernard, 1950. Spain [11, 19]
Tapinoma erraticum (Latreille, 1798). Italy, Spain [18, 21]
Tapinoma israele Forel, 1904. Israel [10]
Tapinoma nigerrimum (Nylander, 1856). Italy, Spain [11, 18-21]
Tapinoma simrothi Krausse, 1911. Cyprus, Italy, Spain, North of Africa, Israel, Jordan, Lebanon [10, 18, 21]
Temnothorax recedens (Nylander, 1856). Italy [18]
Tetramorium brevicorne Bondroit, 1918. Italy [9]
Tetramorium caespitum (Linnaeus, 1758). Mediterranean Basin, Spain, Italy [10, 11, 18, 19]
Tetramorium persignatum Bolton 1995. Israel, Jordan, Lebanon [10]
Tetramorium punicum (Smith F, 1861). Israel, Jordan, Lybia [10]
Tetramorium semilaeve André, 1883. Spain, Italy [18, 20, 21]

3. IDENTIFICATION OF THE CITRUS ANTS

The number of ants that can be collected in citrus groves is variable and depends on the sampling methods used. Many are ground-dwelling and are easily collected by using pit-fall traps. Others wander around looking for honeydew on the trees and can be collected near Hemipterans. There are also cryptic species that are hard to find. It is not possible to present taxonomic keys to identify all ant species of citrus orchards. Herein we present simplified dichotomous keys, adapted from the fundamental work by Bolton [22], for the 4 ant subfamilies and the different genera of Formicidae that are generally found in citrus orchards. These keys apply to the worker caste.

3.1. Key to Subfamilies

1. Body with a single reduced or isolated segment (the petiole) between alitrunk and **2**

- Body with two reduced or isolated segments (the petiole and postpetiole) between alitrunk and gaster .... **Myrmicinae**

2. Apex of gaster with a semicircular to circular acidopore (the orifice of the formic acid appearing as a short nozzle). Sting absent.. **Formicinae**

- Apex of gaster without an acidopore. Sting present or absent... 3

3. Sting vestigial or absent, in any case not visible without dissection .... **Dolichoderinae**

- Sting present. Abdominal segment IV with the sternal and tergal plates rigidly fused .... **Ponerinae**

3.2. Key to Genera

Subfamily Ponerinae

1. Subpetiolar process with a fenestra anteriorly... ***Ponera***

- Subpetiolar process without a fenestra anteriorly... ***Hypoponera***

Subfamily Myrmicinae

1. Postpetiole articulated on dorsal surface of first gastral segment.................................... ***Crematogaster***

- Postpetiole articulated on anterior surface of first gastral segment.. 2

2. Antennae with 10 segments, club with two-segments.. ***Solenopsis***

- Antennae with 12 segments. Tibial spurs of middle and hind legs pectinate .... ***Myrmica***

Antennae with 11-12 segments. Club with three segments or without a distinct club. Tibial spurs usually simple.. 3

3. Propodeum unarmed. Antennae with 12 segments, club with three segments .... ***Monomorium***

- Propodeum armed with a pair of spines or teeth. If unarmed, club with 4-5 segments or indistinct.. 4

4. Antennal club with 4-5 segments or an indistinct club 5

- Antennal club with three segments.. 6

5. Metasternal process minute or absent. Mandibles elongate-triangular and not massively constructed. Monomorphic species... ***Aphaenogaster***

- Metasternal process large. Mandibles short and powerful, massively constructed with their teeth worn down. Polymorphic species .... ***Messor***

6. Very polymorphic with major and minor workers. Major workers with enormous heads, powerful mandibles with two apical teeth and two basal teeth separated by a long diastema .... ***Pheidole***

- Monomorphic species. Mandibles different.. 7

7. Postpetiole much larger than petiole in dorsal view. Alitrunk without erected hairs. Antennae with 12 segments .... ***Cardiocondyla***

- Postpetiole different and alitrunk with erected hairs 8

8. Lateral portions of clypeus raised as a shield wall in front of the antennal insertions .... ***Tetramorium***

- Lateral portions of clypeus not raised as a shield wall in front of the antennal insertions.. 9

9. Antennae with 11 or 12 segments. Clypeus with a middle carina.. ***Temnothorax***

Subfamily Dolichoderinae

1. Petiole in profile usually a simple flattened strip, never with a scale. In dorsal view the gastral segments cover the petiole..***Tapinoma***

- Petiole with scale always present and visible in dorsal view when alitrunk and gaster are in the same plane. Metanotal groove impressed and alitrunk in profile with propodeal declivity convex...***Linepithema***

Subfamily Formicinae

1. Antennae with 11 segments. Propodeum and petiole unarmed. Mesonotum separated from the metanotum by a transverse groove..... ....***Plagiolepis***

- Antennae with 12 segments.. 2

2. Metapleural gland orifice absent. Antennal sockets located far from the posterior margin of the clypeus. Polymorphic species..***Camponotus***

- Metapleural gland orifice present. Antennal sockets located close to the posterior margin of the clypeus...3

3. With the mesosoma in profile view, the propodeal spiracle located very close to the decliveous part of the propodeum. Ocelli not always visible in the workers...................... ....***Lasius***

- With the mesosoma in profile view, the propodeal spiracle located under the horizontal part of the propodeum and at some distance from the decliveous part. Ocelli present and conspicuous in the workers................................. 4

4. Maxillary palps very long, their basal flattened. Propodeal spiracle slit-shaped ....***Cataglyphis***

- With shorter maxillary palps, usually directed downwards. All maxillary segments bear shorter hair. Propodeal spiracle ovoid or circular........ ....***..Formica***

4. MAIN ANT SPECIES OF CITRUS ORCHARDS

The following species are commonly present in citrus orchards in the Mediterranean Region and their importance is universally recognized.

4.1. *Lasius niger* (Linnaeus)

The common black ant has a wide geographical distribution. Seifert [23], in his revision of the genus *Lasius* s.str., stated that the real distribution of *L. niger* may have been overestimated because different Palearctic species morphologically very close, were considered to be the same species by earlier myrmecologists. In many areas of Spain, *L. niger* seems to be replaced by *L. grandis*.

The worker (Fig. **1**) has a total length of 3.0-4.0 mm; the whole head surface with numerous but not very long setae; clypeal carina weak and incomplete, convex in lateral profile; erect setae always present on scapes and hind tibiae; frontal head and clypeus with dense pubescence; head and gaster blackish brown; mesosoma dark brown to medium brown with a yellowish tinge; petiole, coxae, femora and tibiae dark brown; mandibles, scape, trochanter and the tibio-femoral joint region yellowish-reddish brown.

Lasius niger occurs in moderately xerothermous to mesophilic habitats. As a synanthropic species, it lives in cities, gardens and arable land habitats, forming monogynous colonies that contain from 4,000 to 50,000 individuals [24].

Fig. (1). *Lasius niger* (Linnaeus).

Although it easily penetrates dead wood for nest construction, *L. niger* is preferentially a digger and an above-ground constructor, using mineral soil particles. It typically swarms in hot or warm afternoons and evenings with high air humidity. The mating and nuptial flights of *L. niger* can contain thousands of winged males and females and usually take place during the hot summer. Queens have the highest longevity recorded in ants and may live up to 29 years. *Lasius niger* is extremely aggressive towards other ant species, frequently attacking myrmecologists when they open its nests. This ant eats ripe fruits, insects and other small invertebrates but can also attend aphids for their honeydew, bringing them inside the nest and back out again, when necessary. It also attends the pest mealybug *Planococcus citri*, with a significantly negative effect on the activity of its parasitoids and predators [7].

Lasius niger is one of the most frequent species in the citrus orchards of Italy and Spain, where it may constitute over 80% of the entire ant fauna [9, 11].

4.2. *Lasius grandis* Forel

This species is the most abundant of its subgenus in the Iberian peninsula, where it is distributed from sea level up to 2,300 m, in the Sierra Nevada [23].

Lasius grandis is very similar to *L. niger*. Separating its workers from those of *L. niger* is based on the pilosity of the clypeus and morphometric characters.

Lasius grandis lives in open habitats as well as in woodlands, in mesophilic or humid sites. In the Atlas mountains of Morocco, it occurs at 1400-1800 m altitude in deciduous and coniferous woodland and, at 2,200-2,800 m, in open habitats, such as mountain pastures, *Juniperus* stands and more xerothermous grasslands. In Spain its nests are found mainly under stones and in the soil, and it rarely constructs large soil mounds, like those of *L. niger.* The time of nuptial flight and other aspects of its biology are unknown. In Spanish citrus groves *L. grandis* can be common, reaching about 40% of the total number of ground-dwelling ants [19]. This species tends aphids that infest many trees [25] and shrubs and has monogynous colonies. It is as aggressive as *L. niger* after its nest is disturbed.

4.3. *Linepithema humile* (Mayr)

The Argentine ant is native to the drainage basin of the Paraná river in Argentina, Brazil, Paraguay, and Uruguay. It is an invasive species that is now distributed worldwide, including the Mediterranean Region, and is ranked among the 100 worst animal invaders in the world.

The workers of *L. humile* are small, medium to dark brown ants, reaching 2-3 mm in length (Fig. **2**). Body surface is smooth and shiny and lacks hairs on the dorsum of the head and thorax. The petiole is composed of a single, scale-like segment, and the sting is absent. Workers are extremely fast-moving and industrious, being often recruited in high numbers.

Fig. (2). *Linepithema humile* (Mayr).

This species has the unique capacity of forming supercolonies of interconnected nests that lack colonial boundaries and intraspecific aggression, acting as a single cooperative unit, and may spread over hundreds to thousands of kilometers [26]. Three continental supercolonies of *L. humile* have been identified In Southern Europe. The absence of intraspecific aggression over large areas allows non-native populations of *L. humile* to reach high densities and ecological dominance, often displacing native ant species.

This species is polygynous and polydomous, and many of its nests have numerous dealate (wingless) queens. In the introduced populations of *L. humile*, the queens generally do not disperse in the winged form and colonies reproduce by budding off into new units, established by as few as ten workers and a single queen.

This species is omnivorous and strongly attracted to various sugary liquids. As a consequence, *L. humile* tends a variety of Hemiptera that infest citrus, including *Aleurothrixus floccosus Maskell*, *Aphis spiraecola Patch*, *Coccus hesperidum (L.)*, *Planococcus citri (Risso)*, and *Saissetia oleae (Olivier)*, protecting them from natural enemies. The ants provide extra protection by covering mealybugs in shelters made of bits of bark and soil particles, loosely stuck together.

The Argentine ant can also protect insects and mites that do not produce honeydew from parasitoid and predator activity [5, 6]. For example, populations of *Panonychus citri (McGregor),* and *Aonidiella aurantii (Maskell)*, can increase due to the presence of the Argentine ant. In fact, this ant is extremely aggressive against coccinellid predators of mites, like *Stethorus punctillum*, and disrups the activity of the parasitoid *Aphytis melinus (Weise)*.

The Argentine ant is not very frequent in Mediterranean citrus groves. When present, it is not as abundant as native ant species [11, 19, 25] and its pest status is less important than in other parts of the world.

4.4. *Tapinoma simrothi* Krausse

Tapinoma is a large and diverse genus distributed throughout the world. About 90 species are recognized, of which 25 occur in the Palearctic region.

The worker of *T. simrothi* (Fig. **3)** is 3.0-4.0 mm long, with a uniformly dark brown to black body; total fifth gastral tergite ventral; pronotum lacking erect hairs; front border of clypeus with a slit-like cleft; legs including tarsi uniformly brownish.

Fig. (3). *Tapinoma simrothi* Krausse.

Species of *Tapinoma* utilize various sites, such as open soil, soil under rocks, wood or other plant material or preformed cavities. Clay soils, wherein *T. simrothi* nests and forms polygynous colonies of 4,000 to 20,000 workers, are preferred. Species of *Tapinoma* are general scavengers, with a preference for honeydew, often tending aphids or coccids. Nine species were recorded in the Mediterranean Region, but only *T. simrothi, T. erraticum* and *T. nigerrimum* (the synonymy of *nigerrimum* with *erraticum* [1] is not considered here) occur in citrus orchards. They damage the leaves, shoots, twigs and flowers of many plants [16, 27]. *Tapinoma simrothi* is among the most frequent ant species in citrus orchards, where it damages young leaves and shoots of citrus in order to obtain sugary sap. However, this behavior does not occur when numerous honeydew-producing insects are present. The main importance of this ant is due to its positive influence on the growth and survival of the pest aphids *Toxoptera aurantii* (Boyer de Fonscolombe) and *Aphis spiraecola* [28]. It is an extremely aggressive and competitive ant and can easily displace *L. humile* in some environments [29].

4.5. *Plagiolepis pygmaea* (Latreille) and *Plagiolepis schmitzii* Forel

Plagiolepis pygmaea and *P. schmitzii* are commonly found in citrus groves. Both are considered nectarivorous and honeydew foragers but also prey on other arthropods [30]. The workers are relatively small, less than 2.0-mm long, have a brownish and glossy body with little pubescence; and the antennae with 11 segments. The two species can be separated by *P. schmitzii* having the third and fourth funiculus segments elongated, subequal, and twice as long as the second one, whereas in *P. pygmaea* the third funiculus segment is shorter than the fourth and only slightly longer than the second.

Plagiolepis pygmaea is present in southern Europe, North Africa, and western Asia, where it usually occurs in arid habitats with low vegetation. It has polygynous and polydomous colonies with 5-35 queens, in subterranean nests located under stones or in gravel. Queens may found independent new communities, but colony reproduction usually takes place by budding. *Plagiolepis pygmaea* is one of the few native ant species that can coexist with *L. humile* in both natural and agricultural ecosystems [30, 31]. *Plagiolepis pygmaea* is considered a pest of citrus because of its relationships with Hemipterans, such as *Planococcus citri* and *Icerya purchasi Maskell* [21, 30]. Although *P. pygmaea* can transfer young nymphs of the dictyospermum scale *Chrysomphalus dictyospermi Morgan* onto citrus shoots and fruits, the nature of this relationship is unknown. *Plagiolepis schmitzii* is less common than *P. pygmaea* but more frequent in North Africa. It lives in soils with a high humidity level and, like *P. pygmaea*, consists of polygynous colonies. The ant occurs in Spanish citrus groves [11, 21], where it may locally be abundant [19]. The importance and economic impact of *Plagiolepis* spp. in citriculture is not clear.

4.6. *Pheidole pallidula* (Nylander)

The genus *Pheidole* consists of over 1,100 species with a worldwide distribution. *Pheidole pallidula* is very common in the Mediterranean Region. Like all members of this genus, it is characterized by a highly dimorphic worker caste, with the major workers (3.0-4.5 mm in length) having an enormous head with strong mandibles, and the minor workers being smaller (2.5-3.0 mm) (Fig. **4**). Antennae with 12 segments and the club has three segments that are subequal in length. *Pheidole pallidula* is an omnivorous species [29] that finds its food mainly by foraging on the soil surface, rarely visiting trees. Workers can hunt slow or defenseless prey. They also eat seeds or even flower nectar, but are mainly scavengers feeding on dead insects.

Fig. (4). *Pheidole pallidula* (Nylander).

This species occurs mainly on sunny slopes with low vegetation density, where it forms colonies containing from 600 to 6,000 minor individuals per nest [32]. The predatory habit of this species is a desirable trait, but is not well known. Although the role of this ant is controversial, *P. pallidula* has been described as a potential predator of medfly pupae in Spanish citrus groves [13].

4.7. Other Ant Species

Several other ant species, such as *Formica cunicularia, F. gerardi, Tetramorium caespitum, T. semilaeve, T. brevicorne, Camponotus nylandery, C. sylvaticus, C. aethiops and C. foreli,* occur in citrus orchards in the Mediterranean Region. However, their ecological role there has not yet been seriously studied.

5. ANT MANAGEMENT IN CITRUS ORCHARDS

Control of ants that are associated with outbreaks of phloem-feeding Hemipterans is a key component of integrated pest management in citrus orchards. In fact, the control of ants in citrus groves results in reductions of the populations of whiteflies, aphids, mealybugs and scales caused by their natural enemies. An effective ant control program requires knowledge of the biology and behavior of the species present in the agroecosystem. The bio-ethological characteristics of ants (*e.g.* interspecific competition, choice of nest sites and feeding behavior) influence the efficacy of the various available control methods, thus the choice of the best strategy to keep their populations at low levels.

From an applied point of view, the ant control methods that can be used in citrus agroecosystem consist mainly of placing sticky bands around the tree trunk, of residual insecticides (as a barrier on the trunk or on the ground around skirt-pruned trees, or as foliar treatments), and of pesticide-treated baits.

Sticky bands made of polybutene-based glue, which prevent ants from climbing the trees, are used successfully in Australia and South Africa. For this method to be efficient, it is necessary that tree leaves do not touch the ground, obtained by appropriate pruning, and of preventing alternative access routes to tree canopies, by proper weed control. In addition, bands must be kept clean from dust, leaves, dead insects, *etc.* Bands can be phytotoxic, especially to young trees that are exposed to direct sunlight. Although effective, this method is not normally used by growers, because it requires regular maintenance and intensive labor.

Current strategies to control ants in citrus orchards are usually based on the application of residual insecticides, such as diazinon and chlorpyriphos. When applied to the trunk and ground beneath the tree, or to the whole canopy every 1-3 months, these insecticides provide a repellent barrier and/or kills foraging workers. However, the effectiveness of spraying is limited and many applications are required, because the insecticide does not reach the workers, queens, eggs and larvae that reside in the nest [33]. A further disadvantage of contact insecticides is that they have a wide action spectrum, thus interfering with the activities of beneficial insects and increasing the populations of secondary phytophagous arthropods.

Pesticide-treated baits are an effective control method that combines a slow acting insecticide with a feeding attractant. Because foraging workers transfer the toxic baits among members of the community, this method also affects the nest population and, in the case of Argentine ants, has the additional advantage of being distributed among nests of this unicolonial species [34]. To achieve maximum effectiveness of ant control, baits should be distributed in early spring, when the honeydew-producing insect populations are still low and feeding attractants can be the most important resource for foraging workers. Moreover, during this period toxic baits can target developing ant larvae, thus providing a long-term impact. The concentration of active ingredient must be low, so that toxicity is delayed for a few days and foraging workers do not die before distributing the baits to colony members. The optimal percentages of carbohydrate, protein and fat in the feeding attractant vary according to the ant species. Solid baits are generally used against protein-feeding ants, such as *Pheidole* spp. [35]. Liquid baits are more appropriate for species that feed mainly on sugar. Argentine ants can be controlled by using liquid baits made of a syrup (25% sucrose-water) and fipronil, imidacloprid, thiamethoxam or borate, at ultralow concentrations, kept inside a releasing device called ant bait station [34, 36]. The dispenser consists of a central reservoir that slowly releases the bait in response to ant feeding and allows only the entrance of ants. Efficient control of high infestations of Argentine ants requires 55 bait stations per hectare, which can be gradually reduced as the ant population decreases.

REFERENCES

[1] Bolton B, Alpert G, Ward PS, Naskrecki P. Bolton's catalogue of ants of the world: 1758-2005. CD-ROM. Cambridge, Mass.: Harvard University Press 2007.

[2] Hölldobler B, Wilson EO. The ants. Cambridge, Mass.: Harvard University Press 1990

[3] Bernhard S, Dixon AFG. Ecology and evolution of aphid-ant interactions. Annu Rev Ecol Evol, System 2005; 36: 345-72.

[4] Way MJ. Mutualism between ants and honeydew-producing Homoptera. Annu Rev Entomol 1963; 8: 307-44.

[5] Haney PB, Luck RF, Moreno DS. Increases in densities of the citrus red mite, *Panonychus citri* (Acarina: Tetranychidae), in association with the Argentine ant, *Iridomyrmex humilis* (Hymenoptera: Formicidae), in southern California citrus. Biocontrol 1987; 32(1): 49-57.

[6] Pekas A, Tena A, Aguilar A, Garcia-Mari F. Effect of Mediterranean ants (Hymenoptera: Formicidae) on California Red scale (Hemiptera:Diaspididae) populations in citrus orchards. Environ Entomol 2010; 39(3): 827-34.

[7] Campos JM, Martínez-Ferrer MT, Forés V. Parasitism disruption by ants of *Anagyrus pseudococci* (Girault) and *Leptomastix dactylopii* Howard (Hymenoptera: Encyrtidae), two parasitoids of the citrus mealybug *Planococcus citri* (Risso) (Homoptera: Pseudococcidae). Bull IOBC/WPRS 2006; 29(3): 33-46.

[8] Villalba M, Vila N, Marzal C, Garcia Marí F. Influencia en el control biológico del cotonet *Planococcus citri* (Homoptera: Pseudococcidae) de la liberación inoculativa de enemigos naturales y la eliminación de hormigas, en parcelas de cítricos. Bol San Veg Plagas 2006; 32: 203-14.

[9] Marras PM, Sanna F, Pantaleoni RA. Influence of ant-exclusion on *Planococcus citri* density in a citrus orchard. Bull IOBC/WPRS 2008; 38: 104-10.

[10] Haney PB. Identification, Ecology and Control of the Ants in Citrus: A World Survey. In: Goren R, Mendel K, Eds. Proceedings of the Sixth International Citrus Congress; 1988: Tel Aviv, Israel: Balaban Publishers 1988; pp.1227-51.

[11] Cerdá X, Palacios R, Retana J. Ant community structure in citrus orchards in the Mediterranean basin: impoverishment as a consequence of habitat homogeneity. Environ Entomol 2009; 38(2): 317-24.

[12] Xiao Y, Qureshi JA, Stansly PA. Contribution of predation and parasitism to mortality of citrus leafminer *Phyllocnistis citrella* Stainton (Lepidoptera: Gracillaridae) populations in Florida. Biol Control 2007; 40: 396-404.

[13] Urbaneja A, García Marí F, Tortosa D *et al.* Influence of ground predators on the survival of the Mediterranean fruit fly pupae, *Ceratitis capitata*, in Spanish citrus orchards. Biocontrol 2006; 51: 611-26.

[14] Barzman MS, Mills NJ, Thi Thu Cuc N. Traditional knowledge and rationale for weaver ant husbandry in the Mekong delta of Vietnam. Agriculture and human values - Fall 1996; 13(4): 2-9.

[15] Cerdà A, Jurgensen MF. The influence of ants on soil and water losses from an orange orchard in eastern Spain. J Appl Entomol 2008; 132: 306-14.

[16] Martelli GM. Danni del *Tapinoma nigerrimum* Nyl. alle parti aeree di alcune piante del genere Citrus. Riv Patol Veg 1933; 23: 286-9.

[17] Chang VCS, Ota AK. Fire ant damage to Polyethylene tubing used in drip irrigation systems. J Econ Entomol 1976; 69(4): 447-50.

[18] La Pergola A, Alicata A, Longo S. Survey of the ants (Hymenoptera Formicidae) in citrus orchards with different types of crop management in Sicily. Bull IOBC/WPRS 2008; 38: 233- 7.

[19] Palacios RM, Martinez-Ferrer T, Cerdá X. Composiciòn, abundancia y fenologìa de las hormigas (Hymenoptera: Formicidae) en campo de cìtricos de Tarragona. Bol San Veg Plagas 1999; 25(2): 229-40.

[20] Piñol J, Espadaler X, Cañellas N, Peréz N. Effects of the concurrent exclusion of ants and earwigs on aphid abundance in an organic citrus grove. BioControl 2009; 54: 515-27.

[21] Alvis L, Garcia-Marí F. Identification and abundance of ants (Hymenoptera: Formicidae) in citrus tree from Valencia (Spain). Bull IOBC/WPRS 2006; 29(3): 111-6.

[22] Bolton BO. Identification guide to the ant genera of the world. Cambridge, MA: Harvard University Press 1994.

[23] Seifert B. A taxonomic revision of the Palaearctic members of the ant subgenus *Lasius* s. str. (Hymenoptera, Formicidae). Abh Ber Natkdmus Görlitz 1992; 66: 1-67.

[24] Seifert B. *Lasius platythorax* n. sp., a widespread sibling species of *Lasius niger* (Hymenoptera, Formicidae). Entomol Gen 1991; 16: 69-81.

[25] Monzó C, Vanaclocha P, Gómez K, Tortosa D *et al.* Survey of the Ants (Hymenoptera: Formicidae) in soil of citrus orchards in Spain. Bull IOBC/WPRS 2006; 29(3): 283.

[26] Holway DA, Suarez AV. Colony structure variation and interspecific competitive ability in the invasive Argentine ant. Oecologia 2004; 138: 216-22.

[27] Bernard F. Contribution à la connaissance du *Tapinoma simrothi* Krausse, fourmi la plus nuisibles aux cultures du Maghreb. Bulletin de la Société d'Histoire naturelle de l'Afrique du Nord 1978 ; 67(3/4): 87-101.

[28] Dartigues D. Répartition spatio-temporelle des aphides et influence des fourmis, sur orangers en Kabylie. Fruits 1991; 46(4): 461-69.

[29] Bernard F. Les fourmis (Hymenoptera Formicidae) d'Europe occidentale et septentrionale. Paris: Masson et C. Ed. 1968.

[30] Liotta G. La *Plagiolepis pygmaea* Latr. (La formichetta nera) (Hymenoptera - Formicidae). Boll Ist Entomol Agr Osserv Fitopatol Palermo 1965; 5: 1-43.

[31] Abril S, Gomez C. Ascertaining key factors behind the coexistence of the native ant species *Plagiolepis pygmaea* with the invasive Argentine Ant *Linepithema humile* (Hymenoptera: Formicidae). Sociobiology 2009; 53: 559-68.

[32] Bernard F. Rôle du terrain et de la végétation dans la répartition des fourmis méditerranéennes de France. Compte Rendu de la Société de Biogéographie 1980; 485 : 149-54.

[33] Klotz JH, Rust MK, Gonzalez D *et al.* Directed sprays and liquid baits to manage ants in vineyards and citrus groves. J Agric Urban Entomol 2003; 20: 31-40.

[34] Greenberg L, Klotz JH, Rust MK. Liquid borate baits for control of Argentine ant (*Linepithema humile*) in organic citrus (Hymenoptera: Formicidae). Fla Entomol 2006; 89: 469-74.

[35] Taniguchi G, Thompson T, Sipes B. Control of the big-headed ant, *Pheidole megcephala* (Hymenoptera: Formicidae) in pineapple cultivation using Amdro in bait stations. Sociobiology 2005; 45: 1-7.

[36] Klotz JH, Rust MK, Phillips P. Liquid bait delivery systems for controlling Argentine ants in citrus groves (Hymenoptera: Formicidae). Sociobiology 2004; 43: 419-27.

CHAPTER 18

Secondary Pests

Uri Gerson[1*] and Vincenzo Vacante[2]

[1]*Department of Entomology, Faculty of Agricultural, Food and Environmental Quality Sciences, The Hebrew University of Jerusalem, P.O. Box 12, Rehovot, 76-100, Israel and* [2]*Dipartimento Partimonio Architettonico ed Urbanistico (PAU), Mediterranean University of Reggio Calabria, Via Melissari, Località Feo di Vito, 89060 Reggio Calabria, Italy*

Abstract: Herein we treat secondary pests that occasionally infest citrus in the Mediterranean Region. Many of these species occur fortuitously, whereas a few may become pests. Some cause economic-level damage only in one or a few countries, being pests during a few years, later disappearing or becoming almost harmless. Therefore this list is probably incomplete, and will require changes in the future. Brief systematic and biological data are provided for each species, with a minimum of references.

Keywords: Citriculture, Mediterranean Region, citrus pests, secondary pests, bio-ecology, control.

1. INTRODUCTION

A large number of phytophagous arthropods have been recorded from citrus trees in the Mediterranean Region; for instance, over 70 species were listed from Israel alone [1]. Many of these species probably occur on citrus fortuitously, arriving by various means, including being wind-borne. Few, however, may cause some economic damage, and these can be divided into major and minor pests. The former were discussed in earlier parts of this eBook, and the latter will be dealt with below. It must be emphasized that some secondary pests cause discernible damage only in one or a few countries in the Region. Further, they may be economic pests during one or more years, later disappearing or becoming almost harmless. For these reasons the list of species discussed below is probably incomplete, others becoming (or ceasing to be) pests in future years. Their families will be presented in the same order as above, with only brief descriptions and a minimum of references.

2. ACARI: TETRANYCHIDAE

2.1. *Eutetranychus citri* Attiah

Diagnostic characters: The oval body is 0.4-0.5 mm in length, yellow-brownish, with legs as long as the body. The dorsal setae are short, spatulate and placed on small tubercles. The male is reddish, smaller, its caudal part pointed, legs longer than body. This species resembles *Eutetranychus orientalis* (Klein), differing in having only 5 setae on tibia II and the first pair of dorso-central setae (*c*1) being located behind *c*2 (*E. orientalis* has 6-7 setae on tibia II and its *c*1, *c*2 setae are in line).

Life history and economic damage: This mite is a minor pest of limes in Egypt [2], occurring also on cotton in India [3]. Nothing is otherwise known about the extent of its damage or about any control measures undertaken.

3. INSECTA: HEMIPTERA

3.1. *Closterotomus trivialis* (Costa) (Miridae)

Diagnostic characters: Body 6.6-7.9 mm in length, yellow to green in color and with a yellow to black head (Fig. **1**). The first antennal segment is shorter than the width of the head and the second is 1.3 times longer than the width of the first thoracic segment (pronotum), third and fourth segments together. The pronotum has two central and two lateral black spots. Femora III are brown and the tibiae bear black spots and thorns.

***Address correspondence to Uri Gerson:** Department of Entomology, Faculty of Agricultural, Food and Environmental Quality Sciences, The Hebrew University of Jerusalem, P.O. Box 12, Rehovot, 76-100; E-mail: Gerson@agri.huji.ac.il

Fig. (1). *Closterotomus trivialis* (Costa) feeding on young fruit of citrus.

Life history and economic damage: This species, commonly referred to as the plant bug, is known in various Mediterranean countries [4, 5], including Spain, Italy and Greece. It develops on herbaceous and woody cultivated plants, feeding on young leaves, shoots, buds, blossoms and flowers, and migrating from herbaceous to trees. The bug causes leaf and fruit deformation, necrosis of shoots and flower loss. During feeding the pest may transmit virus diseases. *Closterotomus trivialis* has a single annual generation, overwintering as eggs in bark cracks, in untreated wounds and/or within various wooden supports. Young bugs live on weeds under the canopy and then move onto citrus [6], where it may be harmful in years of poor bloom.

Management: On citrus the populations of *C. trivialis* are limited by some Hymenopterous egg parasites, like *Telenomus lopicida* Silvestri (Scelionidae) and *Tetrastichus miridivorus* Domenichini (Eulophidae). Regular flowering suggests that there is no need for any treatment. In late winter and in spring, weeds (like *Urtica* spp.) under the tree canopy are to be inspected and if pest populations are dense the weeds should be treated with a contact insecticide.

3.2. *Hemiberlesia rapax* (Comstock) (Diaspididae)

Diagnostic characters: Shield round, grey, convex, exuvium located at one side; body of adult female bright yellow. The dorsal macroducts are one-barred; anal opening large; without perivulvar pores; median lobes strong, other lobes like small points.

Life history and economic damage: This species, commonly called greedy scale, is very polyphagous, attacking most woody plants. It raises two or more annual generations and a female produces 30-50 eggs. The pest often colonizes the bark, trunk and limbs of its host plants, but in heavy infestations it attacks leaves and fruit also. Infestations by *Hemiberlisia rapax* often cause leaf yellowing, development of necrotic patches and premature leaf drop, dieback of stems and of growing points, fruit discoloration and premature drop. On citrus in Italy it is considered of little economic importance [7], although a serious pest of other fruit trees [8].

Management: No control measures against this pest are usually needed in citrus orchards, where *H. rapax* is often attacked by the Hymenopterous *Aphytis proclia* (Walker) (Aphelinidae) and several coccinellid beetles.

3.3. *Unaspis yanonensis* Kuwana (Diaspididae)

Diagnostic characters: The shield of the female, which bears a central ridge, is blackish-brown with a gray margin. The ventral shield is white. The body is elongate, distinctly segmented; length about 2.5 mm. About two-thirds of the anterior body is strongly sclerotized. Pygidum with three pairs of well developed lobes; the median ones large, not close to each other, diverging and serrated at their inner margins. Shield of male white.

Life history and economic damage: This species, known as the arrowhead scale, occurs on citrus in most citrus-growing areas of the world. In the Mediterranean Region there are two annual generations, in April and in July; the pest overwinters as fertilized females, each producing about 200 eggs [9]. *Unaspis yanonensis* occurs on leaves and twigs. Its damage includes necrotic spots on the foliage followed by leaf drop; irregular stem growth and dieback, external pits and spots on the fruit skin and irregularly shaped fruit; premature ripening, fruit drop and rot and even tree death within a year [9, 10].The economic impact of *U. yanonensis* is however difficult to measure, especially when it occurs in mixed populations with other armored scale insects [10, 11].

Management: The Hymenopterous *Aphytis yanonensis* Rosen and DeBach (Aphelinidae) was introduced into southern Europe, but it does not completely control the pest [12]. It has been suggested that this species cannot survive there from one year to the next, and that renewed releases of *A. yanonensis* are required in order to fully control the pest [13].

4. ORTHOPTERA

4.1. *Anacridium aegyptium* Linnaeus (Acrididae)

Diagnostic characters: The body is pale-reddish, with a pronounced, orange-colored pronotum (Fig. **2**), fore wings gray-brown, with rust-colored veins, hind wings transparent, and veins dark. Female body 75-90 mm in length, male 50-65 mm, larva greenish.

Life history and economic damage: Known as the Egyptian grasshopper, this insect is a sporadic pest of citrus orchards. It lives in trees and shrubs, only seldom feeding on annuals. This grasshopper usually occurs in small numbers, feeding on the leaves. The females undergo a prolonged summer diapause, beginning to lay eggs in May-July of the following year, thus completing only a single annual generation [14]. Although rare, mass infestations may cause serious damage [1], and require application of contact pesticides.

Fig. (2). *Anacridium aegyptium* Linnaeus.

5. COLEOPTERA

5.1. *Otiorrhynchus cribricollis* Gyllenhal (Curculionidae)

Diagnostic characters: The body is 6-8 mm long, brownish-black with reddish brown legs and antennae (Fig. **3**). The rostrum is short, bearing a triangular longitudinal groove that extends toward the apex. The antennae are fine, setose; the scape is clavate, curved and the second segment of the funiculum is slightly longer than the first. The elytra bear 10 striae, dotted and marked, separated by granulose spaces and with short hairs. The femora are smooth and the tibiae are flattened and the tip ends with a pronounced external expansion.

Fig. (3). *Otiorrhynchus cribricollis* Gyllenhal on citrus leaf.

Life history and economic damage: This weevil has a wide geographical distribution, including southern Europe, the Mediterranean Region, USA, Australia and New Zealand. It lives on herbaceous and woody plants and has a single annual generation, overwintering as a larva in the soil, where it feeds on roots and underground organs of wild and cultivated plants [15]. The adult is nocturnal, infesting citrus seedlings or young trees during summer, its feeding causing semicircular wounds on the leaf margins [16]. Mature citrus groves are not significantly damaged by the weevil whose attacks do not require any control.

Management: Weevil infestations of young citrus trees may require control. This can be implemented by tying strips of glass wool around the trunk, and by tillage, which disrupt the development of the weevil's larvae.

5.2. *Apate monachus* Fabricius (Bostrychidae)

Diagnostic characters: Adult body is about 15 mm long, brown-black, pronotum large, hood-like, covering the downwards-placed head. Females bear a pair of caudal projections that do not occur in males. Larvae up to 25 mm in length, white except for brownish head and prothorax.

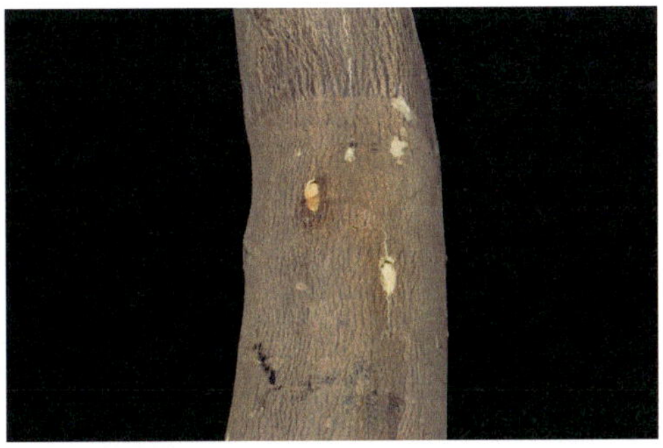

Fig. (4). Branch of citrus attacked by larvae of *Apate monachus* Fabricius.

Life history and economic damage: This widely-distributed species, commonly called the black giant bostrychid, is polyphagous, attacking over 40 shrub and tree species. Adults emerge in spring and immediately begin to burrow galleries in young citrus stems, thin branches or in weakened and dead trees, producing about 80 eggs per female (Fig. **4**). Citron and sour lemons are the preferred citrus host plants [17]. The emerging larvae continue to bore galleries and remain therein throughout their lives; an entire generation may last from one to three years [1]. No external damage is usually seen, but heavily excavated branches may break in storms. This beetle is only seldom regarded as a pest of citrus.

Management: The pest prefers *Calicotome* shrubs, and damage to citrus trees sometimes occurs near these bushes. Heavy infestations can thus be contained by removing these shrubs [17]. The adult or larval beetles may be removed from the galleries by inserting wires, or blocking the gallery entries after inserting a volatile pesticide.

6. LEPIDOPTERA

6.1. Noctuidae

The larvae (caterpillars) of several noctuids (also known as owlet moths), such as *Helicoverpa* (= *Heliothis*) *armigera* (Hübner), *Spodoptera exigua* (Hübner) and *Trichoplusia ni* (Hübner), may occur on citrus trees. The larvae of these moths, which are serious pests of many crops, cause only sporadic damage to citrus. They usually feed on very young leaves, often in nurseries, rarely on young fruit. Because they are unable to develop on citrus trees, their damage is usually minor and transitory, and (except in rare cases) do not require any management measures.

6.2. *Euzopherodes (Ephestia) vapidella* Mannerheim (Pyralidae)

Diagnostic characters: The adult is about 4 mm in length, forewings red-brown, hind wings white. Larvae with a brown head, body whitish, with long setae, length up to 10 mm.

Life history and economic damage: This pest, also known as the citrus stub moth, undergoes three annual generations in Middle Eastern citrus groves [1]. The larvae gnaw beneath the citrus bark, especially around grafts on sweet lime. They also feed on suckers growing from the rootstock and on young buds. When gnawing around the bark they may girdle it, causing subsequent breakage [18].

Management: Damage can be avoided by pruning close to the base as early as possible in the spring, removing rootstock branches and treating wounds with grafting wax.

6.3. *Archips rosanus* (Linnaeus) (Tortricidae)

Diagnostic characters: The adult is 18-22 mm long, forewings brown. The wings of the male have a dark apical spot near the costal margin, which becomes paler towards the outer fringe, and another median dark spot that transverses the wing. In the female this spot is often incomplete, with reticulations, and the remaining spots are attenuated. The hind wings are gray and their apices are sometimes yellowish to orange.

Life history and economic damage: This polyphagous species occurs in Eurasia and North America, infesting herbaceous and woody plants, including citrus. The pest has a single annual generation, overwintering in egg masses on trunks and large branches. The eggs hatch in late winter, the larvae infesting the new growth, where they construct nests by tying leaves and flowers together (Fig. **5**). The adults emerge in May-June. Damage is sporadic, consisting flower and shoot destruction and nibbling on fruits around the pedicles [19].

Fig. (5). Mature larva of *Archips rosanus* (Linnaeus) on citrus leaves.

Management: The presence of eggs on trunks during winter suggests that they can be controlled by winter sprays. Pest numbers may later be monitored with sex pheromone traps. The predatory mite *Allothrombium fuliginosum* (Hermann) (Trombidiidae) and the Hymenopterous egg parasite *Trichogramma embryophagum* Hartig var. *cacoeciae* (Trichogrammatidae) limit the pest's numbers. Large populations can also be controlled with applications of *Bacillus thuringiensis* Berliner [20].

6.4. *Cacoecimorpha pronubana* (Hübner) (Tortricidae)

Diagnostic characters: The adult has a wingspan of 14-18 mm. The fore wings of the female are cinnamon in color, and those of the male have pronounced median and basal bands that form a triangle with the costal margin. The median band of the female is less pronounced, the marginal band being replaced by a dark spot located at the apical corner.

Life history and economic damage: This species is widely distributed, occurring in the Mediterranean Region, several European countries and in South Africa. It infests herbaceous and woody plants, including olive and citrus [21]. The pest may complete 2-6 annual generations, depending on environmental conditions, overwintering as larvae. The latter infest citrus fruits, causing deformations [22] or tunneling within.

Management: Control measures are usually unnecessary, because several natural enemies keep pest numbers below their economic damage level. The enemies include the Hymenopterous *Trichogramma* spp. (Trichogrammatidae), *Elachertus affinis* Masi (Eulophidae), *Apanteles xanthostigma* (Haliday) (Braconidae), *Itoplectis alternans*

Gravenhorst and *Phaeogenes nigridens* (Wesmael) (Ichneumonidae), and the Dipterous *Actia pilipennis* (Fallén) (Tachinidae) [20].

7. DIPTERA

7.1. *Contarinia citri* Barnes (Cecidomyiidae)

Diagnostic characters: The adult body is dark brown, about 1.5 mm in length, forewings gray. Antennae with 2+12 segments, most with circumfila. Larvae with a sternal spatula, about 0.15 mm long. Young larvae initially white, later turning yellow, length up to 2.0 mm. Some authors [23] believe that *C. citri* may be one of a number of species that attack citrus blossoms in different parts of the world.

Life history and economic damage: In the Mediterranean Region (Italy, Turkey, Cyprus, and Israel) the pest, known as the citrus blossom midge, completes 2-4 annual generations [24, 25], the number probably dependent on the availability of lemon blossoms, its preferred host-plant [26]. A generation may be completed in about one month at 20°C. The larvae live and bore in the blossoms and the flowers, feeding on the petals, stamens and ovaries, which are then vulnerable to bacterial rot. Heavily infested blossoms, which may contain as many as 100-200 larvae, drop, losses coming up to 50% [27].

The larvae spend the spring and summer in the soil, where their populations enter short or prolonged periods of diapause [26]. Prolonged (lasting several years) diapause is well known in gall midges [28]. This probably accounts for the sporadic occurrence of this pest, which may disappear from citrus orchards for several years and then suddenly occur in large numbers.

Management: Due to the sporadic occurrence of this pest, no special management measures can be recommended. It is attacked by several Hymenopterous parasitoids [27], but not to a large extent.

REFERENCES

[1] Avidov Z, Harpaz I. Plant Pests of Israel. Jerusalem: Israel Universities Press 1969.

[2] Attiah HH. The Genus *Eutetranychus* in U.A.R., with description of three new species. Bull Soc Entomol Egypte 1967; 51: 11-6.

[3] Nasser OA, Ghai S. Taxonomic studies on tetranychid mites infesting vegetable and fruit crops in Delhi and surrounding areas. Orient Insects 1981; 15: 333-96.

[4] Wagner E. Die Miridae Hahn, 1831, des Mitelmeerraumes und der Makaronesischen Inseln (Hemiptera, Heteroptera). Teil. 1. Entomologische Abhandlungen 1974; 37 (Supplement): 1-484.

[5] Tamanini L. Gli eterotteri della Basilicata e della Calabria (Italia meridionale) (Hemiptera, Heteroptera). Memorie del Museo civico di storia naturale di Verona 1981; 3: 1-164.

[6] Tremblay E. Entomologia applicata. Vol. II, parte 1. Napoli: Liguori Editore 1995.

[7] Bianchi A, Pacchiacucchi A, Guarino L, Maffeo E. Segnalata una nuova cocciniglia su actinidia nel Lazio. Informatore Agrario 1994; 50(46): 73-5.

[8] Danzig EM, Pellizzari G. Diaspididae. In: Kozár F., Ed. Catalogue of Palaearctic Coccoidea. Hungarian Academy of Sciences. Budapest: Akaprint Nyomdaipari Kft. 1998; pp. 172-370.

[9] Davis EE, French S., Venette RC. Mini Risk Assessment Arrowhead Scale: *Unaspis yanonensis* (Kuwana) [Hemiptera: Diaspididae]. 2005. http://aphisweb.aphis.usda.gov/plant_health/plant_pest_info/pest_detection/downloads/pra/uyanonensispra.pdf.

[10] EPPO. *Unaspis citri*: diagnostic protocols for regulated pests. Bull OEPP/EPPO 2004; 34: 299-301.

[11] Onillon JC. Lutte biologique et intégrée dans les vergers de citrus en zone Méditerranéenne. Entomophaga 1988; 33: 481-94.

[12] Bénassy C. Use in France of *Aphytis yanonensis* Rosen and De B. (Hym.: Aphelinidae), a specific parasite of the arrowhead scale, *Unaspis yanonensis* Kuw. (Hom.: Diaspidinae). Fruits 1988; 43: 287-91.

[13] Kreiter P, Marro JP, Dijoux L, Tourniaire R. Japanese citrus fruit scale, *Unaspis yanonensis,* is still causing problems on the French Riviera. Phytoma 1998; 508: 32-3.

[14] Uvarov B. Grasshoppers and Locusts, A Handbook of General Acridology. Vol. 2. London: Centre for Overseas Pest Research 1977.

[15] Vacante V. I coleotteri Curculionidi della fragola in Sicilia. I Contributo. La difesa delle piante 1989; 12: 107-18.

[16] Di Martino E. L'*Otiorrhynchus cribricollis* Gyll. dannos oagli agrumi della Sicilia orientale. Riv Agrum Acireale 1956; I: 239-51.

[17] Bodenheimer FS. Citrus Entomology in the Middle East. The Hague: W. Junk 1951.

[18] Lopez-Colon, J.I. 1996. Presence of *Apate monachus* Fabricius, 1775, in Andalusia (Coleoptera, Bostrichidae). Nouv Rev Entomol 13: 294-7.

[19] Di Martino E. In: Spina P, Di Martino E, Eds. Trattato di agrumicoltura. Bologna: Edagricole 1985; pp. 113-95.

[20] Trembaly E. Entomologia applicata. Vol. II, parte seconda. Napoli: Liguori Editore 1986.

[21] Ragusa S. Osservazioni su *Cacoecia pronubana* Hb. (Verme del Garofano) in Sicilia. Boll Ist Ent Univ Oss Fitop Palermo 1967; 7: 41-62.

[22] Zocchi R. *Cacoecimorpha pronubana* (Hb.). Agric Ricerca 1980; 3 (12): 51-4.

[23] Harris KM, Yukawa J. The taxonomic status of the Japanese citrus flower-bud midge, *Contarinia okadai* (Miyoshi) comb. n. and of the citrus blossom midge, *C. citri* Barnes (Diptera: Cecidomyiidae). Bull Entomol Res 1980; 70: 277-85.

[24] Genduso P. Un cecidomide dannoso alla zagara dei limoni, nuovo per la Sicilia e per la fauna Mediterranea [*Contarinia citri* (?) Barnes]. Boll Ist Entomol Agrar Oss Fitopatol Palermo 1963; 5: 65-73.

[25] Uygun N, Şekeroglu E. Doğu Akdeniz Bölgesi turunçgillerinde zararlı *Contarinia citri* Barnes (Diptera: Cecidomyiidae) üzerinde gözlemler (Observations on *Contarinia citri* Barnes (Diptera: Cecidomyiidae), a citrus pest in the Eastern Mediterranean Region). Türkiye Bitki Koruma Dergisi 1983; 7: 193-8.

[26] Gerson U, Neubaeur I. The citrus blossom midge, *Contarinia citri* Barnes (Diptera: Cecidomyiidae) in Israel. Phytoparasitica 1976; 4: 163-72.

[27] Rubin A. *Contarinia citri* Barnes - a pest of citrus new to Israel. Israel J Agric Res 1965; 15: 103-4.

[28] Barnes HF. Gall Midges of Economic Importance. Vol VII. Gall Midges of Cereal Crops. London: Crosby Lockwood 1958

CHAPTER 19

Conclusions

Uri Gerson[1*] and Vincenzo Vacante[2]

[1]*Department of Entomology, Faculty of Agricultural, Food and Environmental Quality Sciences, The Hebrew University of Jerusalem, P.O. Box 12, Rehovot, 76-100, Israel;* [2]*Dipartimento Patrimonio Architettonico ed Urbanistico (PAU), Mediterranean University of Reggio Calabria, Via Melissari, Località Feo di Vito, 89060 Reggio Calabria, Italy*

Abstract: In this concluding Chapter we discuss the importance of several basic steps that can optimize the outcome of a citrus IPM program. In particular, we treat the need for stricter quarantine services and determining the Economic Injury Level (EIL) and Action Threshold (AT) for each pest. We emphasize the need to know more about the direct and indirect effects of the various citrus hosts on the pests and on their natural enemies, about the indirect effects of soil conditions and horticultural practices on pests and their enemies, quantitative estimates of the impact of the natural enemies (or any other IPM program) on pest numbers and (even more important) on yield quantity and quality. We also advocate initiating further in-depth studies on the use of indigenous natural enemies, exploring interactions between suites, or guilds, of natural enemies, and a far more extensive application of molecular tools, all of which could provide support for tracking the results of IPM projects. Finally, we suggest the setting up of a centralized database about citrus pests and their natural enemies in the Mediterranean Region.

Keywords: Citriculture, Mediterranean Region, citrus pests, optimization IPM.

Many facets of citriculture in the Mediterranean Region, including horticultural and economic aspects, were reviewed in this essay (Part I, Chapter **1**), as preludes to an extensive discussion of the various arthropods that attack citrus in this Region, their natural enemies and their Integrated Pest Management (IPM) (Part II, Chapters 2 and 3). Three main threads run through this essay: the need to identify the various pests and to study their life histories; the need to identify their efficient natural enemies and to demonstrate their efficacy, and finally the need to understand the concept, mechanics and application of IPM.

With these data, *e.g.* the state of our current knowledge at hand, and in order to conclude, we now sum up what we believe should be done in order to advance and promote the further use of IPM in the Mediterranean Region.

1. There is a strong need for stricter quarantine services, based on better identification facilities. Such facilities are dependent not only on well-trained personnel, but also on comprehensive identification methods, whether conventional binary keys, which mostly use morphological characters, molecular methods using unique markers [1-4], or precise behavioral observations (*e.g.* Ben-David *et al.*) [5]. While such methodology is of paramount importance in efforts to restrict the invasions of exotic pests, especially if they transmit citrus virus diseases, like aphids (Chapters **9,10**), it is also essential for the correct identification of the natural enemies. Furthermore, quarantine facilities serve as obligatory checkpoints for the introduction of exotic biological control agents, screening out undesired organisms (like unknown races of such agents, hyperparasites and various diseases). Quarantine facilities should also study and rear the quarantined exotic organisms in closed facilities.

Standards for diagnostic processes are being developed and draft diagnostic protocols for the identification of various pests are being sent for international consultation [6].

The many cases of exotic citrus pests that had invaded, and could invade, the Mediterranean Region, as noted in the foregoing Chapters, indicate that stricter quarantine services are a very real and urgent necessity. We do however note that much useful information is made available for the interested public through EPPO [7], but it is very general in nature and not specifically targeted at citrus pests.

***Address correspondence to Uri Gerson:** Department of Entomology, Faculty of Agricultural, Food and Environmental Quality Sciences, The Hebrew University of Jerusalem, P.O. Box 12, Rehovot, 76-100, Israel; E-mail: Gerson@agri.huji.ac.il

The data from such facilities could also serve as a mechanism for record-keeping of historical significance. As noted by Warner *et al.* [8], record keeping in the past was flawed due to the absence of regulatory requirements. Such records are essential for understanding the spread of pests, and their natural enemies, throughout the Region.

2. Determining the Economic Injury Level (EIL) and the Action Threshold (AT) for each pest, which are fundamental elements in any IPM program. Determination of individual EIL levels will enable growers to avoid unnecessary pesticide applications, thereby improving any conservation biological control that is due to natural enemies. This topic was thoroughly discussed in Chapter **3**, above.

3. Obtaining a better and fuller understanding about the direct and indirect effects of the various citrus hosts, their ages, the in-between cover crops and the surrounding vegetation on the pests and on their natural enemies. As noted in Chapter **6.2.3**, the citrus rust mite, *Phyllocoptruta oleivora* (Ashmead), appears to prefer lemons, oranges, tangerines and rough lemon, whereas pummelo and sour orange seem to be resistant [9]. The citrus red mite, *Panonychus citri* (McGregor), developed more rapidly and raised larger populations on lemons than on orange trees [10]. Another example of differential pest attack due to host species is the California red scale, *Aonidiella aurantii* (Maskell) (Chapter **4.4.1**). It prefers to settle on citrus species and varieties that have relatively few oil glands (*e.g.* sour orange) as compared to those with abundant oil glands (mandarin), a factor that affects the damage caused to these fruits. As to natural enemies, the phytoseiid *Euseius tularensis* Congdon, a predator of citrus thrips in California, prefers young citrus foliage on which it raises larger populations [11]. To this should be added various nutrients of citrus origin, like nectar and pollen, which encourage the populations of predatory mites. In addition, the presence of other pests and their excretions (*e.g.* honeydew), and various phytopathogenic fungi (*e.g.* mildews) may affect the numbers and activities of pests as well as of their natural enemies.

Habitat management is a form of conservation biological control that includes, among other methodologies, the provision of nutrients for natural enemies and shelter from adverse conditions, thus enhancing biological control in agricultural systems [12]. A further step is habitat manipulation [13], carried out to mitigate the impacts of invasive arthropod pests as well as to conserve natural enemies. One common method is to encourage the numbers of natural enemies by providing various plants with floral resources, such as nectar and pollen. An example is growing legumes as cover crops between the rows of citrus trees in California, which increased the numbers of *E. tularensis* [14]. Ground cover plants may serve as overwintering reservoirs for predatory mites [15, 16]. Another approach in habitat manipulation is to cover the ground in the citrus orchard with polyester/polyolefin mulch, which would prevent root-feeding pests from entering the soil and could hinder their adults from exiting and continuing their life cycles. In addition, such mulches also eliminate the need for herbicides [17]. Taking into account such factors may affect and determine the outcome of IPM programs.

An aspect sometimes overlooked in citrus IPM is the succession of pests and their natural enemies as the orchards age. An example pertains to the armored scale insects that attack citrus and to their natural enemies (Chapter **4**). The California red scale is especially harmful to young trees, later settling mostly on their inner, shaded parts and inflicting less damage. On the other hand, injury caused by the chaff scale, *Parlatoria pergandii* Comstock, or by the citrus mussel scale, *Lepidosaphes beckii* (Newman), begins to be noticed only when the trees reach their tenth to twelfth years. These two pests are parasitized by different species of the hymenopterous aphelinid *Aphytis* [18], which thus also show a succession in occurrence in the citrus orchard. Such succession indicates that somewhat different monitoring and control methods should be used in orchards of different ages.

4. Obtaining a better and fuller understanding about the indirect effects of soil conditions and horticultural practices on pests and their natural enemies. For instance, several environmental causes, bolstered by experimental work, have been proposed to explain outbreaks of the two-spotted spider mite, *Tetranychus urticae* Koch, in eastern Spain. Low to moderate saline stress in the soil brought about an increase in the reproduction of *T. urticae* infesting clementines near Valencia, Spain [19]. Bruessow *et al.* [20] have argued that replacing sour orange rootstock susceptible to *Citrus tristeza viruS* (CTV) by a tolerant rootstock may have rendered the trees more susceptible to *T. urticae* in Spanish citrus orchards. The aggregation, and thus damage, of this pest on Spanish clementines at the end of summer may have been due to the mechanical removal of the summer and fall flushes. Practices that promote abundant normal summer and fall vegetative growth could thus minimize mite damage to fruit quality [21].

The indirect effects of horticultural chemicals (fertilizers, herbicides, fungicides and insecticides) on pests and their natural enemies should also be better understood. Fertilizers and herbicides promote plant growth and thus, indirectly, larger pest numbers. Fungicides and insecticides, while applied to reduce pest populations, could cause spider mite outbreaks [22] or changes in the pest and natural enemy fauna, due to resistance or insensitivity to these chemicals. A separate issue, little studied in citrus pest control, is the effect of sublethal doses of pesticides [23]. Of special interest for the further application of IPM would be studying the effects of the judicial use of "soft" pesticides, like essential oils [24, 25] or the petroleum extracts of *Citrus aurantium* peels [26].

5. Better (quantitative) estimates of the impact of the natural enemies (or any other IPM program) on pest numbers and (even more important) on yield quantity and quality. Our primary purpose in citrus IPM is not to kill as many pests as possible, but to obtain more and better fruits. Many methods for estimating and demonstrating the effects of natural enemies on pest numbers were discussed by Gerson *et al.* [27]. Briefly, these include direct observations (*e.g.* pest numbers and extent of their damage before and after releases of natural enemies), using of cages and barriers, removal of the natural enemy, whether mechanically or by pesticides (the "pesticide check method") and through statistical inference methods.

Such quantitative data are essential for evaluating the success, or failure, of any control program. Without them it would be difficult to substantiate claims for the success of any such efforts. Solid quantitative documentation and a parallel economic analysis of citrus IPM efforts would bolster the confidence of growers and could lead to further investments in IPM projects.

6. Initiating further in-depth studies on the use of indigenous natural enemies, a point which we made many years ago [28]. Using such enemies has been somewhat neglected in the past due to successes with introduced parasites and predators. However, the more recent constraints on the introduction of exotic natural enemies have made studies on indigenous enemies that could control citrus pests more urgent. For instance, Abad-Moyano *et al.* [29] reported that the indigenous phytoseiid *Euseius stipulatus* (Athias Henriot) is the most important natural enemy of the citrus red mite, *Panonychus citri*. The use of indigenous acaropathogenic fungi [30] is another possible option.

7. Exploring interactions between suites, or guilds, of natural enemies. These interactions are often discussed as intraguild predation (IGP) or competitive displacement. The former takes place when two different natural enemies share a host (or prey) and at least one also feeds on the other. The outcome may increase, or decrease, overall pest control. Competitive displacement occurs when one of two ecological homologues, species that exist in the same niche, eliminates the other without (or almost without) any contact between them. IGP among phytoseiid predators of citrus mites was described by Abad-Moyano *et al.* [31]. In that case the superiority of *E. stipulatus* over other phytoseiids in intraguild interactions reduced the control of *T. urticae* infesting clementines in Spain. Competitive displacement of one citrus armored scale insect [*Aonidiella citrina* (Coquillett)] by another, closely-related species (*Aonidiella aurantii*) has been studied in California, as has the gradual displacement of their parasites (*Aphytis* spp.) in the same habitat [32].

8. A far more extensive application of molecular tools can provide much support for tracking the results of IPM projects. This includes (as noted above) determining the correct systematic status of pests and their natural enemies, but also discovering the reason for failures in biocontrol, following food chains and determining the effects of certain symbiotic bacteria on pest and natural enemy ecology.

9. Setting up a centralized database about citrus pests and their natural enemies in the Mediterranean Region. We envision an information center that will be available to all plant protection practitioners, whether growers, extension workers or scientists who are active in the crop protection disciplines. Such a center would encourage the interdisciplinary exchange of information and participation in national programs, and should increase the application of any new technologies in citrus IPM and biological control research. Data stored in the system should include historical information, successful and (not less important) unsuccessful biological control projects, up-to-date information about the biology and phenology of pests, natural enemies and host plants. This proposal follows Warner *et al.* [8], who recently proposed setting up a centralized database system for all cases of classical biocontrol of arthropods in the USA.

Finally, we conclude in the hope that our essay will contribute to the promotion, application and success of IPM in the Mediterranean citriculture.

REFERENCES

[1] Bailly X, Migeon A, Navajas M. Analysis of microsatellite variation in the spider mite pest *Tetranychus turkestani* (Acari: Tetranychidae) reveals population genetic structure and raises questions about related ecological factors. Biol J Linnean *Soc* 2003; 82: 69-78.

[2] Ben-David T, Melamed S, Gerson U, Morin S. ITS2 sequences as barcodes for identifying and analyzing spider mites (Acari: Tetranychidae). Exp Appl Acarol 2007 41**:** 169-81.

[3] Hurtado MA, Ansaloni T, Cros-Arteil S, Jacas JA, Navajas M. Sequence analysis of the ribosomal internal transcribed spacers region in spider mites (Prostigmata: Tetranychidae) occurring in citrus orchards in Eastern Spain: use for species discrimination. Ann Appl Biol 2008; 153: 167-74.

[4] Knapp M, Wagener B, Navajas M. Molecular discrimination between the spider mite *Tetranychus evansi* Baker & Pritchard, an important pest of tomatoes in southern Africa, and the closely related species *T. urticae* Koch (Acarina: Tetranychidae). Afz Entomol 2003; 11: 300-04.

[5] Ben-David T, Gerson U, Morin, S. Asymmetric reproductive interference between two closely related spider mites: *Tetranychus urticae* and *T. turkestani* (Acari: Tetranychidae). Exp Appl Acarol 2009; 48**:** 213-27.

[6] Clover G, Hammons S, Unger JG. .International diagnostic protocols for regulated plant pests. Bulletin OEPP 2010; 40: 24-9.

[7] EPPO/PQR 46. EPPO's plant quarantine data retrieval system, vers. 4.6. http://www.eppo.org./quarantine/Pest_Risk /PRA_documents.htm (accessed 2007).

[8] Warner KD, Getz C, Maurano S, Powers K. An analysis of historical trends in classical biological control of arthropods suggests need for a new centralized database in the USA. Biocontrol Sc Tech 2009; 19: 675-88.

[9] Yue B, Sun M, Qiu A. A comparison of citrus resistance to citrus rust mite (Acari: Eriophyidae) in the laboratory. Syst Appl Acarol 1998; 3: 59-62.

[10] Fleschner CA. Field approach to population studies of tetranychid mites on citrus and avocado in California. In: Becker EC, Ed. Proc. 10th Int. Cong. Entomol. Vol. 2. Montreal 1958. Ottawa: Mortimer 1958; pp. 669-74.

[11] Grafton-Cardwell EE, Ouyang Y. Augmentation of *Euseius tularensis* (Acari: Phytoseiidae) in citrus. Environ Entomol 1995; 24: 738-47.

[12] Landis DA, Wratten SD, Gurr GM. Habitat management to conserve natural enemies of arthropod pests in agriculture. Annu Rev Entomol 2000: 45: 175-201.

[13] Jonsson M, Wratten SD, Landis DA, Tompkins JML, Cullen R. Habitat manipulation to mitigate the impacts of invasive arthropod pests. Biol Invas 2010; 12: 2933-45.

[14] Grafton-Cardwell EE, Ouyang Y, Bugg RL. Leguminous cover crops to enhance population development of *Euseius tularensis* (Acari: Phytoseiidae) in citrus. Biol Cont 1999; 16: 73-80.

[15] Fadamiro HY, Xiao Y, Nesbitt M, Childers CC. Diversity and seasonal abundance of predacious mites in Alabama Satsuma citrus. Ann Entomol Soc Am 2009; 102: 617-28.

[16] Mailloux J, Le Bellec F, Kreiter S, Tixier M-S, Dubois P. Influence of ground cover management on diversity and density of phytoseiid mites (Acari: Phytoseiidae) in Guadeloupean citrus orchards. Exp Appl Acarol 2010; 52: 275-90.

[17] Duncan LW, Stuart RJ, Gmitter FG, Lapointe. Use of landscape fabric to manage *Diaprepes* root weevil in citrus groves. Fla Entomol 2009: 92: 74-9.

[18] Avidov Z, Harpaz I. Plant Pests of Israel. Jerusalem: Israel Universities Press, 1969.

[19] Aucejo-Romero S, Gomez-Cadenas A, Jacas-Miret JA. Effects of NaCl-stressed citrus plants on life-history parameters of *Tetranychus urticae* (Acari: Tetranychidae). Exp Appl Acarol 2004; 33: 55-67.

[20] Bruessow F, Asins MJ, Jacas JA, Urbaneja A. Replacement of CTV-susceptible sour orange rootstock by CTV-tolerant ones may have triggered outbreaks of *Tetranychus urticae* in Spanish citrus. Agric Ecosys Environ 2010; 13: 93-8.

[21] Ansaloni T, Pascual-Ruiz S, Hurtado MA, Jacas JA. Can summer and fall vegetative growth regulate the incidence of *Tetranychus urticae* Koch on clementine fruit? Crop Prot 2008; 27: 459-64.

[22] Gerson U, Cohen E. Resurgences of spider mites (Acari: Tetranychidae) induced by synthetic pyrethroids. Exp Appl Acarol 1989; 6: 29-46.

[23] Desneux N, Decourtye A, Delpuech J-M. The sublethal effects of pesticides on beneficial arthropods. Annu Rev Entomol 2007; 52: 81-106.

[24] Cloyd RA, Galle CL, Keith SR, Kalscheur NA, Kemp KE. Effect of commercially available plant-derived essential oil products on arthropod pests. J Econ Entomol 2009; 102: 1567-79.

[25] Han J, Choi B-R, Lee S-G, Kim SI, Ahn Y-J. Toxicity of plant essential oils to acaricide-susceptible and -resistant *Tetranychus urticae* (Acari: Tetranychidae) and *Neoseiulus californicus* (Acari: Phytoseiidae). J Econ Entomol 2010; 103: 1293-8.

[26] Siskos EP, Konstantopoulou MA, Mazomenos BE. Insecticidal activity of *Citrus aurantium* peel extract against *Bactrocera oleae* and *Ceratitis capitata* adults (Diptera: Tephritidae). J Appl Entomol 2009; 133:108-16.

[27] Gerson U, Smiley RL, Ochoa R. Mites (Acari) for Pest Control. Oxford: Blackwell Publishing, 2003.

[28] Gerson U, Vacante V. The use of indigenous acarine predators to control citrus mite pests. Bull OILB/SROP 1993: 16(7): 115-19.

[29] Abad-Moyano R, Urbaneja A, Hoffmann D, Schausberger P. Effects of *Euseius stipulatus* on establishment and efficacy in spider mite suppression of *Neoseiulus californicus* and *Phytoseiulus persimilis* in Clementine. Exp Appl Acarol 2010; 50: 329-41.

[30] Paz Z, Gerson U, Sztejnberg A. Assaying three new fungi against citrus mites in the laboratory, and a field trial. BioControl 2007; 52: 855-62.

[31] Abad-Moyano R, Pina T, Dembilio O, Ferragut F, Urbaneja A. Survey of natural enemies of spider mites (Acari: Tetranychidae) in citrus orchards in eastern Spain. Exp Appl Acarol 2009; 47: 49-61.

[32] DeBach P. The competitive displacement and coexistence principles. Annu Rev Entomol 1966; 11: 183-212.

INDEX

A

C

D

E

G

H

I

J

K

L

M

N

O

P

Q

R

S

V

W

X

Y

Z

Made in the USA
Las Vegas, NV
03 November 2026

57424417R00171